DOT-VNTSC-FRA-02-13

U.S. Department
of Transportation
**Federal Railroad
Administration**

Noise Characteristics of the Transrapid TR08 Maglev System

Office of Railroad
Development
Washington, DC 20590

U.S. Department of Transportation
Research and Special Programs Administration
John A. Volpe National Transportation Systems Center
Cambridge, MA 02142

July 2002

REPORT DOCUMENTATION PAGE		Form Approved OMB No. 0704-0188
Public reporting burden for this collection of information is estimated to average 1 hour per response, including the time for reviewing instructions, searching existing data sources, gathering and maintaining the data needed, and completing and reviewing the collection of information. Send comments regarding this burden estimate or any other aspect of this collection of information, including suggestions for reducing this burden, to Washington Headquarters Services, Directorate for Information Operations and Reports, 1215 Jefferson Davis Highway, Suite 1204, Arlington, VA 22202-4302, and to the Office of Management and Budget, Paperwork Reduction Project (0704-0188), Washington, DC 20503.		
1. AGENCY USE ONLY (Leave blank)	2. REPORT DATE July 2002	3. REPORT TYPE AND DATES COVERED April 2001 – July 2002
4. TITLE AND SUBTITLE Noise Characteristics of the Transrapid TR08 Maglev System		5. FUNDING NUMBERS RR294/R2130
6. AUTHOR(S): Bernd Barsikow[5], Daniel R. Disk[3], Carl E. Hanson[4], Michael Hellmig[5], Amishi Joshi[2], Arnold Kupferman[1], Ronald Mauri[2], Christopher J. Roof[2], Paul Valihura[2]		
7. PERFORMING ORGANIZATION NAME(S) AND ADDRESS(ES) US Department of Transportation Research and Special Programs Administration John A. Volpe National Transportation Systems Center 55 Broadway Cambridge, MA 02142-1093		8. PERFORMING ORGANIZATION REPORT NUMBER DOT-VNTSC-FRA-02-13
9. SPONSORING/MONITORING AGENCY NAME(S) AND ADDRESS(ES) U.S. Department of Transportation Federal Railroad Administration Office of Railroad Development 1120 Vermont Avenue, NW (Mail Stop 20) Washington, D.C. 20590	Port Authority of Allegheny County 345 Sixth Avenue, 3rd Floor Pittsburgh, PA 15222-2527	10. SPONSORING/MONITORING AGENCY REPORT NUMBER
11. SUPPLEMENTARY NOTES [1]Federal Railroad Administration; [2]Volpe Center; [3]MAGLEV, Inc., [4]Harris Miller Miller and Hanson, [5]akustik-data Engineering		
12a. DISTRIBUTION/AVAILABILITY STATEMENT		12b. DISTRIBUTION CODE
13. ABSTRACT (Maximum 200 words) As part of the Federal Railroad Administration's (FRA) Magnetic Levitation Transportation Technology Deployment Program, this technical report has been prepared to characterize the noise associated with the operation of the Transrapid International (TRI) TR08 Maglev System. The TRI TR08 Maglev System is an advanced transportation technology in which magnetic forces levitate, propel, and guide a vehicle over a specially-designed guideway. The TR08 Maglev System is the technology that is being considered for deployment in the U.S., and potential noise impacts of the technology were not known. This document presents noise data collected during measurements of the TRI TR08 Maglev System in August 2001, and May 2002, at the TRI Test Facility in the Emsland region of Germany. Noise data presented for the TR08 include single-microphone and array data for various guideway types, as well as speed- and distance-based noise relationships. Noise data are also compared with other transportation technologies.		
14. SUBJECT TERMS magnetic levitation, Maglev, noise, environment, high-speed ground transportation, data, measurements		15. NUMBER OF PAGES 338
		16. PRICE CODE
17. SECURITY CLASSFICATION OF REPORT Unclassified	18. SECURITY CLASSIFICATION OF THIS PAGE Unclassified	19. SECURITY CLASSIFICATION OF ABSTRACT Unclassified	20. LIMITATION OF ABSTRACT Unlimited

TABLE OF CONTENTS

Table Of Contents iii
List Of Figures vi
List Of Tables xxv
Preface xxxi
Executive Summary ES-1
CHAPTER 1   Introduction 1-1
   1.1   Purpose 1-1
   1.2   Background 1-1
   1.3   Roles And Responsibilities 1-2
   1.4   TR08 Maglev System Technology Information 1-3
   1.5   Test Facility 1-8
   1.6   Noise Measurements 1-10
      1.6.1   General 1-10
      1.6.2   Applicability of Data 1-11
   1.7   Report Organization 1-11
CHAPTER 2   Noise Measurements 2-1
   2.1   Introduction 2-1
   2.2   Measuring Sites and Weather Conditions 2-1
      2.2.1   Single-Microphone Measurements 2-3
      2.2.2   Microphone-Array Measurements 2-7
   2.3   Instrumentation and Procedures 2-12
      2.3.1   Measuring Equipment 2-12
      2.3.2   Principles and Application of Microphone-Array Technology 2-13
   2.4   Measurement Program 2-18
      2.4.1   Single-Microphone Measurements 2-18
      2.4.2   Microphone-Array Measurements 2-19
   2.5   Analyses and Results 2-25
      2.5.1   Single-Microphone Measurements 2-25
      2.5.2   Microphone-Array Measurements 2-32
   2.6   Model Calculations 2-60
      2.6.1   The AD-PRO Prediction Model 2-60
      2.6.2   Predictions of the Wayside Noise Generated by the TR08 Travelling on the Reference Concrete Guideway 2-61
      2.6.3   Source Reference Sound-Exposure Levels at 50 ft Distance from Guideway 2-67
   2.7   Summary of Results 2-68
CHAPTER 3   Validation and Verification 3-1
   3.1   Description of Validation and Verification Effort 3-1
   3.2   Test Protocol 3-1
      3.2.1   Measurement Proposal 3-1
      3.2.2   Test Plan 3-1
   3.3   Field Verification of Measurements and Protocol 3-2
   3.4   Validation of Contractor Data and Analyses 3-3
      3.4.1   Measurements Alongside Guideway 3-3
      3.4.2   Measurements Beneath Guideway 3-7

3.5 Findings and conclusions 3-7
CHAPTER 4   Supplemental Data and Discussion 4-1
  4.1 Comparison of TR08 Noise with That from Other High-Speed Transportation Systems 4-1
  4.2 TR08 Reference Data 4-1
  4.3 Regulatory Framework 4-2
CHAPTER 5   References 5-1
APPENDIX A. Noise Test Plans A-1
APPENDIX B. Results of single-microphone measurements at the various guideway types B-1
  B.1 Introduction B-1
  B.2 Reference Concrete Guideway B-2
    B.2.1 Microphone at 30.5 m (100.0 ft) distance from track centerline B-2
    B.2.2 Microphone at 25.0 m (82.0 ft) distance from track centerline B-6
    B.2.3 Microphone at 15.2 m (50.0 ft) distance from track centerline B-10
    B.2.4 Microphone at 6.5 m (21.3 ft) distance from track centerline (high position) B-14
    B.2.5 Microphone at 6.5 m (21.3 ft) distance from track centerline (low position) B-18
    B.2.6 Microphone beneath guideway centerline B-21
  B.3 Prototype Steel Guideway B-25
    B.3.1 Microphone at 30.5 m (100.0 ft) distance from track centerline ... B-25
    B.3.2 Microphone at 25.0 m (82.0 ft) distance from track centerline B-29
    B.3.3 Microphone at 15.2 m (50.0 ft) distance from track centerline B-33
    B.3.4 Microphone at 6.5 m (21.3 ft) distance from track centerline (high position) B-37
    B.3.5 Microphone at 6.5 m (21.3 ft) distance from track centerline (low position) B-41
    B.3.6 Microphone beneath guideway centerline B-45
  B.4 Prototype Concrete Guideway B-49
    B.4.1 Microphone at 30.5 m (100.0 ft) distance from track centerline ... B-49
    B.4.2 Microphone at 25.0 m (82.0 ft) distance from track centerline B-53
    B.4.3 Microphone at 15.2 m (50.0 ft) distance from track centerline B-57
    B.4.4 Microphone at 6.5 m (21.3 ft) distance from track centerline (high position) B-61
    B.4.5 Microphone at 6.5 m (21.3 ft) distance from track centerline (low position) B-65
    B.4.6 Microphone beneath guideway centerline B-69
  B.5 Hybrid Guideway B-73
    B.5.1 Microphone at 30.5 m (100.0 ft) distance from track centerline ... B-73
    B.5.2 Microphone at 25.0 m (82.0 ft) distance from track centerline B-77
    B.5.3 Microphone at 15.2 m (50.0 ft) distance from track centerline B-81
    B.5.4 Microphone at 6.5 m (21.3 ft) distance from track centerline (high position) B-85
    B.5.5 Microphone at 6.5 m (21.3 ft) distance from track centerline (low position) B-89

- B.5.6 Microphone beneath guideway centerline B-93
- B.6 North switch B-97
  - B.6.1 Microphone at 30.5 m (100.0 ft) distance from track centerline ... B-97
  - B.6.2 Microphone at 25.0 m (82.0 ft) distance from track centerline ... B-101
  - B.6.3 Microphone at 15.2 m (50.0 ft) distance from track centerline ... B-105
  - B.6.4 Microphone at 6.5 m (21.3 ft) distance from track centerline (high position) B-109
  - B.6.5 Microphone at 6.5 m (21.3 ft) distance from track centerline (low position) B-113
  - B.6.6 Microphone beneath guideway centerline B-117
- B.7 At-Grade Guideway B-121
  - B.7.1 Beam 340 (Steel) B-121
  - B.7.2 Beam 341 (Concrete) B-126
- APPENDIX C. Sound Source Distribution Graphics C-1
- APPENDIX D. Tabular One-Third Octave Band Data D-1

LIST OF FIGURES

Figure ES-1. Sound-Exposure Level as a Function of Vehicle Speed ES-2

Figure 1-1. The TRI TR08 Maglev System 1-4

Figure 1-2. Typical TRI TR08 Interior Plans for (a) medium-density intercity seating, (b) high-density commuter type seating, and (c) first-class intercity seating 1-5

Figure 1-3. TRI TR08 Guideway (can be either elevated on columns or mounted at grade) 1-6

Figure 1.4. TRI TR08 Support and Guidance Systems 1-7

Figure 1-5. TRI TR08 Propulsion 1-7

Figure 1-6. Transrapid Test Facility (TVE) 1-9

Figure 1-7. Typical speed profile of TR08 at the Transrapid Test Facility 1-10

Figure 2-1. General location of noise measurement sites. 2-2

Figure 2-2. The TR08 passing by the measuring site with the prototype steel guideway (column 213) and the WV array (in high position) mounted on the hydraulic tower; the 6.5 m-microphone (in low position) can be seen in the foreground. 2-9

Figure 2-3. The TR08 passing by the measuring site with the reference concrete guideway and the hydraulic tower positioned between column 228 and 229 with the attached WX32 array. 2-10

Figure 2-4. The WH array mounted on the hydraulic tower in the area of the reference concrete guideway. 2-11

Figure 2-5. Width of main lobe as a function of frequency for the WV and WH array, viz., for the three sub-arrays whose working frequency ranges are indicated; results are with 25 dB Dolph-shading. 2-16

Figure 2-6. Width of main lobe in the vertical direction as a function of frequency for the WX32 array; results are with 25 dB Dolph-shading. 2-18

Figure 2-7. Sound-exposure level as a function of vehicle speed measured at 30.5 m (100.0 ft) distance from track centerline and 1.2 m (4.0 ft) above the ground for various guideway types. 2-28

Figure 2-8. Sound-exposure level as a function of vehicle speed measured at 25.0 m (82.0 ft) distance from track centerline and 3.5 m (11.5 ft) above the ground for various guideway types. 2-29

Figure 2-9. Sound-exposure level as a function of vehicle speed measured at 15.2 m (50.0 ft) distance from track centerline and 1.5 m (5.0 ft) above the ground for various guideway types. 2-29

Figure 2-10. Sound-exposure level as a function of vehicle speed measured at 6.5 m (21.3 ft) distance from track centerline and the height of the upper surface of the guideway for various guideway types. 2-30

Figure 2-11. Sound-exposure level as a function of vehicle speed measured at 6.5 m (21.3 ft) distance from track centerline and 1.5 m (5.0 ft) below the upper surface of the guideway for various guideway types. 2-30

Figure 2-12. Sound-exposure level as a function of vehicle speed measured beneath guideway centerline at a height of 1.5 m (5.0 ft) above the ground for various guideway types. 2-31

Figure 2-13. Averaged sound-source distributions measured with the WV array during passbys of the TR08 on the reference concrete guideway at speeds between 150 and 400 km/h (93 and 249 mph). 2-34

Figure 2-14. Averaged sound-source distributions measured with the WV array during passbys of the TR08 on the prototype steel guideway at speeds between 150 and 400 km/h (93 and 249 mph)... 2-35

Figure 2-15. Averaged sound-source distributions measured with the WV array during passbys of the TR08 on the prototype concrete guideway at speeds between 150 and 400 km/h (93 and 249 mph). 2-36

Figure 2-16. Averaged sound source distributions measured with the WV array during passbys of the TR08 on the hybrid guideway at speeds between 150 and 400 km/h (93 and 249 mph) 2-37

Figure 2-17. Vertical sound-source distributions measured with the WV array during passbys of the TR08 on the different guideway types at a speed of 150 km/h (93 mph). 2-40

Figure 2-18. Vertical sound-source distributions measured with the WV array during passbys of the TR08 on the different guideway types at a speed of 200 km/h (124 mph). 2-40

Figure 2-19. Vertical sound-source distributions measured with the WV array during passbys of the TR08 on the different guideway types at a speed of 300 km/h (186 mph). 2-41

Figure 2-20. Vertical sound-source distributions measured with the WV array during passbys of the TR08 on the different guideway types at a speed of 400 km/h (249 mph). 2-41

Figure 2-21. Horizontal sound-source distributions measured with the WH array during passbys of the TR08 on the reference concrete guideway at speeds between 150 and 400 km/h (93 and 249 mph). 2-43

Figure 2-22. Two-dimensional sound-source distributions measured with the WX32 array during passbys of the TR08 on the reference concrete guideway at speeds between 150 and 400 km/h (93 and 249 mph).................................... 2-45

Figure 2-23. Two-dimensional sound-source distributions measured with the WX16 array during passbys of the TR08 on the reference concrete guideway at speeds between 150 and 400 km/h (93 and 249 mph). 2-46

Figure 2-24. Simultaneous presentation of one- and two-dimensional sound-source distributions measured with the WH08 and the WX16 array during passbys of the TR08 on the reference concrete guideway at speeds between 150 and 400 km/h (93 and 249 mph). 2-47

Figure 2-25. Unweighted narrow-band spectra of the sound-pressure level (SPL) measured with the WV array during passbys of the TR08 on the reference concrete guideway at speeds between 150 and 400 km/h (93 and 249 mph); $\Delta f = 11.6$ Hz. 2-52

Figure 2-26. Unweighted one-third octave-band spectra of the sound-pressure level (SPL) measured with the WV array during passbys of the TR08 on the reference concrete guideway at speeds between 150 and 400 km/h (93 and 249 mph). 2-53

Figure 2-27. Unweighted narrow-band spectra of the sound-pressure level (SPL) measured with the WV array during passbys of the TR08 on the prototype steel guideway at speeds between 150 and 400 km/h (93 and 249 mph); $\Delta f = 11.6$ Hz. 2-54

Figure 2-28. Unweighted one-third octave-band spectra of the sound-pressure level (SPL) measured with the WV array during passbys of the TR08 on the prototype steel guideway at speeds between 150 and 400 km/h (93 and 249 mph). 2-55

Figure 2-29. Unweighted narrow-band spectra of the sound-pressure level (SPL) measured with the WV array during passbys of the TR08 on the prototype concrete guideway at speeds between 150 and 400 km/h (93 and 249 mph); $\Delta f = 11.6$ Hz. 2-56

Figure 2-30. Unweighted one-third octave-band spectra of the sound-pressure level (SPL) measured with the WV array during passbys of the TR08 on the prototype concrete guideway at speeds between 150 and 400 km/h (93 and 249 mph). 2-57

Figure 2-31. Unweighted narrow-band spectra of the sound-pressure level (SPL) measured with the WV array during the passbys of the TR08 on the hybrid guideway at speeds between 150 and 400 km/h (93 and 249 mph); $\Delta f=11.6$ Hz. 2-58

Figure 2-32. Unweighted one-third octave-band spectra of the sound-pressure level (SPL) measured with the WV array during passbys of the TR08 on the hybrid guideway at speeds between 150 and 400 km/h (93 and 249 mph)..... 2-59

Figure 2-33. Measured (———) and predicted (- · - · -) A-weighted time histories during passbys of the TR08 on the reference concrete guideway at speeds of 150 km/h (93 mph); measuring distances are 30.5 m (100.0 ft) (top), 25.0 m (82.0 ft) (mid), and 15.2 m (50.0 ft) (bottom). 2-62

Figure 2-34. Measured (——) and predicted (- · - · -) A-weighted time histories during passbys of the TR08 on the reference concrete guideway at a speed of 200 km/h (124 mph); measuring distances are 30.5 m (100.0 ft) (top), 25.0 m (82.0 ft) (mid), and 15.2 m (50.0 ft) (bottom) 2-63

Figure 2-35. Measured (——) and predicted (- · - · -) A-weighted time histories during passbys of the TR08 on the reference concrete guideway at a speed of 300 km/h (186 mph); measuring distances are 30.5 m (100.0 ft) (top), 25.0 m (82.0 ft) (mid), and 15.2 m (50.0 ft) (bottom) 2-64

Figure 2-36. Measured (——) and predicted (- · - · -) A-weighted time histories during passbys of the TR08 on the reference concrete guideway at a speed of 400 km/h (249 mph); measuring distances are 30.5 m (100.0 ft) (top), 25.0 m (82.0 ft) (mid), and 15.2 m (50.0 ft) (bottom) 2-65

Figure 3-1 Comparison of Maximum Sound Level Data 3-4

Figure 3-2 Example Sound Level and Onset Rate Time History Data 3-5

Figure 3-3 TR08 Sound Level Onset Rate Data 3-6

Figure 3-4. Comparison of Sound Level Data Beneath Guideway 3-7

Figure B-1. Time history of the A-weighted SPL during a passby of the TR08 travelling on the reference concrete guideway at about 100 km/h (62 mph) measured at 30.5 m *(100 ft) distance from track centerline and 1.2 m (4.0 ft) above the grund B-2

Figure B-2. Time history of the A-weighted SPL during a passby of the TR08 travelling on the reference concrete guideway at about 150 km/h (93 mph) measured at 30.5 m (100.0 ft) distance from track centerline and 1.2 m (4.0 ft) above the ground B-2

Figure B-3. Time history of the A-weighted SPL during a passby of the TR08 travelling on the reference concrete guideway at about 200 km/h (124 mph) measured at 30.5 m (100.0 ft) distance from track centerline and 1.2 m (4.0 ft) above the ground B-3

Figure B-4. Time history of the A-weighted SPL during a passby of the TR08 travelling on the reference concrete guideway at about 300 km/h (186 mph) measured at 30.5 m (100.0 ft) distance from track centerline and 1.2 m (4.0 ft) above the ground B-3

Figure B-5. Time history of the A-weighted SPL during a passby of the TR08 travelling on the reference concrete guideway at about 400 km/h (249 mph) measured at 30.5 m (100.0 ft) distance from track centerline and 1.2 m (4.0 ft) above the ground B-4

Figure B-6. Time history of the A-weighted SPL during a passby of the TR08 travelling on the reference concrete guideway at about 100 km/h (62 mph) measured at 25.0 m (82 ft) distance from track centerline and 3.5 m (11.5 ft) above the ground B-6

Figure B-7. Time history of the A-weighted SPL during a passby of the TR08 travelling on the reference concrete guideway at about 150 km/h (93 mph) measured at 25.0 m (82.0 ft) distance from track centerline and 3.5 m (11.5 ft) above the ground. B-6

Figure B-8. Time history of the A-weighted SPL during a passby of the TR08 travelling on the reference concrete guideway at about 200 km/h (124 mph) measured at 25.0 m (82.0 ft) distance from track centerline and 3.5 m (11.5 ft) above the ground. B-7

Figure B-9. Time history of the A-weighted SPL during a passby of the TR08 travelling on the reference concrete guideway at about 300 km/h (186 mph) measured at 25.0 m (82.0 ft) distance from track centerline and 3.5 m (11.5 ft) above the ground. B-7

Figure B-10. Time history of the A-weighted SPL during a passby of the TR08 travelling on the reference concrete guideway at about 400 km/h (249 mph) measured at 25.0 m (82.0 ft) distance from track centerline and 3.5 m (11.5 ft) above the ground. B-8

Figure B-11. Time history of the A-weighted SPL during a passby of the Tr08 travelling on the reference concrete guideway at about 100 km/h (62 mph) measured at 15.2 m (50.0 ft) distance from track centerline and 1.5 m (5.0 ft) above the ground. B-10

Figure B-12. Time history of the A-weighted SPL during a passby of the TR08 travelling on the reference concrete guideway at about 150 km/h (93 mph) measured at 15.2 m (50.0 ft) distance from track centerline and 1.5 m (5.0 ft) above the ground. B-10

Figure B-13. Time history of the A-weighted SPL during a passby of the TR08 travelling on the reference concrete guideway at about 200 km/h (124 mph) measured at 15.2 m (50.0 ft) distance from track centerline and 1.5 m (5.0 ft) above the ground. B-11

Figure B-14. Time history of the A-weighted SPL during a passby of the TR08 travelling on the reference concrete guideway at about 300 km/h (186 mph) measured at 15.2 m (50.0 ft) distance from track centerline and 1.5 m (5.0 ft) above the ground. B-11

Figure B-15. Time history of the A-weighted SPL during a passby of the TR08 travelling on the reference concrete guideway at about 400 km/h (249 mph) measured at 15.2 m (50.0 ft) distance from track centerline and 1.5 m (5.0 ft) above the ground. B-12

Figure B-16. Time history of the A-weighted SPL during a passby of the TR08 travelling on the reference concrete guideway at about 100 km/h (62 mph) measured at 6.5 m (21.3 ft) distance from track centerline and the height of the upper surface of the guideway. B-14

Figure B-17. Time history of the A-weighted SPL during a passby of the TR08 travelling on the reference concrete guideway at about 150 km/h (93 mph) measured at 6.5 m (21.3 ft) distance from track centerline and the height of the upper surface of the guideway. B-14

Figure B-18. Time history of the A-weighted SPL during a passby of the TR08 travelling on the reference concrete guideway at about 200 km/h (124 mph) measured at 6.5 m (21.3 ft) distance from track centerline and the height of the upper surface of the guideway. B-15

Figure B-19. Time history of the A-weighted SPL during a passby of the TR08 travelling on the reference concrete guideway at about 300 km/h (186 mph) measured at 6.5 m (21.3 ft) distance from track centerline and the height of the upper surface of the guideway. B-15

Figure B-20. Time history of the A-weighted SPL during a passby of the TR08 travelling on the reference concrete guideway at about 400 km/h (249 mph) measured at 6.5 m (21.3 ft) distance from track centerline and the height of the upper surface of the guideway. B-16

Figure B-21. Time history of the A-weighted SPL during a passby of the TR08 travelling on the reference concrete guideway at about 100 km/h (62 mph) measured at 6.5 m (21.3 ft) distance from track centerline and 1.5 m (5.0 ft) below the upper surface of the guideway. B-18

Figure B-22. Time history of the A-weighted SPL during a passby of the TR08 travelling on the reference concrete guideway at about 150 km/h (93 mph) measured at 6.5 m (21.3 ft) distance from track centerline and 1.5 m (5.0 ft) below the upper surface of the guideway. B-18

Figure B-23. Time history of the A-weighted SPL during a passby of the TR08 travelling on the reference concrete guideway at about 200 km/h (124 mph) measured at 6.5 m (21.3 ft) distance from track centerline and 1.5 m (5.0 ft) below the upper surface of the guideway. B-19

Figure B-24. Time history of the A-weighted SPL during a passby of the TR08 travelling on the reference concrete guideway at about 400 km/h (249 mph) measured at 6.5 m (21.3 ft) distance from track centerline and 1.5 m (5.0 ft) below the upper surface of the guideway. B-19

Figure B-25. Time history of the A-weighted SPL during a passby of the TR08 travelling on the reference concrete guideway at about 100 km/h (62 mph) measured beneath the guidway centerline at a height of 1.5 m (5.0 ft) above the ground. B-21

Figure B-26. Time history of the A-weighted SPL during a passby of the TR08 travelling on the reference concrete guideway at about 150 km/h (93 mph) measured beneath guideway centerline at a height of 1.5 m (5.0 ft) above the ground. B-21

Figure B-27. Time history of the A-weighted SPL during a passby of the TR08 travelling on the reference concrete guideway at about 200 km/h (124 mph) measured beneath guideway centerline at a height of 1.5 m (5.0 ft) above the ground. B-22

Figure B-28. Time history of the A-weighted SPL during a passby of the TR08 travelling on the reference concrete guideway at about 300 km/h (186 mph) measured beneath guideway centerline at a height of 1.5 m (5.0 ft) above the ground. B-22

Figure B-29. Time history of the A-weighted SPL during a passby of the TR08 travelling on the reference concrete guideway at about 400 km/h (249 mph) measured beneath guideway centerline at a height of 1.5 m (5.0 ft) above the ground. B-23

Figure B-30. Time history of the A-weighted SPL during a passby of the TR08 travelling on the prototype steel guideway at about 100 km/h (62 mph) measured at 30.5 m (100.0 ft) distanace from track centerline and 1.2 m (4.0 ft) above the ground. B-25

Figure B-31. Time history of the A-weighted SPL during a passby of the TR08 travelling on the prototype steel guideway at about 150 km/h (93 mph) measured at 30.5 m (100.0 ft) distance from track centerline and 1.2 m (4.0 ft) above the ground. B-25

Figure B-32. Time history of the A-weighted SPL during a passby of the TR08 travelling on the prototype steel guideway at about 200 km/h (124 mph) measured at 30.5 m (100.0 ft) distance from track centerline and 1.2 m (4.0 ft) above the ground. B-26

Figure B-33. Time history of the A-weighted SPL during a passby of the TR08 travelling on the prototype steel guideway at about 300 km/h (186 mph) measured at 30.5 m (100.0 ft) distance from track centerline and 1.2 m (4.0 ft) above the ground. B-26

Figure B-34. Time history of the A-weighted SPL during a passby of the TR08 travelling on the prototype steel guideway at about 400 km/h (249 mph) measured at 30.5 m (100.0 ft) distance from track centerline and 1.2 m (4.0 ft) above the ground. B-27

Figure B-35. Time history of the A-weighted SPL during a passby of the TR08 travelling on the prototype steel guideway at aobut 100 km/h (62 mph) measured at 25.0 m (82.0 ft) distance from track centerline and 3.5 m (11.5 ft) above the ground. B-29

Figure B-36. Time history of the A-weighted SPL during a passby of the TR08 travelling on the prototype steel guideway at about 150 km/h (93 mph) measured at 25.0 m (82.0 ft) distance from track centerline and 3.5 m (11.5 ft) above the ground. B-29

Figure B-37. Time history of the A-weighted SPL during a passby of the TR08 travelling on the prototype steel guideway at about 200 km/h (124 mph) measured at 25.0 m (82.0 ft) distance from track centerline and 3.5 m (11.5 ft) above the ground. B-30

Figure B-38. Time history of the A-weighted SPL during a passby of the TR08 travelling on the prototype steel guideway at about 300 km/h (186 mph) measured at 25.0 m (82.0 ft) distance from track centerline and 3.5 m (11.5 ft) above the ground. B-30

Figure B-39. Time history of the A-weighted SPL during a passby of the TR08 travelling on the prototype steel guideway at about 400 km/h (249 mph) measured at 25.0 m (82.0 ft) distance from track centerline and 3.5 m (11.5 ft) above the ground. B-31

Figure B-40. Time history of the A-weighted SPL during a pssby of the TR08 travelling on the prototype steel guideway at about 100 km/h (62 mph) measured at 15.2 m (50.0 ft) distance from track centerline and 1.5 m (5.0 ft) above the ground. B-33

Figure B-41. Time history of the A-weighted SPL during a passby of the TR08 travelling on the prototype steel guideway at about 150 km/h (93 mph) measured at 15.2 m (50.0 ft) distance from track centerline and 1.5 m (5.0 ft) above the ground. B-33

Figure B-42. Time history of the A-weighted SPL during a passby of the TR08 travelling on the prototype steel guideway at about 200 km/h (124 mph) measured at 15.2 m (50.0 ft) distance from track centerline and 1.5 m (5.0 ft) above the ground..B-34

Figure B-43. Time history of the A-weighted SPL during a passby of the TR08 travelling on the prototype steel guideway at about 300 km/h (186 mph) measured at 15.2 m (50.0 ft) distance from track centerline and 1.5 m (5.0 ft) above the ground..B-34

Figure B-44. Time history of the A-weighted SPL during a passby of the TR08 travelling on the prototype steel guideway at about 400 km/h (249 mph) measured at 15.2 m (50.0 ft) distance from track centerline and 1.5 m (5.0 ft) above the ground..B-35

Figure B-45. Time history of the A-weighted SPL during a passby of the TR08 travelling on the prototype steel guideway at about 100 km/h (62 mph) measured at 6.5 m (21.3 ft) distance from track centerline and the height of the upper surface of the guideway...B-37

Figure B-46. Time history of the A-weighted SPL during a passby of the TR08 travelling on the prototype steel guideway at about 150 km/h (93 mph) measured at 6.5 m (21.3 ft) distance from track centerline and the height of the upper surface of the guideway...B-37

Figure B-47. Time history of the A-weighted SPL during a passby of the TR08 travelling on the prototype steel guideway at about 200 km/h (124 mph) measured at 6.5 m (21.3 ft) distance from track centerline and the height of the upper surface of the guideway...B-38

Figure B-48. Time history of the A-weighted SPL during a passby of the TR08 travelling on the prototype steel guideway at about 300 km/h (186 mph) measured at 6.5 m (21.3 ft) distance from track centerline and the height of the upper surface of the guideway...B-38

Figure B-49. Time history of the A-weighted SPL during a passby of the TR08 travelling on the prototype steel guideway at about 400 km/h (249 mph) measured at 6.5 m (21.3 ft) distance from track centerline and the height of the upper surface of the guideway. B-39

Figure B-50. Time history of the A-weighted SPL during a passby of the TR08 travelling on the prototype steel guideway at about 100 km/h (62 mph) measured at 6.5 m (21.3 ft) distance from track centerline and 1.5 m (5.0 ft) below the upper surface of the guideway. B-41

Figure B-51. Time history of the A-weighted SPL during a passby of the TR08 travelling on the prototype steel guideway at about 150 km/h (93 mph) measured at 6.5 m (21.3 ft) distance from track centerline and 1.5 m (5.0 ft) below the upper surface of the guideway. B-41

Figure B-52. Time history of the A-weighted SPL during a passby of the TR08 travelling on the prototype steel guideway at about 200 km/h (124 mph) measured at 6.5 m (21.3 ft) distance from track centerline and 1.5 m (5.0 ft) below the upper surface of the guideway. B-42

Figure B-53. Time history of the A-weighted SPL during a passby of the TR08 travelling on the prototype steel guideway at about 300 km/h (186 mph) measured at 6.5 m (21.3 ft) distance from track centerline and 1.5 m (5.0 ft) below the upper surface of the guideway. B-42

Figure B-54. Time history of the A-weighted SPL during a passby of the TR08 travelling on the prototype steel guideway at about 400 km/h (249 mph) measured at 6.5 m (21.3 ft) distance from track centerline and 1.5 m (5.0 ft) below the upper surface of the guideway. B-43

Figure B-55. Time history of the A-weighted SPL during a passby of the TR08 travelling on the prototype steel guideway at about 100 km/h (62 mph) measured beneath the guideway centerline at a height of 1.5 m (5.0 ft) above the ground. B-45

Figure B-56. Time history of the A-weighted SPL during a passby of the TR08 travelling on the prototype steel guideway at about 150 km/h (93 mph) measured beneath guideway centerline at a height of 1.5 m (5.0 ft) above the ground. ...B-45

Figure B-57. Time history of the A-weighted SPL during a passby of the TR08 travelling on the prototype steel guideway at about 200 km/h (124 mph) measured beneath guideway centerline at a height of 1.5 m (5.0 ft) above the ground. B-46

Figure B-58. Time history of the A-weighted SPL during a passby of the TR08 travelling on the prototype steel guideway at about 300 km/h (186 mph) measured beneath guideway centerline at a height of 1.5 m (5.0 ft) above the ground. B-46

Figure B-59. Time history of the A-weighted SPL during a passby of the TR08 travelling on the prototype steel guideway at about 400 km/h (249 mph) measured beneath guideway centerline at a height of 1.5 m (5.0 ft) above the ground. B-47

Figure B-60. Time history of the A-weighted SPL during a passby of the TR08 travelling on the prototype concrete guideway at about 100 km/h (62 mph) measured at 30.5 m (100.0 ft) distance from track centeriline and 1.2 m (4.0 ft) above the ground..B-49

Figure B-61. Time history of the A-weighted SPL during a passby of the TR08 travelling on the prototype concrete guideway at about 150 km/h (93 mph) measured at 30.5 m (100.0 ft) distance from track centerline and 1.2 m (4.0 ft) above the ground..B-49

Figure B-62. Time history of the A-weighted SPL during a passby of the TR08 travelling on the prototype concrete guideway at about 200 km/h (124 mph) measured at 30.5 m (100.0 ft) distance from track centerline and 1.2 m (4.0 ft) above the ground..B-50

Figure B-63. Time history of the A-weighted SPL during a passby of the TR08 travelling on the prototype concrete guideway at about 300 km/h (186 mph) measured at 30.5 m (100.0 ft) distance from track centerline and 1.2 m (4.0 ft) above the ground..B-50

Figure B-64. Time history of the A-weighted SPL during a passby of the TR08 travelling on the prototype concrete guideway at about 400 km/h (249 mph) measured at 30.5 m (100.0 ft) distance from track centerline and 1.2 m (4.0 ft) above the ground..B-51

Figure B-65. Time history of the A-weighted SPL during a passby of the TR08 travelling on the prototype concrete guideway at about 100 km/h (62 mph) measured at 25.0 m (82.0 ft) distance from track centerline and 3.5 m (11.5 ft) above the ground..B-53

Figure B-66. Time history of the A-weighted SPL during a passby of the TR08 travelling on the prototype concrete guideway at about 150 km/h (93 mph) measured at 25.0 m (82.0 ft) distance from track centerline and 3.5 m (11.5 ft) above the ground..B-53

Figure B-67. Time history of the A-weighted SPL during a passby of the TR08 travelling on the prototype concrete guideway at about 200 km/h (124 mph) measured at 25.0 m (82.0 ft) distance from track centerline and 3.5 m (11.5 ft) above the ground..B-54

Figure B-68. Time history of the A-weighted SPL during a passby of the TR08 travelling on the prototype concrete guideway at about 300 km/h (186 mph) measured at 25.0 m (82.0 ft) distance from track centerline and 3.5 m (11.5 ft) above the ground..B-54

Figure B-69. Time history of the A-weighted SPL during a passby of the TR08 travelling on the prototype concrete guideway at about 400 km/h (249 mph) measured at 25.0 m (82.0 ft) distance from track centerline and 3.5 m (11.5 ft) above the ground..B-55

Figure B-70. Time history of the A-weighted SPL during a passby of the the TR08 travelling on the prototype concrete guideway at about 100 km/h (62 mph) measured at 15.2 m (50.0 ft) distance from track centerline and 1.5 m (5.0 ft) above the ground. B-57

Figure B-71. Time history of the A-weighted SPL during a passby of the TR08 travelling on the prototype concrete guideway at about 150 km/h (93 mph) measured at 15.2 m (50.0 ft) distance from track centerline and 1.5 m (5.0 ft) above the ground. B-57

Figure B-72. Time history of the A-weighted SPL during a passby of the TR08 travelling on the prototype concrete guideway at about 200 km/h (124 mph) measured at 15.2 m (50.0 ft) distance from track centerline and 1.5 m (5.0 ft) above the ground. B-58

Figure B-73. Time history of the A-weighted SPL during a passby of the TR08 travelling on the prototype concrete guideway at about 300 km/h (186 mph) measured at 15.2 m (50.0 ft) distance from track centerline and 1.5 m (5.0 ft) above the ground. B-58

Figure B-74. Time history of the A-weighted SPL during a passby of the TR08 travelling on the prototype concrete guideway at about 400 km/h (249 mph) measured at 15.2 m (50.0 ft) distance from track centerline and 1.5 m (5.0 ft) above the ground. B-59

Figure B-75. Time history of the A-weighted SPL during a passby of the TR08 travelling on the prototype concrete guideway at about 100 km/h (62 mph) measured at 6.5 m (21.3 ft) distanace from track centerline and the height of the upper surface of the guideway. B-61

Figure B-76. Time history of the A-weighted SPL during a passby of the TR08 travelling on the prototype concrete guideway at about 150 km/h (93 mph) measured at 6.5 m (21.3 ft) distance from track centerline and the height of the upper surface of the guideway. B-61

Figure B-77. Time history of the A-weighted SPL during a passby of the TR08 travelling on the prototype concrete guideway at about 200 km/h (124 mph) measured at 6.5 m (21.3 ft) distance from track centerline and the height of the upper surface of the guideway. B-62

Figure B-78. Time history of the A-weighted SPL during a passby of the TR08 travelling on the prototype concrete guideway at about 300 km/h (186 mph) measured at 6.5 m (21.3 ft) distance from track centerline and the height of the upper surface of the guideway. B-62

Figure B-79. Time history of the A-weighted SPL during a passby of the TR08 travelling on the prototype concrete guideway at about 400 km/h (249 mph) measured at 6.5 m (21.3 ft) distance from track centerline and the height of the upper surface of the guideway. B-63

Figure B-80. Time history of the A-weighted SPL during a passby of the TR08 travelling on the prototype concrete guideway at about 100 km/h (62 mph) measured at 6.5 m(21.3 ft) distance from track centerline and 1.5 m (5.0 ft) below the upper surface of the guideway...B-65

Figure B-81. Time history of the A-weighted SPL during a passby of the TR08 travelling on the prototype concrete guideway at about 150 km/h (93 mph) measured at 6.5 m (21.3 ft) distance from track centerline and 1.5 m (5.0 ft) below the upper surface of the guideway. B-65

Figure B-82. Time history of the A-weighted SPL during a passby of the TR08 travelling on the prototype concrete guideway at about 200 km/h (124 mph) measured at 6.5 m (21.3 ft) distance from track centerline and 1.5 m (5.0 ft) below the upper surface of the guideway. B-66

Figure B-83. Time history of the A-weighted SPL during a passby of the TR08 travelling on the prototype concrete guideway at about 300 km/h (186 mph) measured at 6.5 m (21.3 ft) distance from track centerline and 1.5 m (5.0 ft) below the upper surface of the guideway. B-66

Figure B-84. Time history of the A-weighted SPL during a passby of the TR08 travelling on the prototype concrete guideway at about 400 km/h (249 mph) measured at 6.5 m (21.3 ft) distance from track centerline and 1.5 m (5.0 ft) below the upper surface of the guideway. B-67

Figure B-85. Time history of the A-weighted SPL during a passby of the TR08 travelling on the prototype concrete guideway at about 100 km/h (62 mph) measured beneath the guideway centerline at a height of 1.5 m (5.0 ft) above the ground. B-69

Figure B-86. Time history of the A-weighted SPL during a passby of the TR08 travelling on the prototype concrete guideway at about 150 km/h (93 mph) measured beneath guideway centerline at a height of 1.5 m (5.0 ft) above the ground. B-69

Figure B-87. Time history of the A-weighted SPL during a passby of the TR08 travelling on the prototype concrete guideway at about 200 km/h (124 mph) measured beneath guideway centerline at a height of 1.5 m (5.0 ft) above the ground. B-70

Figure B-88. Time history of the A-weighted SPL during a passby of the TR08 travelling on the prototype concrete guideway at about 300 km/h (186 mph) measured beneath guideway centerline at a height of 1.5 m (5.0 ft) above the ground. B-70

Figure B-89. Time history of the A-weighted SPL during a passby of the TR08 travelling on the prototype concrete guideway at about 400 km/h (249 mph) measured beneath guideway centerline at a height of 1.5 m (5.0 ft) above the ground. B-71

Figure B-90. Time history of the A-weighted SPL during a passby of the TR08 travelling on the hybrid guideway at about 100 km/h (62 mph) measured at 30.5 m (100.0 ft) distance from track centerline and 1.2 m (4.0 ft) above the ground. B-73

Figure B-91. Time history of the A-weighted SPL during a passby of the TR08 travelling on the hybrid guideway at about 150 km/h (93 mph) measured at 30.5 m (100.0 ft) distance from track centerline and 1.2 m (4.0 ft) above the ground. B-73

Figure B-92. Time history of the A-weighted SPL during a passby of the TR08 travelling on the hybrid guideway at about 200 km/h (124 mph) measured at 30.5 m (100.0 ft) distance from track centerline and 1.2 m (4.0 ft) above the ground. B-74

Figure B-93. Time history of the A-weighted SPL during a passby of the TR08 travelling on the hybrid guideway at about 300 km/h (186 mph) measured at 30.5 m (100.0 ft) distance from track centerline and 1.2 m (4.0 ft) above the ground. B-74

Figure B-94. Time history of the A-weighted SPL during a passby of the TR08 travelling on the hybrid guideway at about 400 km/h (249 mph) measured at 30.5 m (100.0 ft) distance from track centerline and 1.2 m (4.0 ft) above the ground. B-75

Figure B-95. Time history of the A-weighted SPL during a passby of the TR08 travelling on the hybrid guideway at about 100 km/h (62 mph) measured at 25.0 m (82.0 ft) distance from track centerline and 3.5 m (11.5 ft) above the ground. B-77

Figure B-96. Time history of the A-weighted SPL during a passby of the TR08 travelling on the hybrid guideway at about 150 km/h (93 mph) measured at 25.0 m (82.0 ft) distance from track centerline and 3.5 m (11.5 ft) above the ground. B-77

Figure B-97. Time history of the A-weighted SPL during a passby of the TR08 travelling on the hybrid guideway at about 200 km/h (124 mph) measured at 25.0 m (82.0 ft) distance from track centerline and 3.5 m (11.5 ft) above the ground. B-78

Figure B-98. Time history of the A-weighted SPL during a passby of the TR08 travelling on the hybrid guideway at about 300 km/h (186 mph) measured at 25.0 m (82.0 ft) distance from track centerline and 3.5 m (11.5 ft) above the ground. B-78

Figure B-99. Time history of the A-weighted SPL during a passby of the TR08 travelling on the hybrid guideway at about 400 km/h (249 mph) measured at 25.0 m (82.0 ft) distance from track centerline and 3.5 m (11.5 ft) above the ground. B-79

Figure B-100. Time history of the A-weighted SPL during a passby of the TR08 travelling on the hybrid guideway at about 100 km/h (62 mph) measured at 15.2 m (50.0 ft) distance from track centerline and 1.5 m (5.0 ft) above the ground. B-81

Figure B-101. Time history of the A-weighted SPL during a passby of the TR08 travelling on the hybrid guideway at about 150 km/h (93 mph) measured at

15.2 m (50.0 ft) distance from track centerline and 1.5 m (5.0 ft) above the ground. B-81

Figure B-102. Time history of the A-weighted SPL during a passby of the TR08 travelling on the hybrid guideway at about 200 km/h (124 mph) measured at 15.2 m (50.0 ft) distance from track centerline and 1.5 m (5.0 ft) above the ground. B-82

Figure B-103. Time history of the A-weighted SPL during a passby of the TR08 travelling on the hybrid guideway at about 300 km/h (186 mph) measured at 15.2 m (50.0 ft) distance from track centerline and 1.5 m (5.0 ft) above the ground. B-82

Figure B-104. Time history of the A-weighted SPL during a passby of the TR08 travelling on the hybrid guideway at about 400 km/h (249 mph) measured at 15.2 m (50.0 ft) distance from track centerline and 1.5 m (5.0 ft) above the ground. B-83

Figure B-105. Time history of the A-weighted SPL during a passby of the TR08 travelling on the hybrid guideway at about 100 km/h (62 mph) measured at 6.5 m (21.3 ft) distance from track centerline and the height of the upper surface of the guideway. B-85

Figure B-106. Time history of the A-weighted SPL during a passby of the TR08 travelling on the hybrid guideway at about 150 km/h (93 mph) measured at 6.5 m (21.3 ft) distance from track centerline and the height of the upper surface of the guideway. B-85

Figure B-107. Time history of the A-weighted SPL during a passby of the TR08 travelling on the hybrid guideway at about 200 km/h (124 mph) measured at 6.5 m (21.3 ft) distance from track centerline and the height of the upper surface of the guideway. B-86

Figure B-108. Time history of the A-weighted SPL during a passby of the TR08 travelling on the hybrid guideway at about 300 km/h (186 mph) measured at 6.5 m (21.3 ft) distance from track centerline and the height of the upper surface of the guideway. B-86

Figure B-109. Time history of the A-weighted SPL during a passby of the TR08 travelling on the hybrid guideway at about 400 km/h (249 mph) measured at 6.5 m (21.3 ft) distance from track centerline and the height of the upper surface of the guideway. B-87

Figure B-110. Time history of the A-weighted SPL during a passby of the TR08 travelling on the hybrid guideway at about 100 km/h (62 mph) measured at 6.5 m (21.3 ft) distance from track centerline and 1.5 m (5.0 ft) below the upper surface of the guideway. B-89

Figure B-111. Time history of the A-weighted SPL during a passby of the TR08 travelling on the hybrid guideway at about 150 km/h (93 mph) measured at 6.5 m (21.3 ft) distance from track centerline and 1.5 m (5.0 ft) below the upper surface of the guideway. B-89

Figure B-112. Time history of the A-weighted SPL during a passby of the TR08 travelling on the hybrid guideway at about 200 km/h (124 mph) measured at 6.5 m (21.3 ft) distance from track centerline and 1.5 m (5.0 ft) below the upper surface of the guideway. B-90

Figure B-113. Time history of the A-weighted SPL during a passby of the TR08 travelling on the hybrid guideway at about 300 km/h (186 mph) measured at 6.5 m (21.3 ft) distance from track centerline and 1.5 m (5.0 ft) below the upper surface of the guideway. B-90

Figure B-114. Time history of the A-weighted SPL during a passby of the TR08 travelling on the hybrid guideway at about 400 km/h (249 mph) measured at 6.5 m (21.3 ft) distance from track centerline and 1.5 m (5.0 ft) below the upper surface of the guideway. B-91

Figure B-115. Time history of the A-weighted SPL during a passby of the TR08 travelling on the hybrid guideway at about 100 km/h (62 mph) measured beneath guideway centerline at a height of 1.5 m (5.0 ft) above the ground. B-93

Figure B-116. Time history of the A-weighted SPL during a passby of the TR08 travelling on the hybrid guideway at about 150 km/h (93 mph) measured beneath guideway centerline at a height of 1.5 m (5.0 ft) above the ground. B-93

Figure B-117. Time history of the A-weighted SPL during a passby of the TR08 travelling on the hybrid guideway at about 200 km/h (124 mph) measured beneath guideway centerline at a height of 1.5 m (5.0 ft) above the ground. ... B-94

Figure B-118. Time history of the A-weighted SPL during a passby of the TR08 travelling on the hybrid guideway at about 300 km/h (186 mph) measured beneath guideway centerline at a height of 1.5 m (5.0 ft) above the ground. ... B-94

Figure B-119. Time history of the A-weighted SPL during a passby of the TR08 travelling on the hybrid guideway at about 400 km/h (249 mph) measured beneath guideway centerline at a height of 1.5 m (5.0 ft) above the ground. ... B-95

Figure B-120. Time history of the A-weighted SPL during a passby of the TR08 travelling on the North switch at about 100 km/h (62 mph) measured at 30.5 m (100.0 ft) distance from track centerline and 1.2 m (4.0 ft) above the ground. B-97

Figure B-121. Time history of the A-weighted SPL during a passby of the TR08 travelling on the North switch at about 150 km/h (93 mph) measured at 30.5 m (100.0 ft) distance from track centerline and 1.2 m (4.0 ft) above the ground. B-97

Figure B-122. Time history of the A-weighted SPL during a passby of the TR08 travelling on the North switch at about 200 km/h (124 mph) measured at 30.5 m (100.0 ft) distance from track centerline and 1.2 m (4.0 ft) above the ground. B-98

Figure B-123. Time history of the A-weighted SPL during a passby of the TR08 travelling on the North switch at about 300 km/h (186 mph) measured at 30.5 m (100.0 ft) distance from track centerline and 1.2 m (4.0 ft) above the ground. B-98

Figure B-124. Time history of the A-weighted SPL during a passby of the TR08 travelling on the North switch at about 385 km/h (239 mph) measured at 30.5 m (100.0 ft) distance from track centerline and 1.2 m (4.0 ft) above the ground. B-99

Figure B-125. Time history of the A-weighted SPL during a passby of the TR08 travelling on the North switch at about 100 km/h (62 mph) measured at 25.0 m (82.0 ft) distance from track centerline and 3.5 m (11.5 ft) above the ground. B-101

Figure B-126. Time history of the A-weighted SPL during a passby of the TR08 travelling on the North switch at about 150 km/h (93 mph) measured at 25.0 m (82.0 ft) distance from track centerline and 3.5 m (11.5 ft) above the ground. B-101

Figure B-127. Time history of the A-weighted SPL during a passby of the TR08 travelling on the North switch at about 200 km/h (124 mph) measured at 25.0 m (82.0 ft) distance from track centerline and 3.5 m (11.5 ft) above the ground. B-102

Figure B-128. Time history of the A-weighted SPL during a passby of the TR08 travelling on the North switch at about 300 km/h (186 mph) measured at 25.0 m (82.0 ft) distance from track centerline and 3.5 m (11.5 ft) above the ground. B-102

Figure B-129. Time history of the A-weighted SPL during a passby of the TR08 travelling on the North switch at about 385 km/h (239 mph) measured at 25.0 m (82.0 ft) distance from track centerline and 3.5 m (11.5 ft) above the ground. B-103

Figure B-130. Time history of the A-weighted SPL during a passby of the TR08 travelling on the North switch at about 100 km/h (62 mph) measured at 15.2 m (50.0 ft) distance from track centerline and 1.5 m (5.0 ft) above the ground. B-105

Figure B-131. Time history of the A-weighted SPL during a passby of the TR08 travelling on the North switch at about 150 km/h (93 mph) measured at 15.2 m (50.0 ft) distance from track centerline and 1.5 m (5.0 ft) above the ground. B-105

Figure B-132. Time history of the A-weighted SPL during a passby of the TR08 travelling on the North switch at about 200 km/h (124 mph) measured at 15.2 m (50.0 ft) distance from track centerline and 1.5 m (5.0 ft) above the ground. B-106

Figure B-133. Time history of the A-weighted SPL during a passby of the TR08 travelling on the North switch at about 300 km/h (186 mph) measured at 15.2 m (50.0 ft) distance from track centerline and 1.5 m (5.0 ft) above the ground. B-106

Figure B-134. Time history of the A-weighted SPL during a passby of the TR08 travelling on the North switch at about 385 km/h (239 mph) measured at 15.2 m (50.0 ft) distance from track centerline and 1.5 m (5.0 ft) above the ground. B-107

Figure B-135. Time history of the A-weighted SPL during a passby of the TR08 travelling on the North switch at about 100 km/h (62 mph) measured at 6.5 m (21.3 ft) distance from track centerline and the height of the upper surface of the guideway. B-109

Figure B-136. Time history of the A-weighted SPL during a passby of the TR08 travelling on the North switch at about 150 km/h (93 mph) measured at 6.5 m (21.3 ft) distance from track centerline and the height of the upper surface of the guideway. B-109

Figure B-137. Time history of the A-weighted SPL during a passby of the TR08 travelling on the North switch at about 200 km/h (124 mph) measured at 6.5 m (21.3 ft) distance from track centerline and the height of the upper surface of the guideway. B-110

Figure B-138. Time history of the A-weighted SPL during a passby of the TR08 travelling on the North switch at about 300 km/h (186 mph) measured at 6.5 m (21.3 ft) distance from track centerline and the height of the upper surface of the guideway. B-110

Figure B-139. Time history of the A-weighted SPL during a passby of the TR08 travelling on the North switch at about 385 km/h (239 mph) measured at 6.5 m (21.3 ft) distance from track centerline and the height of the upper surface of the guideway. B-111

Figure B-140. Time history of the A-weighted SPL during a passby of the TR08 travelling on the North switch at about 100 km/h (62 mph) measured at 6.5 m (21.3 ft) distance from track centerline and 1.5 m (5.0 ft) below the upper surface of the guideway. B-113

Figure B-141. Time history of the A-weighted SPL during a passby of the TR08 travelling on the North switch at about 150 km/h (93 mph) measured at 6.5 m (21.3 ft) distance from track centerline and 1.5 m (5.0 ft) below the upper surface of the guideway. B-113

Figure B-142. Time history of the A-weighted SPL during a passby of the TR08 travelling on the North switch at about 200 km/h (124 mph) measured at 6.5 m (21.3 ft) distance from track centerline and 1.5 m (5.0 ft) below the upper surface of the guideway. B-114

Figure B-143. Time history of the A-weighted SPL during a passby of the TR08 travelling on the North switch at about 300 km/h (186 mph) measured at 6.5 m (21.3 ft) distance from track centerline and 1.5 m (5.0 ft) below the upper surface of the guideway. B-114

Figure B-144. Time history of the A-weighted SPL during a passby of the TR08 travelling on the North switch at about 385 km/h (239 mph) measured at 6.5 m (21.3 ft) distance from track centerline and 1.5 m (5.0 ft) below the upper surface of the guideway. B-115

Figure B-145. Time history of the A-weighted SPL during a passby of the TR08 travelling on the North switch at about 100 km/h (62 mph) measured beneath guideway centerline at a height of 1.5 m (5.0 ft) above the ground. B-117

Figure B-146. Time history of the A-weighted SPL during a passby of the TR08 travelling on the North switch at about 150 km/h (93 mph) measured beneath guideway centerline at a height of 1.5 m (5.0 ft) above the ground. B-117

Figure B-147. Time history of the A-weighted SPL during a passby of the TR08 travelling on the North switch at about 200 km/h (124 mph) measured beneath guideway centerline at a height of 1.5 m (5.0 ft) above the ground. B-118

Figure B-148. Time history of the A-weighted SPL during a passby of the TR08 travelling on the North switch at about 300 km/h (186 mph) measured beneath guideway centerline at a height of 1.5 m (5.0 ft) above the ground. B-118

Figure B-149. Time history of the A-weighted SPL during a passby of the TR08 travelling on the North switch at about 385 km/h (239 mph) measured beneath guideway centerline at a height of 1.5 m (5.0 ft) above the ground. B-119

Figure B-150. Time history of the A-weighted SPL during a passby of the TR08 travelling on the at-grade steel guideway at about 100 km/h (62 mph) measured at 6.5 m (21.3 ft) distance from track centerline and the height of the upper surface of the guideway...B-121

Figure B-151. Time history of the A-weighted SPL during a passby of the TR08 travelling on the at-grade steel guideway at about 100 km/h (62 mph) measured at 6.5 m (21.3 ft) distance from track centerline and 1.5 m (5.0 ft) below the upper surface of the guideway...B-121

Figure B-152. Time history of the A-weighted SPL during a passby of the TR08 travelling on the at-grade steel guideway at about 300 km/h (186 mph) measured at 6.5 m (21.3 ft) distance from track centerline and the height of the upper surface of the guideway...B-122

Figure B-153. Time history of the A-weighted SPL during a passby of the TR08 travelling on the at-grade steel guideway at about 300 km/h (186 mph) measured at 6.5 m (21.3 ft) distance from track centerline and 1.5 m (5.0 ft) below the upper surface of the guideway...B-122

Figure B-154. Time history of the A-weighted SPL during a passby of the TR08 travelling on the at-grade steel guideway at about 370 km/h (230 mph) measured at 6.5 m (21.3 ft) distance from track centerline and the height of the upper surface of the guideway...B-123

Figure B-155. Time history of the A-weighted SPL during a passby of the TR08 travelling on the at-grade steel guideway at about 370 km/h (230 mph) measured at 6.5 m (21.3 ft) distance from track centerline and 1.5 m (5.0 ft) below the upper surface of the guideway...B-123

Figure B-156. Time history of the A-weighted SPL during a passby of the TR08 travelling on the at-grade concrete guideway at about 100 km/h (62 mph) measured at 6.5 m (21.3 ft) distance from track centerline and the height of the upper surface of the guideway...B-126

Figure B-157. Time history of the A-weighted SPL during a passby of the TR08 travelling on the at-grade concrete guideway at about 100 km/h (62 mph) measured at 6.5 m (21.3 ft) distance from track centerline and 1.5 m (5.0 ft) below the upper surface of the guideway. B-126

Figure B-158. Time history of the A-weighted SPL during a passby of the TR08 travelling on the at-grade concrete guideway at about 300 km/h (186 mph) measured at 6.5 m (21.3 ft) distance from track centerline and the height of the upper surface of the guideway. B-127

Figure B-159. Time history of the A-weighted SPL during a passby of the TR08 travelling on the at-grade concrete guideway at about 300 km/h (186 mph) measured at 6.5 m (21.3 ft) distance from track centerline and 1.5 m (5.0 ft) below the upper surface of the guideway. B-127

Figure B-160. Time history of the A-weighted SPL during a passby of the TR08 travelling on the at-grade concrete guideway at about 370 km/h (230 mph) measured at 6.5 m (21.3 ft) distance from track centerline and the height of the upper surface of the guideway. B-128

Figure B-161. Time history of the A-weighted SPL during a passby of the TR08 travelling on the at-grade concrete guideway at about 370 km/h (230 mph) measured at 6.5 m (21.3 ft) distance from track centerline and 1.5 m (5.0 ft) below the upper surface of the guideway. B-128

Figure C-1 Averaged sound-source distributions measured with the WV array during passbys of the TR08 on the reference concrete guideway at speeds between 150 and 400 km/h (93 and 249 mph). C-2

Figure C-2 Averaged sound-source distributions measured with the WV array during passbys of the TR08 on the prototype steel guideway at speeds between 150 and 400 km/h (93 and 249 mph). C-3

Figure C-3 Averaged sound-source distributions measured with the WV array during passbys of the TR08 on the prototype concrete guideway at speeds between 150 and 400 km/h (93 and 249 mph). C-4

Figure C-4 Averaged sound-source distributions measured with the WV array during passbys of the TR08 on the hybrid guideway at speeds between 150 and 400 km/h (93 and 249 mph). C-5

LIST OF TABLES

Table ES-1. Comparison of TR08 Sound Exposure Levels with those of other High-Speed Ground Transportation Systems ES-3

Table ES-2. Japanese Shinkansen Noise Limits ES-4

Table ES-3. German Magnetic Levitation Noise Standards ES-4

Table 1-1: Specifications* of the Transrapid TR08 Maglev System 1-5

Table 2-1. Heights of the single microphones above ground or guideway 2-3

Table 2-2. Number, date, and time of vehicle passby, array type, vehicle speed, temperature, and wind speed. 2-20

Table 2-3. Parameters of localized sound sources 2-49

Table 2-4. Measured and predicted sound-exposure levels during passbys of the TR08 on the reference concrete guideway. 2-66

Table 2-5. Reference sound exposure levels of individual sound sources of the TR08 system 2-67

Table 4-1. Comparison of TR08 sound exposure levels with those of other high-speed ground transportation systems. 4-1

Table 4-2. Source Reference SELs of the TR08 system at 15 m (50 ft) for application in FRA Detailed Noise Analysis 4-2

Table 4-3. Japanese Shinkansen Noise Limits 4-3

Table 4-4. German Magnetic Levitation Noise Standards 4-3

Table B-1. Results of the microphone positioned close to the reference concrete guideway at 30.5 m (100.0 ft) distance from track centerline and 1.2 m (4.0 ft) above the ground (measuring series A/B). B-5

Table B-2. Results of the microphone positioned close to the reference concrete guideway at 25.0 m (82.0 ft) distance from track centerline and 3.5 m (11.5 ft) above the ground (measuring series A/B). B-9

Table B-3. Results of the microphone positioned close to the reference concrete guideway at 15.2 m (50.0 ft) distance from track centerline and 1.5 m (5.0 ft) above the ground (measuring series A/B). B-13

Table B-4. Results of the microphone positioned close to the reference concrete guideway at 6.5 m (21.3 ft) distance from track centerline and the height of the upper surface of the guideway (measuring series A) B-17

Table B-5. Results of the microphone positioned close to the reference concrete guideway at 6.5 m (21.3 ft) distance from track centerline and 1.5 m (5.0 ft) below the upper surface of the guideway (measuring series B) B-20

Table B-6. Results of the microphone positioned beneath the centerline of the reference concrete guideway at a height of 1.5 m (5.0 ft) above the ground (measuring series A/B). B-24

Table B-7. Results of the microphone positioned close to the prototype steel guideway at 30.5 m (100.0 ft) distance from track centerline and 1.2 m (4.0 ft) above the ground (measuring series H). B-28

Table B-8. Results of the microphone positioned close to the prototype steel guideway at 25.0 m (82.0 ft) distance from track centerline and 3.5 m (11.5 ft) above the ground (measuring series H). B-32

Table B-9. Results of the microphone positioned close to the prototype steel guideway at 15.2 m (50.0 ft) distance from track centerline and 1.5 m (5.0 ft) above the ground (measuring series H). B-36

Table B-10. Results of the microphone positioned close to the prototype steel guideway at 6.5 m (21.3 ft) distance from track centerline and the height of the upper surface of the guideway (measuring series G/H). B-40

Table B-11. Results of the microphone positioned close to the prototype steel guideway at 6.5 m (21.3 ft) distance from track centerline and 1.5 m (5.0 ft) below the upper surface of the guideway (measuring series H). B-44

Table B-12. Results of the microphone positioned beneath the centerline of the prototype steel guideway at a height of 1.5 m (5.0 ft) above the ground (measuring series H). B-48

Table B-13. Results of the microphone positioned close to the prototype concrete guideway at 30.5 m (100.0 ft) distance from track centerline and 1.2 m (4.0 ft) above the ground (measuring series L) B-52

Table B-14. Results of the microphone positioned close to the prototype concrete guideway at 25.0 m (82.0 ft) distance from track centerline and 3.5 m (11.5 ft) above the ground (measuring series L). B-56

Table B-15. Results of the microphone positioned close to the prototype concrete guideway at 15.2 m (50.0 ft) distance from track centerline and 1.5 m (5.0 ft) above the ground (measuring series L). B-60

Table B-16. Results of the microphone positioned close to the prototype concrete guideway at 6.5 m (21.3 ft) distance from track centerline and the height of the upper surface of the guideway (measuring series K/L). B-64

Table B-17. Results of the microphone positioned close to the prototype concrete guideway at 6.5 m (21.3 ft) distance from track centerline and 1.5 m (5.0 ft) below the upper surface of the guideway (measuring series L). B-68

Table B-18. Results of the microphone positioned beneath the centerline of the prototype concrete guideway at a height of 1.5 m (5.0 ft) above the ground (measuring series L)... B-72

Table B-19. Results of the microphone positioned close to the hybrid guideway at 30.5 m (100.0 ft) distance from track centerline and 1.2 m (4.0 ft) above the ground (measuring series J).. B-76

Table B-20. Results of the microphone positioned close to the hybrid guideway at 25.0 m (82.0 ft) distance from track centerline and 3.5 m (11.5 ft) above the ground (measuring series J). B-80

Table B-21. Results of the microphone positioned close to the hybrid guideway at 15.2 m (50.0 ft) distance from track centerline and 1.5 m (5.0 ft) above the ground (measuring series J). B-84

Table B-22. Results of the microphone positioned close to the hybrid guideway at 6.5 m (21.3 ft) distance from track centerline and the height of the upper surface of the guideway (measuring series I/J). B-88

Table B-23. Results of the microphone positioned close to the hybrid guideway at 6.5 m (21.3 ft) distance from track centerline and 1.5 m (5.0 ft) below the upper surface of the guideway (measuring series J). B-92

Table B-24. Results of the microphone positioned beneath the centerline of the hybrid guideway at a height of 1.5 m (5.0 ft) above the ground (measuring series J). B-96

Table B-25. Results of the microphone positioned close to the North switch at 30.5 m (100.0 ft) distance from track centerline and 1.2 m (4.0 ft) above the ground (measuring series Z)...B-100

Table B-26. Results of the microphone positioned close to the North switch at 25.0 m (82.0 ft) distance from track centerline and 3.5 m (11.5 ft) above the ground (measuring series Z)...B-104

Table B-27. Results of the microphone positioned close to the North switch at 15.2 m (50.0 ft) distance from track centerline and 1.5 m (5.0 ft) above the ground (measuring series Z)...B-108

Table B-28. Results of the microphone positioned close to the North switch at 6.5 m (21.3 ft) distance from track centerline and the height of the upper surface of the guideway (measuring series Z)..B-112

Table B-29. Results of the microphone positioned close to the North switch at 6.5 m (21.3 ft) distance from track centerline and 1.5 m (5.0 ft) below the upper surface of the guideway (measuring series Z). B-116

Table B-30. Results of the microphone positioned beneath the centerline of the North switch at a height of 1.5 m (5.0 ft) above the ground (measuring series Z). B-120

Table B-31. Results of the microphone positioned close to the at-grade steel guideway at 6.5 m (21.3 ft) distance from track centerline and the height of the upper surface of the guideway (measuring series Y)....................................B-124

Table B-32. Results of the microphone positioned close to the at-grade steel guideway at 6.5 m (21.3 ft) distance from track centerline and 1.5 m (5.0 ft) below the upper surface of the guideway (measuring series Y)...............................B-125

Table B-33. Results of the microphone positioned close to the at-grade concrete guideway at 6.5 m (21.3 ft) distance from track centerline and the height of the upper surface of the guideway (measuring series Y). B-129

Table B-34. Results of the microphone positioned close to the at-grade concrete guideway at 6.5 m (21.3 ft) distance from track centerline and 1.5 m (5.0 ft) below the upper surface of the guideway (measuring series Y). B-130

Table D-1. Sound-pressure levels of the unweighted one-third octave-band spectra measured with the WV array during passbys of the TR08 on the reference concrete guideway at speeds between 150 and 400 km/h (93 and 249 mph) at a height of 2.1 m (6.9 ft) with reference to the upper surface of the guideway. D-1

Table D-2. Sound-pressure levels of the unweighted one-third octave-band spectra measured with the WV array during passbys of the TR08 on the reference concrete guideway at speeds between 150 and 400 km/h (93 and 249 mph) at a height of 0.7 m (2.3 ft) with reference to the upper surface of the guideway. D-2

Table D-3. Sound-pressure levels of the unweighted one-third octave-band spectra measured with the WV array during passbys of the TR08 on the reference concrete guideway at speeds between 150 and 400 km/h (93 and 249 mph) at a height of 0 m (0 ft) with reference to the upper surface of the guideway. D-3

Table D-4. Sound-pressure levels of the unweighted one-third octave-band spectra measured with the WV array during passbys of the TR08 on the reference concrete guideway at speeds between 150 and 400 km/h (93 and 249 mph) at a height of -0.7 m (-2.3 ft) with reference to the upper surface of the guideway. D-4

Table D-5. Sound-pressure levels of the unweighted one-third octave-band spectra measured with the WV array during passbys of the TR08 on the reference concrete guideway at speeds between 150 and 400 km/h (93 and 249 mph) at a height of -1.4 m (-4.6 ft) with reference to the upper surface of the guideway. D-5

Table D-6. Sound-pressure levels of the unweighted one-third octave-band spectra measured with the WV array during passbys of the TR08 on the prototype steel guideway at speeds between 150 and 400 km/h (93 and 249 mph) at a height of 2.1 m (6.9 ft) with reference to the upper surface of the guideway. D-6

Table D-7. Sound-pressure levels of the unweighted one-third octave-band spectra measured with the WV array during passbys of the TR08 on the prototype steel guideway at speeds between 150 and 400 km/h (93 and 249 mph) at a height of 0.7 m (2.3 ft) with reference to the upper surface of the guideway. D-7

Table D-8. Sound-pressure levels of the unweighted one-third octave-band spectra measured with the WV array during passbys of the TR08 on the prototype steel guideway at speeds between 150 and 400 km/h (93 and 249 mph) at a height of 0 m (0 ft) with reference to the upper surface of the guideway. D-8

Table D-9. Sound-pressure levels of the unweighted one-third octave-band spectra measured with the VV array during passbys of the TR08 on the prototype steel guideway at speeds between 150 and 400 km/h (93 and 249 mph) at a height of -0.7 m (-2.3 ft) with reference to the upper surface of the guideway. D-9

Table D-10. Sound-pressure levels of the unweighted one-third octave-band spectra measured with the VV array during passbys of the TR08 on the prototype steel guideway at speeds between 150 and 400 km/h (93 and 249 mph) at a height of -1.4 m (-4.6 ft) with reference to the upper surface of the guideway. D-10

Table D-11. Sound-pressure levels of the unweighted one-third octave-band spectra measured with the VV array during passbys of the TR08 on the prototype concrete guideway at speeds between 150 and 400 km/h (93 and 249 mph) at a height of 2.1 m (6.9 ft) with reference to the upper surface of the guideway. D-11

Table D-12. Sound-pressure levels of the unweighted one-third octave-band spectra measured with the VV array during passbys of the TR08 on the prototype concrete guideway at speeds between 150 and 400 km/h (93 and 249 mph) at a height of 0.7 m (2.3 ft) with reference to the upper surface of the guideway. D-12

Table D-13. Sound-pressure levels of the unweighted one-third octave-band spectra measured with the VV array during passbys of the TR08 on the prototype concrete guideway at speeds between 150 and 400 km/h (93 and 249 mph) at a height of 0 m (0 ft) with reference to the upper surface of the guideway. D-13

Table D-14. Sound-pressure levels of the unweighted one-third octave-band spectra measured with the VV array during passbys of the TR08 on the prototype concrete guideway at speeds between 150 and 400 km/h (93 and 249 mph) at a height of -0.7 m (-2.3 ft) with reference to the upper surface of the guideway. D-14

Table D-15. Sound-pressure levels of the unweighted one-third octave-band spectra measured with the VV array during passbys of the TR08 on the prototype concrete guideway at speeds between 150 and 400 km/h (93 and 249 mph) at a height of -1.4 m (-4.6 ft) with reference to the upper surface of the guideway. D-15

Table D-16. Sound-pressure levels of the unweighted one-third octave-band spectra measured with the VV array during passbys of the TR08 on the hybrid guideway at speeds between 150 and 400 km/h (93 and 249 mph) at a height of 2.1 m (6.9 ft) with reference to the upper surface of the guideway. D-16

Table D-17. Sound-pressure levels of the unweighted one-third octave-band spectra measured with the WV array during passbys of the TR08 on the hybrid guideway at speeds between 150 and 400 km/h (93 and 249 mph) at a height of 0.7 m (2.3 ft) with reference to the upper surface of the guideway...................D-17

Table D-18. Sound-pressure levels of the unweighted one-third octave-band spectra measured with the WV array during passbys of the TR08 on the hybrid guideway at speeds between 150 and 400 km/h (93 and 249 mph) at a height of 0 m (0 ft) with reference to the upper surface of the guideway.........................D-18

Table D-19. Sound-pressure levels of the unweighted one-third octave-band spectra measured with the WV array during passbys of the TR08 on the hybrid guideway at speeds between 150 and 400 km/h (93 and 249 mph) at a height of -0.7 m (-2.3 ft) with reference to the upper surface of the guideway. D-19

Table D-20. Sound-pressure levels of the unweighted one-third octave-band spectra measured with the WV array during passbys of the TR08 on the hybrid guideway at speeds between 150 and 400 km/h (93 and 249 mph) at a height of -1.4 m (-4.6 ft) with reference to the upper surface of the guideway. D-20

PREFACE

As part of the Federal Railroad Administration's (FRA) Magnetic Levitation Transportation Technology Deployment Program, this technical report characterizes the noise associated with the operation of the Transrapid International TR08 Maglev System, a transportation system employing magnetic levitation (maglev). This report presents measurements of noise associated with the TR08 Maglev System; these measurements were taken at the Transrapid Test Facility (TVE), in the Emsland region of Germany, in August, 2001, and May, 2002. The data presented and analyzed herein can be utilized to support the required environmental planning and deployment activities for any Transrapid Maglev project in the U.S.

This report is the product of a combined international effort. The authors wish to thank the many individuals and organizations whose contributions were instrumental in the coordination, test plan development, field measurements, analysis, documentation, and supporting functions for this report. The participation, cooperation, and forbearance of the operators and staff of TVE during the measurement campaign were especially appreciated. The valuable assistance of Mr. Robert Budell, Mr. Frank Litzmann and Mr. Christian Rausch of Transrapid International, and Mr. Laurence Blow and Mr. Reed Tanger of Transrapid USA, are also acknowledged, as their contributions were critical to the success of this effort. The staff of IABG (Industrieanlagen Betriebsgesellschaft), including Mr. Gerold Snieders and Mr. Hans-Gert Runde, provided important information to the measurement team; special thanks is extended to Dr. Klaus-Peter Schmitz (IABG), who, despite his many other responsibilities at TVE, gave his attention to our needs during the measurement program. In the coordination of the measurement campaign and in the production of this report, we also acknowledge the support of staff at akustik-data Engineering, Harris Miller Miller & Hanson, MAGLEV, Inc., and the John A. Volpe National Transportation Systems Center. Particularly, the assistance of Mr. Edd Manges from MAGLEV, Inc., and Mr. Albrecht Ebner of akustik-data Engineering was useful.

EXECUTIVE SUMMARY

INTRODUCTION

In the Transportation Equity Act for the 21st Century (TEA-21), Congress authorized the Magnetic Levitation Transportation Technology Deployment Program (Maglev Deployment Program) to demonstrate the benefits of an operating transportation system employing magnetic levitation (maglev). Maglev is an advanced transportation technology in which magnetic forces lift, propel, and guide a vehicle over a specially designed guideway at variable speeds, including those in excess of 386 kilometers per hour (km/h) (240 miles per hour (mph)). The U.S. Department of Transportation (USDOT) assigned the Maglev Deployment Program to the Federal Railroad Administration (FRA) for implementation.

To satisfy the requirements of the National Environmental Policy Act, and to fulfill the directives of the Programmatic Environmental Impact Statement and associated Record of Decision published in 2001 (FRA), the potential environmental and related impacts associated with the operation of the proposed technology must be analyzed. The maglev technology that is being considered for deployment in the U.S. is the Transrapid International (TRI) TR08 Maglev System. This technical report characterizes the noise associated with the operation of the TRI TR08 Maglev System based on measurements made at the Transrapid Test Facility (TVE) in Germany in August, 2001. The data presented and analyzed herein can be utilized to support the required environmental planning and deployment activities for any TRI Maglev project in the U.S. that uses the TR08 technology.

The FRA is the lead agency for the Maglev Deployment Program. The project sponsor of the Pennsylvania Project (the Pennsylvania Maglev Alternative), the Port Authority of Allegheny County, and that organization's private partner MAGLEV, Inc., coordinated, managed, and had technical oversight of the measurement and reporting of the noise characteristics of the TR08. The actual noise data were collected, analyzed, and reported by akustik-data Engineering under subcontract to Harris Miller Miller & Hanson, Inc. (HMMH). Technical staff from the John A. Volpe National Transportation Systems Center (Volpe Center) provided active oversight both in the development of the test plans and during the measurements made by HMMH and akustik-data. The Volpe Center also independently collected noise data which were utilized during a detailed validation and verification of the final data provided by akustik-data and presented herein.

MEASUREMENTS

The environmental measurements described in this report took place at the TVE in the Emsland region of Germany, in August, 2001, and May, 2002. The TVE consists of a 31.5 km (19.5 mile) guideway, with 2 loops at either end of a straight section. Various guideway types (steel, concrete, hybrid, switch) and configurations (at-grade, elevated) exist at the test track.

Noise measurements were undertaken to obtain wayside noise data during the operation of the TR08 vehicle on the different guideway types at the test track. Acoustic data were obtained for several vehicle speeds and guideway configurations. Data include both single-microphone measurements at various distances from the guideway and at multiple heights relative to the local ground surface, as well as microphone array measurements. In addition to being useful for overall characterization of noise associated with the TR08, the single-microphone data are also used for comparison with similar data for other technologies. The array data are useful for sound-source localization and more detailed environmental planning. Results are intended to provide reference levels for use during the environmental impact assessments of the Maglev Deployment Program.

Data are generally presented in A-weighted decibels (dBA) which best represent the response of the human ear.

RESULTS

Figure ES-1 presents a summary of sound exposure level (SEL) data for the TR08 as a function of vehicle speed. The data shown in Figure ES-1 represent measurements for 5 elevated guideway types measured at a distance of 30.5 m (100.0 ft) from track centerline and at a height of 1.2 m (4.0 ft) above the local ground surface. Figure ES-1 shows sound level data collected at five speeds and includes trend lines spanning those speeds.

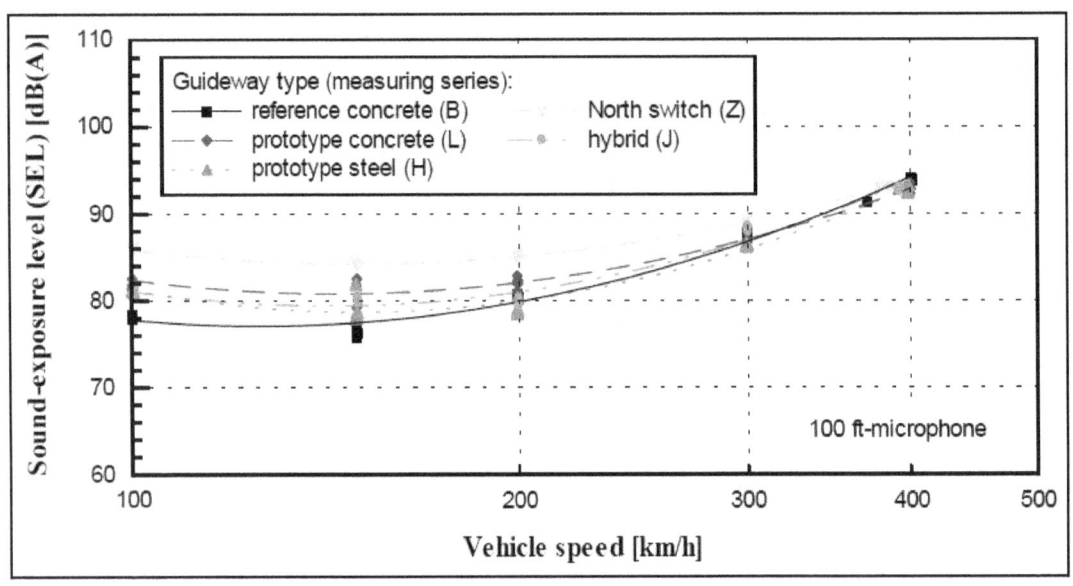

Figure ES-1. Sound-Exposure Level as a Function of Vehicle Speed

COMPARISON OF TR08 SOUND LEVELS WITH OTHER RAIL SOURCES

The results of the noise measurements of the TR08 Maglev System may be compared with similar data, documented by the FRA (1998), for other high-speed ground transportation systems. Table ES-1 presents such a comparison in terms of SEL at a reference distance of 30.5 m (100 ft) for representative speeds. In the table, all trains have been normalized to the same length for direct comparison; as a result of the data normalization, SEL values for the TR08 in Table ES-1 are greater than those presented elsewhere in this report.

Table ES-1. Comparison of TR08 Sound Exposure Levels with those of other High-Speed Ground Transportation Systems

Speed [km/h (mph)]	SEL (dBA) at 30.5 m (100 ft)*						
	Maglev Technology					Wheel-on-Rail Technology	
	TR08				TR07		
	reference concrete guideway	prototype concrete guideway	prototype steel guideway	hybrid beam	reference concrete guideway	Acela	TGV
100 (62)	83	86	85	85	(NA)	(NA)	(NA)
150 (93)	81	82	84	85	80	87	88
200 (124)	86	87	88	85	83	92	92
240 (150)	(NA)	(NA)	(NA)	(NA)	85	94	93
300 (186)	93	94	95	92	90	(NA)	97
400 (249)	99	99	100	98	93	(NA)	(NA)

*Trains normalized to 225 m (740 ft) in length.

REGULATORY CONTEXT

The purpose of this document is to measure and report the noise characteristics of the Transrapid TR08 Maglev System. The commercial transportation application of magnetic levitation system technology is new to the United States. As such, no maglev-specific noise regulations currently exist. Accordingly, either existing standards originally developed for conventional rail technologies will be applied to maglev applications within the U.S. or new regulations and/or exceptions for the unique technology will be developed. A brief review of high-speed rail standards in Japan and Germany helps to provide context for the U.S.

The Japanese regulations were developed for the Shinkansen, a traditional high-speed rail system. Conversely, the German regulations were developed specifically for maglev technology. In both cases, sound level limits are specified for specific land use categories. By defining the noise level limits in terms of land use, the regulations specify maximum sound levels received at *receptors*, as opposed to specific sound levels emitted by the vehicle noise source. In neither case are measurement distances specified for these levels.

Table ES-2 presents the Japanese noise limits specific to the Shinkansen high-speed rail system. While the Shinkansen was originally opened in 1964, recognition of its unique capabilities and potential prompted the specification that noise limits would be instituted in several steps through the year 2001 (Ono 2000).

Table ES-2. Japanese Shinkansen Noise Limits

Land Use Category	Maximum Sound Level, L_{ASmx} (dB)
I: Residential	70
II: Non-Residential	75

Source: Japanese Ministry of the Environment 2002

Table ES-3 presents the German noise limits specific to maglev technology. Whereas the Japanese noise limits are based on maximum sound levels (L_{AFmx}), the German limits are based on the hourly equivalent sound level ($L_{Aeq,1h}$).

Table ES-3. German Magnetic Levitation Noise Standards

Land Use Category	Hourly Equivalent Sound Level, $L_{Aeq,1h}$ (dBA)	
	Day	Night
I: Hospitals, Schools, Spas, Retirement Homes	57	47
II: Residential	59	49
III: Core, Village, Mixed-Use	64	54
IV: Industrial	69	59

Source: Bundesrat 2001

While no maglev-specific noise regulations have been developed for the U.S., guidance materials for high-speed rail in general have been developed. The FRA presented guidance for the determination of potential noise impacts in the *High-Speed Ground Transportation Noise and Vibration Impact Assessment* document (FRA 1998). Based in part on Federal Highway Administration (FHWA) methodologies, and similar to the Japanese and German regulations, this document also recommends the application of impact criteria based on land use categories.

The introduction of a new transportation technology, specifically the supersonic commercial jet, may also serve as a precedent with respect to the application of current noise regulations to maglev technology. When the Concorde SST was first introduced in the U.S., it did not meet existing aircraft noise regulations.

Recognizing the unique nature of the technology, operation of the Concorde was allowed given that its noise emissions were "reduced to the lowest levels that are economically reasonable, technologically practicable, and appropriate for the Concorde type design" (FAA 1978).

Throughout the TR08 noise characterization process, every effort has been made by the Volpe Center to ensure that the project adheres to appropriate guidance materials and regulatory documentation. In particular, the Environmental Protection Agency (EPA) regulations at Title 40, Part 201 of the Code of Federal Regulations (CFR), provide a regulatory framework for noise emissions from the operation of rail equipment. In part, these regulations preclude the operation of rail cars moving at speeds greater than 72 km/h (45 mph) which produce sounds levels in excess of 93 dBA measured 100 feet from the centerline of the track (see 40 CFR 201.13). Contractor data, validated and verified by the Volpe Center, illustrate that TR08 sound levels, measured at a distance of 100 feet are below 93 dBA at speeds of approximately 300 km/h (186 mph), whereas these sound levels measured for speeds of approximately 380 km/h (236 mph) exceed 93 dBA. Even if it is determined that the EPA standards do not apply to maglev technology, it is likely that maglev noise emissions will be addressed either in a regulatory process or as part of the environmental compliance process.

CHAPTER 1 INTRODUCTION

1.1 PURPOSE

As part of the Federal Railroad Administration's (FRA) Magnetic Levitation Transportation Technology Deployment Program, this technical report characterizes the noise associated with the operation of the Transrapid International TR08 Maglev System, a transportation system employing magnetic levitation (maglev). The TR08 Maglev System is the technology that is being considered for deployment in the U.S. (FRA 2001) by maglev programs in Pennsylvania and Maryland. Under the National Environmental Policy Act, potential environmental impacts, including noise impacts, of a maglev transportation system must be assessed.

The environmental measurements described in this report took place at the Transrapid Test Facility in the Emsland region of Germany, in August, 2001, and May, 2002. The data presented and analyzed herein can be utilized to support the required environmental planning and deployment activities for any Transrapid Maglev project in the U.S. This report also includes a limited comparison of the data to similar data from other modes of transportation as well as to existing national standards originally developed for standard rail transportation and other international standards.

1.2 BACKGROUND

In the Transportation Equity Act for the 21st Century (TEA-21), Congress authorized the Magnetic Levitation Transportation Technology Deployment Program (Maglev Deployment Program) to demonstrate the transportation, economic, environmental, energy, and other benefits of an operating transportation system employing magnetic levitation (maglev). Maglev is an advanced transportation technology in which magnetic forces lift, propel, and guide a vehicle over a specially designed guideway at variable speeds, including those in excess of 386 kilometers per hour (km/h) (240 miles per hour (mph)). The U.S. Department of Transportation (USDOT) assigned the Maglev Deployment Program to the Federal Railroad Administration (FRA) for implementation.

In order to comply with the TEA-21, FRA conducted a competitive award and selection process to demonstrate maglev in a U.S. transportation application. In May 1999, the Secretary of Transportation selected, from a pool of eleven applicants, seven states (or state-designated authorities) to receive grants for pre-construction planning of their maglev programs. To satisfy the requirements of the National Environmental Policy Act (NEPA), FRA, as the lead agency, determined that the Maglev Deployment Program constituted a major Federal action with the potential to have a significant effect on the environment, and accordingly published a Programmatic Environmental Impact Statement (PEIS)

(FRA 2001). The purpose of the PEIS was to describe the Maglev Deployment Program and the potential environmental impacts associated with its possible implementation, as well as to encourage public involvement and to address agency and public concerns (FRA 2001). In developing the PEIS, FRA required each of the seven state participants to prepare environmental assessments, which became the source of baseline data in the Draft PEIS (DPEIS), approved in June 2000. After collecting and incorporating appropriate public and agency comments on the DPEIS, the John A. Volpe National Transportations Systems Center (Volpe Center), on behalf of the FRA, refined the document and prepared the Final PEIS (PEIS), approved in March 2001 (see text at http://www.fra.dot.gov/rdv/maglev/mag_peis.htm). A Record Of Decision (ROD) was executed with the PEIS (FRA 2001).

The PEIS (FRA 2001) analyzed the potential environmental and related impacts associated with the Maglev Deployment Program using the maglev technology available at the time the environmental impacts were being analyzed, and at a level of detail commensurate with the program-level decisions being made at the time of PEIS publication (April 2001). A selection process administered by FRA led to the authorization of further funding for two project-specific environmental impact statements (EIS) for the Maryland and Pennsylvania Maglev Alternatives. Operational parameters of the Transrapid International TR07 Maglev System were used for the PEIS analysis of these two Alternatives, since no direct technical information was available on the proposed vehicle, the TR08 Maglev System. The information about TR07 was based on electromagnetic field measurements made in 1990 and published in 1992 (FRA 1992), and on noise measurements presented in a 1993 publication (FRA 1993). However, the ROD executed with the PEIS (FRA 2001) specified that in order to fully address environmental impacts of the maglev technology, the (1) electromagnetic fields (EMF) and electromagnetic radiation (EMR), (2) noise impacts, and (3) vibration impacts associated with the proposed vehicle, the Transrapid International TR08 Maglev System (see Section 1.4), needed to be identified.

To fulfill the directives of the ROD and its requirements in its grant agreement with FRA, the Port Authority of Allegheny County (the project sponsor of the Pennsylvania Maglev Alternative) contracted with MAGLEV, Inc., to measure and report the noise characteristics of the TR08 Maglev System and co-author this report. The research on noise impacts associated with the TR08 vehicle reported in this technical characterization report will be included by reference in the project-specific EISs, which tier from the PEIS (FRA 2001).

1.3 ROLES AND RESPONSIBILITIES

The Maglev Deployment Program constitutes a major Federal action with the potential to have a significant effect on the environment, and therefore requires compliance with NEPA. The FRA is the lead agency for the Maglev Deployment Program and is responsible for the NEPA compliance process.

The sponsor of the Pennsylvania Project (the Pennsylvania Maglev Alternative) is the Port Authority of Allegheny County. The Port Authority of Allegheny County's private partner, MAGLEV, Inc., coordinated, managed, and had technical oversight of the measurement and reporting of the noise characteristics of the TR08 Maglev System. The actual noise data were collected, analyzed, and reported by akustik-data Engineering Office (akustik-data), a German company under subcontract to Harris Miller Miller & Hanson, Inc. (HMMH).

Technical staff from the Volpe Center provided active oversight and validation both in the development of the test plans and during the measurements made by the contractors. In addition to collecting independent acoustic data, Volpe Center activities included examining techniques and test protocol, reviewing all data and analyses, and documenting these findings.

Assistance to the measurement team, both prior to testing (during development, review, and approval of test plans) and during the measurements at the test facility in the Emsland region of Germany, was also provided by technical staff of the FRA, the Volpe Center, MAGLEV, Inc., Transrapid International, and IABG (Industrieanlagen Betriebsgesellschaft, a European scientific-technical services company). The Test Plans (Appendix A) were reviewed and the measurements were observed by representatives of the Baltimore-Washington Maglev Project (the Maryland Maglev Alternative), including an environmental planning staff person from the Maryland Transit Administration and a representative from Parsons-Engineering Science, a noise and vibration consultant for the Baltimore-Washington Maglev Project.

The Volpe Center has, on behalf of the FRA and the Port Authority of Allegheny County, assembled and co-authored this TR08 noise characterization report along with MAGLEV, Inc., akustik-data, and HMMH.

1.4 TR08 MAGLEV SYSTEM TECHNOLOGY INFORMATION

Maglev is a transportation technology that uses non-contact electromagnetic systems to lift, guide, and propel the vehicle over a specially-designed guideway. Without wheels or other mechanical parts to cause resistance, cruising speeds of 320 to 480 km/h (200-300 mph) can be reached (TRI 2001).

Maglev technology has been researched since the 1960s, and development programs have been conducted by several countries, most notably Japan and Germany. Both these countries have test tracks on which they have conducted extensive testing and refinement of their maglev concepts and of different prototype vehicles. The German technology, the Transrapid International (TRI) Maglev System, has a design based on a long stator linear synchronous motor with conventional electromagnets in an attractive magnetic force configuration. The Japanese technology, the Railway Technical Research Institute's (RTRI) MLU-series system, has a design based on superconducting magnets in an electrodynamic repulsive system (RTRI 2001). The German TRI TR08 Maglev System is the technology that has been selected for deployment in both the Pennsylvania and Maryland maglev projects (FRA 2001).

TRI has been investigating high-speed rail systems utilizing electromagnetic levitation systems since 1969, and commissioned the TR02 in 1971. The eighth generation vehicle, the TR08 (Figure 1-1), and some of its precursor prototype vehicles, the TR07 and TR06, have been demonstrated and tested at the Transrapid Test Facility (Transrapid Versuchsanlage Emsland (TVE)), in the Emsland region of Germany, for more than 15 years (TRI 2001). A significant number of tests and simulated revenue-service operations were conducted with the TR07 from 1989-1999. The TR08 was delivered to the TVE in August 1999. Although significant differences in measured information/data between the TR07 and TR08 are not expected, some design upgrades and changes (e.g., power rail DC segments near stations for battery charging) necessitated new measurements so that the most recent technical and operational environmental performance data may be used for site-specific EIS work for the planned U.S. projects.

Source: Transrapid International (TRI)

Figure 1-1. The TRI TR08 Maglev System

According to the manufacturer, the TR08 is more aerodynamic and more economical than its predecessor, the TR07 (TRI 2001). A hybrid design, using aluminum hollow profiles and aluminum-clad foam sandwich panels, provides a light and stiff structure for the carriage body. A TR08 vehicle consist is made up of two end sections and zero to eight middle sections (Table 1-1). The consist would not be separated in normal operations. The TR08 used at the TVE is a 3-section, pre-production consist that is 79.70 m (261.5 ft) long, weighs 188.50 metric tons (t) (415,571 lbs), has seating for 190+ passengers (Figure 1-2), and is designed for peak 550 km/h (342 mph) operation. The TR08 at TVE is operated as a shuttle on a single guideway with one station. (TRI 2001)

Table 1-1: Specifications* of the Transrapid TR08 Maglev System

System Features	End Section	Middle Section
Vehicle Size:	2	0-8
Section Length	26.99 m (88.6 ft)	24.77 m (81.27 ft)
Section Width	3.70 m (12.14 ft)	3.70 m (12.14 ft)
Section Height	4.16 m (13.65 ft)	4.16 m (13.65 ft)
Payload per Section:		
Passenger Vehicle	10.3 t (22,708 lbs)	13.9 t (30,644 lbs)
Cargo Vehicle	14.0 t (30,865 lbs)	17.5 t (38,581 lbs)
Seats per Section	62-92	84-126
Floor Space per Section	70 m^2 (754 ft^2)	77 m^2 (818 ft^2)

Source: Transrapid International (TRI)
*TRI offers clients multiple configuration options, thus certain specifications may vary slightly among vehicles.

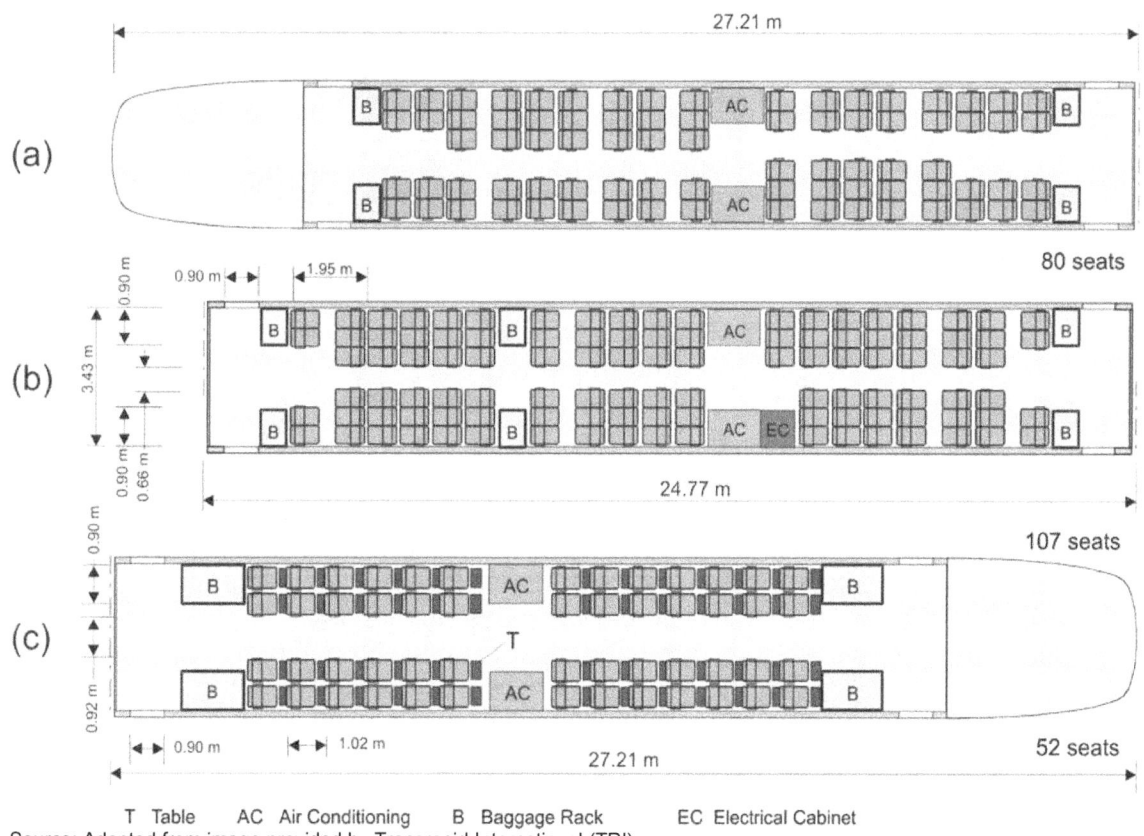

T Table   AC Air Conditioning   B Baggage Rack   EC Electrical Cabinet
Source: Adapted from image provided by Transrapid International (TRI)

Figure 1-2. Typical TRI TR08 Interior Plans for (a) medium-density intercity seating, (b) high-density commuter type seating, and (c) first-class intercity seating.

The guideway is the physical structure along which the maglev vehicles are levitated, guided, and propelled. The guideway can be elevated on bridge-style columns, be mounted at grade on a continuous foundation, or can use other configurations (Figure 1-3). The guideway beam structure can be fabricated from steel, concrete, or in a hybrid of steel and concrete; a flexible steel guideway crossover switch section is used for switching.

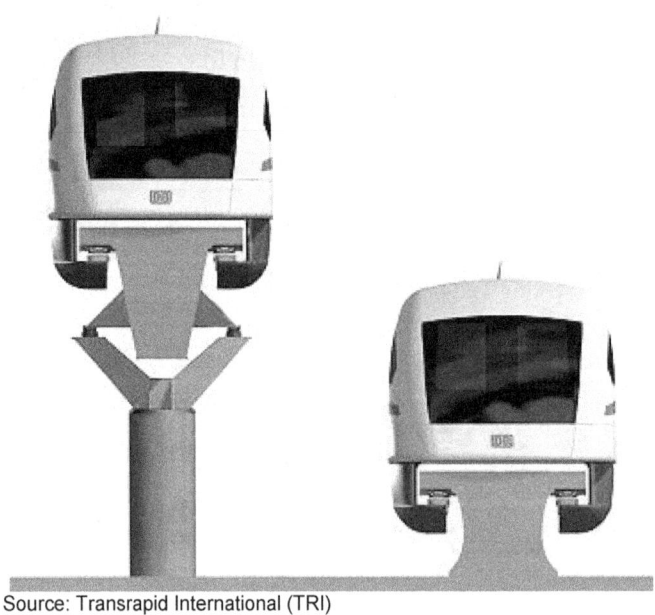

Source: Transrapid International (TRI)

Figure 1-3. TRI TR08 Guideway (can be either elevated on columns or mounted at grade)

The TRI Maglev vehicle levitation frame wraps around the "T-shaped" guideway to securely hold and guide the vehicle. Attractive forces between the electromagnets located in the vehicle levitation frame that surrounds the guideway and the stator packs installed on the underside of the guideway allow the vehicle to levitate. The guidance magnets located on the interior sides of the vehicle frame hold the vehicle laterally in place (Figure 1-4). The levitation and guidance magnets are separated from the guideway by a gap of about 1 cm (0.4 in) to allow for levitation and minor vertical and lateral movement.

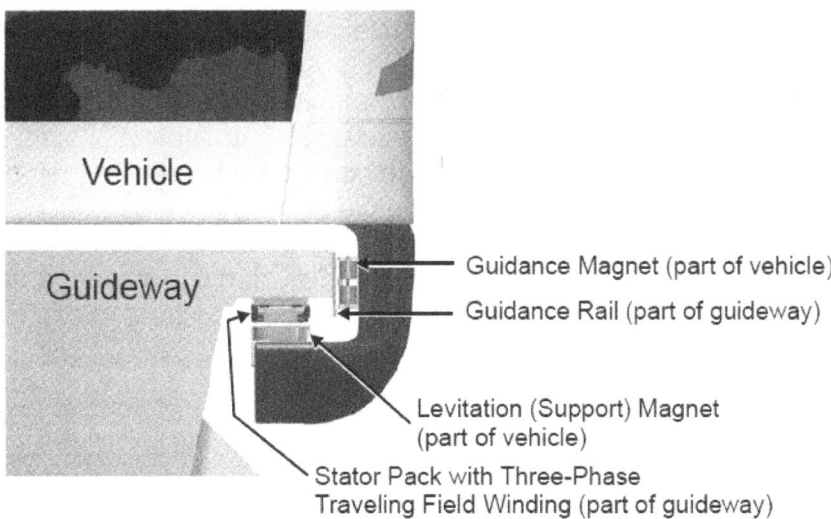

Figure 1.4. TRI TR08 Support and Guidance Systems

The power to propel maglev vehicles is provided via the powered guideway (Figure 1-5). An electric current through the guideway windings generates a traveling electromagnetic field along the guideway. The interaction between the traveling electromagnetic field in the guideway and electromagnetic fields in the vehicle pulls the vehicle along. Adjusting the frequency (0 Hz to approximately 300 Hz) of the alternating electric current can accelerate or decelerate the vehicle – the higher the frequency of the current, the higher the vehicle's speed.

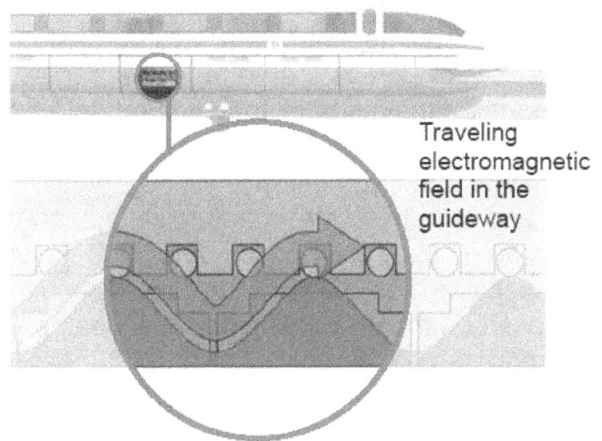

Figure 1-5. TRI TR08 Propulsion

Maglev vehicles are controlled and monitored from a central operations center, and the system runs in automatic mode with pre-programmed speed profiles for revenue operations. Communication between the vehicle and control center is via directional radio data transmission (to date, 38 to 40 gigaHertz (GHz) line-of-sight data radio links have been used for the data transmission line).

Power to the maglev system is supplied by electrical power substations that, in turn, supply several individual switching sections. The substations are located at various points along the guideway and are configured to receive their power from the commercial power grid. For reliability, the current is fed separately and redundantly into each side of the guideway motor. The long-stator linear synchronous motor installed in the guideway is divided into individual segments ("blocks"); only those segments in which the vehicle is located at that moment are switched on and supplied with current.

1.5 TEST FACILITY

The TVE was completed in 1984 with the purpose of simulating long-term operation of vehicles under conditions similar to actual applications. The Versuchs- und Planungsgesellschaft für Magnetbahnsysteme (MVP), a consortium of German companies, is the owner and operator of the TVE, with management sub-contracted to IABG. Revenue operations began at the TVE in 1995, and more than 70,000 visitors have ridden in the Transrapid maglev vehicles (TRI 2001).

The TVE test facility, located in a generally flat, agricultural, lowland landscape in the Emsland region of Germany, consists of a 31.5 km (19.5 mile) guideway, with two loops (one with constant radius and one without a constant radius) at either end of a straight section (Figure 1-6). Various guideway types (steel, concrete, hybrid) and configurations (at-grade, elevated) exist at the test track. The concrete, steel, and hybrid beam sections are elevated on concrete columns. A flexible steel high-speed switch diverts the northbound maglev vehicle on a turnout to the north loop, then moves back to accommodate the through movement onto the high-speed straight section. The switch consists of a continuous steel beam, anchored at the end adjoining the straight section and moveable at the other end with an electro-mechanical actuated drive system. The at-grade section is an embankment of soil built up to the level of the guideway in the high-speed section of the north loop. Two different at-grade guideway types are placed on the embankment, steel beams and concrete beams. Each type is represented by four 6.2 m (20.3 ft) beams for a total of 25 m (82 ft) of concrete at-grade beams and 25 m of steel at-grade beams.

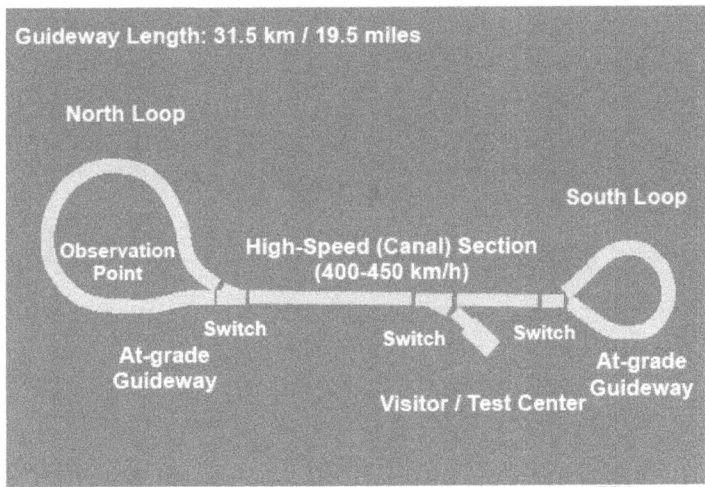

Source: Adapted from image provided by Transrapid International (TRI)

Figure 1-6. Transrapid Test Facility (TVE)

Noise measurements were taken for representative guideway types. The test facility typically operates 3-5 days a week and undergoes maintenance at least 1-2 days a week. Normally, the vehicle runs 5-7 times a day depending on weather and guideway conditions. Each trip consists of two complete runs of the entire test track length. The normal operating sequence provides speeds of 150 km/h, 400 km/h, 200 km/h, and 300 km/h on the straight section (Figure 1-7).

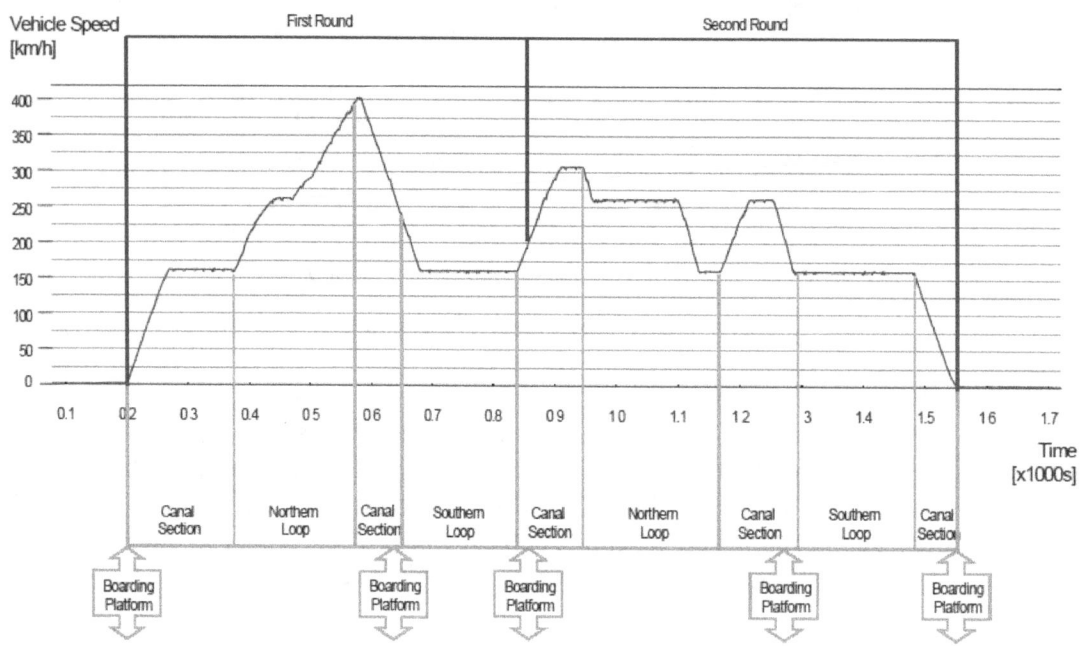

Source: Transrapid International (TRI)

Figure 1-7. Typical speed profile of TR08 at the Transrapid Test Facility

1.6 NOISE MEASUREMENTS

1.6.1 General

During the PEIS analysis, limited technical information was available for the TR08 Maglev System. As such, TR07 operational parameters were used to derive potential noise and vibration effects for the Transrapid Maglev System. In order to assure that technical concerns are addressed accurately, in 2001 the FRA sponsored a noise and vibration measurement program carried out at the TVE on the TR08 vehicle as part of pre-construction, site-specific environmental impact analysis. Test plans are presented in Appendix A, and briefly summarized below. Conditions and events at the time of actual measurement forced some deviations from the original test plan; measurement and analytical procedures actually employed are described in detail in Section 2.3.

For comparative purposes, the noise measurements attempted to replicate, where possible, previous testing programs on the TR07 vehicle. Vibration testing had not been previously conducted by the FRA for the TR07.

Noise measurements were made using microphone arrays placed close to the test track and single microphones placed at European and FRA standard reference distances. A microphone array is a group of closely-spaced microphones which may be used as a highly directional sound-measuring instrument; a microphone array can be used to locate individual sound sources by properly combining the microphone output signals. Vertical arrays were used to determine noise source heights on both the vehicle and the guideway while horizontal arrays were used to pinpoint noise source locations on the TR08 vehicle. For each of the 11 array configurations, the Test Plan (Appendix A) called for 3 passby noise measurements of the TR08 vehicle at each of 4 speeds - 150, 200, 300, and 400 km/h.

1.6.2 Applicability of Data

The measurements, data analyses, and conclusions presented in this report are representative of the TR08 Maglev System as constructed in Germany and tested at the TVE in August 2001, and May, 2002. Any specific applications, such as the construction of a maglev system elsewhere outside of TVE, would probably modify both guideway and vehicle consist specifications (e.g., by modifying the beam design and support structures or by employing different speed profiles) to suit the site-specific operational needs as well as any environmental and/or financial constraints. Such modifications may produce somewhat different noise levels that would therefore require further characterization during acceptance and safety qualification testing of the TR08 Maglev System. The information in this report will be helpful in any such future endeavor.

1.7 REPORT ORGANIZATION

This report is organized into an Executive Summary and five chapters. The Executive Summary presents an overview of the effort to characterize the noise associated with the TR08, and a summary of the data and major findings. Chapter 1 provides background information on the Maglev Deployment Program, states the objectives and approach of the field measurement effort, describes the TR08 technology, and presents the roles and responsibilities of the various co-authors as well as other people involved in the measurement effort. Chapter 2 presents the noise measurements, analyses, and results; results are summarized in Section 2.7. Chapter 3 describes the validation and verification process and results. The data are compared with similar data from other high-speed rail systems and set in a regulatory context in Chapter 4. References are listed in Chapter 5.

CHAPTER 2 NOISE MEASUREMENTS

2.1 INTRODUCTION

This report summarizes the methods and results of noise measurements carried out on August 14-24, 2001, and May 15-17, 2002, at the Transrapid Test Facility (TVE) in the Emsland region of Germany. The measurement program was carried out in two phases because all of the tasks proposed in the original test plan could not be completed in August, 2001. The testing resumed in May, 2002, under nearly identical weather conditions as in the first measurement period.

All measurements were performed by akustik-data Engineering on behalf of the FRA, under subcontract to HMMH. The objective of the measurements was to determine the wayside noise produced during the operation of the Transrapid TR08 Maglev System. Data were obtained with different geometrical arrangements of microphone arrays and up to six single microphones for various vehicle speeds and guideway configurations.

The noise measurements were carried out at four different elevated guideway types, the North switch, and two types of simulated at-grade guideways. Subsequent analyses resulted in sound-source distributions of the vehicle obtained from the microphone-array measurements, time-histories of the sound pressure level from the single-microphone measurements, and different noise descriptors in order to characterize the overall sound emission of the TR08 traveling on the various guideway configurations.

2.2 MEASURING SITES AND WEATHER CONDITIONS

The wayside noise during passbys of the TR08 was measured in areas adjacent to the following TVE track locations, shown in Figure 2-1 (from South to North): beam 213 (prototype steel, ①), beam 215 (prototype concrete, ②), beam 229 and 233 (reference concrete, ③), beam 267 (hybrid, ④), column 291-298 (North switch, ⑤), and beam 340 and 341 (at grade[1], ⑥ and ⑦). An enlarged version of Figure 2-1 is reproduced in Appendix A (page A-27). The measurements that were performed at each of these sites are documented in this chapter.

All measuring sites showed unhampered sound-propagation conditions on the east side of the track, where the single microphones and the microphone arrays were located. This means that the measuring sites were well suited for effective measurements of airborne sound. Only in the area of the hybrid beam, there

---

[1] The term "at-grade" is used loosely with respect to the guideway at beams 340 and 341. Rather than truly being "at-grade" in the traditional sense of the term, the guideway is the typical (elevated) construction with land back-filled in around it to create an "at-grade" section of the guideway. As such, noise characteristics of a true at-grade guideway may vary from those measured at beams 340 and 341. This portion of the guideway, however, is the only area considered somewhat representative of at-grade conditions, as they may be built in the U.S.

was a wooded area on the east side of the track, located at some distance and at a diagonal angle to the direction of the track, so that no major impact by reflected sound was to be expected.

On the other (West) side of the track, however, there were rows of trees extending along all measuring sites. In some cases, an embankment planted with bushes was located between the rows of trees and the track. In principle, reflections might occur in such regions; however, the impact of such reflections on the sound-pressure level on the East side of the track can be regarded as insignificant.

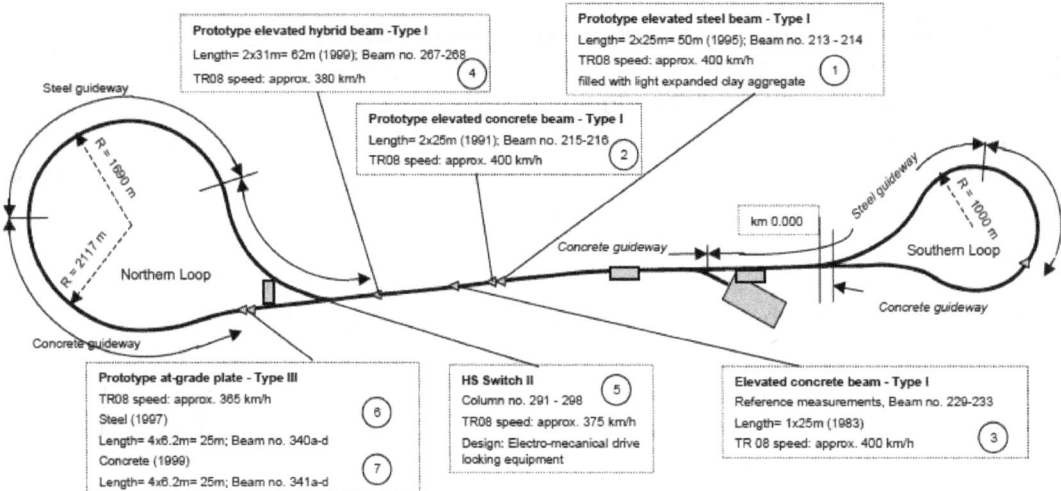

Figure 2-1. General location of noise measurement sites.

Weather data have been recorded during all passbys of the TR08 by means of a weather station, which was always located close to the measuring site with the microphone array. As single microphone measurements were either carried out in close proximity to or within about 30 m (98 ft) of the array, all data recorded by the weather station are also applicable for the sites with the single microphones. The weather station was used to measure the temperature, the wind speed and its direction, the air pressure, and the humidity.

Depending on the time of the day, temperatures during the measurement campaign in August, 2001, were between 16° and 28°C (61° and 82°F), which is typical or even high for a summer day in northern Germany. During this period, the wind was slight to moderate at around 4 meters per second (m/s) (8 knots (kts)), sometimes gusty up to 10 m/s (20 kts), which is quite usual for the Emsland region due to its proximity to the North Sea. There were also isolated periods with almost no wind at all. The air pressure throughout the entire measurement campaign in August, 2001, was relatively constant between 1010 and 1020 hectopascals (hPa). In general, the weather was dry. Only on August 16, 2001, at the beginning of the measurements, there were some slight showers which did not have any impact on the microphones.

During the measurements in May, 2002, temperatures varied between 16° and 23°C (61° and 73°F); i.e., the average temperature was slightly below that recorded in August, 2001. Furthermore, the wind was not as gusty in May, 2002, as compared to August, 2001; maximum wind speed was only 5 m/s in May, 2002. The air pressure was similar to that in August, 2001 (i.e., between 1012 and 1022 hPa). No rainfall occurred over the period of measurements. Thus, the weather conditions for both measurement campaigns can be regarded as good, even if the wind speed at some passbys of the TR08 was on the verge of exceeding recommended wind speed limits for conducting the noise measurements. Therefore, the use of windscreens on the microphones was absolutely necessary.

2.2.1 Single-Microphone Measurements

Measurements with single microphones were made at up to six different measuring positions at each measuring site. Five of the six microphones were placed in a line perpendicular to the track at distances of 6.5, 15.2, 25.0, and 30.5 m (21.3, 50.0, 82.0, and 100.0 ft) from the centerline of the guideway. The 6.5 m-microphone was configured in both a low and a high position. The sixth microphone was located beneath the centerline of the guideway. The heights of the microphones are listed in Table 2-1.

Table 2-1. Heights of the single microphones above ground or guideway

Microphone distance from guideway centerline [m (ft)]	Height above ground [m (ft)]	Height above guideway [m (ft)]
0 (microphone beneath guideway)	1.5 (5.0)	-
6.5 (21.3), low position	-	-1.5 (-5.0)
6.5 (21.3), high position	-	0
15.2 (50.0)	1.5 (5.0)	-
25.0 (82.0)	3.5 (11.5)	-
30.5 (100.0)	1.2 (4.0)	-

Data for the single microphones were recorded at vehicle speeds of about 100, 150, 200, 300, and 400 kilometers per hour (km/h) (62, 93, 124, 186, and 249 miles per hour (mph)). For the single-microphone measurements, the actual vehicle speed during each individual passby was taken from the data that were documented by the control system of the IABG (Industrieanlagen Betriebsgesellschaft). For each measuring site, the following details are given:

Prototype steel guideway (①)

    Location: between column 212 and 214,

    beam length: 50 m (164 ft), so-called "double-span" beam,

    measuring date: August 21 and 22, 2001 (6.5 m-microphones),
August 23, 2001 (microphone beneath guideway, 15.2, 25.0, and 30.5 m-microphones),
May 15.2002 (microphone beneath guideway, 6.5, 15.2, 25.0, and 30.5m microphones),

    measuring plane: at column 213 (6.5, 15.2, 25.0, and 30.5 m-microphones), between column 213 and 214 (microphone beneath guideway),

    vehicle speeds: from 99.8 to 399.7 km/h (62.0 to 248.4 mph), see Tables B-7 to B-12,

    weather conditions: see general remarks in Section 2.2, for August 21 to 23, 2001, see Table 2-2,

    remark: for noise-reduction reasons, the prototype steel beam is filled with light expanded clay aggregate.

Prototype concrete guideway (②)

    Location: between column 214 and 216,

    beam length: 50 m (164 ft), so-called "double-span" beam,

    measuring date: August 23 and 24, 2001 (6.5 m-microphones),
August 21, 2001 (microphone beneath guideway, 15.2, 25.0, and 30.5 m-microphones),
May 15, 2002 (microphones beneath guideway, 6.5, 15.2, 25.0, and 30.5 m microphones),

    measuring plane: at column 215 (6.5, 15.2, 25.0, and 30.5 m-microphones), between column 215 and 216 (microphone beneath guideway),

    vehicle speeds: from 99.9 to 400.1 km/h (62.1 to 248.6 mph), see Tables B-13 to B-18,

    weather conditions: see general remarks in Section 2.2, for August 21, 23, and 24, 2001, see Table 2-2.

Reference concrete guideway (③)

    Location: between column 216 and 266,

    beam length: 25 m (82 ft),

    measuring date: August 14 and 15, 2001 (6.5 m-microphones),
August 15 and 16, 2001 (microphone beneath guideway, 15.2, 25.0, and 30.5 m-microphones),

	May 17. 2002 (microphone beneath guideway, 6.5, 15.2, 25.0, and 30.5 m microphones),
measuring plane:	between column 228 and 229 (6.5 m-microphones), at column 233 (15.2, 25.0, and 30.5 m-microphones), between column 233 and 234 (microphone beneath guideway),
vehicle speeds:	from 99.9 to 401.5 km/h (62.1 to 249.5 mph), see Tables B-1 to B-6,
weather conditions:	see general ramarks in Section 2.2 for August 14 to 16, 2001, see Table 2-2.

Hybrid guideway (④)

Location:	between column 266 and 268,
beam length:	62 m (203 ft), so-called "double-span" beam,
measuring date:	May 16, 2002 (microphone beneath guideway, 6.5, 15.2, 25.0, and 30.5 m-microphones),
measuring plane:	at column 267 (6.5, 15.2, 25.0, and 30.5 m-microphones), between column 266 and 267 (microphone beneath guideway),
vehicle speeds:	from 100.1 to 393.6 km/h (62.2 to 244.6 mph), see Tables B-19 to B-24,
weather conditions:	see general remarks in Section 2.2 for May 16, 2002, see Table 2-2.

North switch (⑤)

Location:	between column 291 and 298,
switch length:	132 m (433 ft),
measuring date:	August 24, 2001 (microphone beneath guideway, 6.5, 15.2, 25.0, and 30.5 m-microphones), May 17, 2002 (microphone beneath guideway 6.5, 15.2, 25.0, and 30.5 m microphones)
measuring plane:	between column 294 and 295,
vehicle speeds:	from 99.9 to 385.3 km/h (62.1 to 239.4 mph), see Tables B-25 to B-30,
weather conditions:	see general remarks in Section 2.2, for August 24, 2001, see Table 2-2.

At-grade guideway (⑥ and ⑦)

Location:	beam 340 a-d (steel), beam 341 a-d (concrete),

beam length:	4 x 6.2 m = 25 m (82 ft) each,
measuring date:	August 17 and 22, 2001 (6.5 m-microphones), May 16, 2002 (6.5 m microphones)
measuring plane:	middle of beam 340, middle of beam 341,
vehicle speeds:	from 100.2 to 372.0 km/h (62.3 to 231.2 mph), see Tables B-31 to B-34,
weather conditions:	see general remarks in Section 2.2 for August 17 and 22, 2001, see Table 2-2.

The above list of details at the various measuring sites reveals that the lengths of the associated guideway types are generally relatively short. For instance, the prototype steel and the prototype concrete guideway each have a length of only 50 m (164 ft), the hybrid guideway is 62 m (203 ft) long, and the at-grade steel and at-grade concrete guideways are each 25 m (82 ft) long. These guideways are always positioned adjacent to other guideway types. For measurements using single microphones with their omnidirectional characteristic, this means that an investigation of the guideway's sound radiation that is unhampered by side effects can only take place if the distance between microphone and guideway is relatively small, in particular, since the TR08 with its length of approximately 79 m (259 ft) always passes different guideway types at the same time.

If one assumes that an adequate measurement can only be achieved if both on the right-hand and left-hand side of the measuring plane at least three times the measuring distance between microphone and track centerline must be of the same guideway type, only measurements beneath the guideway and at a distance of 6.5 m (21.3 ft) can be regarded as fully valid (due to the length of only 25 m (82 ft) of the two at-grade guideway types, even the measurements from the microphone at 6.5 m are, unfortunately, not fully valid). For this reason, for measuring distances of 15.2, 25.0, and 30.5 m (50, 82, and 100 ft), the single-microphone results at the prototype steel, prototype concrete, and hybrid guideway have only limited relevance. Nevertheless, it is safe to assume that the results reproduce the sound-level differences between the various guideway types more or less correctly.

These side effects, however, do not apply to the single-microphone measurements in the region of the North switch and the reference concrete guideway. With a length of 132 m (433 ft), the switch is sufficiently long even for the most remote microphone—a particularly important measurement location, since the switch (acoustically not treated steel beam) has a much higher sound emission compared with the adjacent guideway type (reference concrete guideway). For the reference concrete guideway in the region of column 233, there are no limitations in terms of relevance for the single-microphone results, since both on the right-hand and on the left-hand side of the measuring plane, there are large areas of identical guideway types.

2.2.2 Microphone-Array Measurements

Measurements with differently configured microphone arrays were made during passbys of the TR08 at four measuring sites; i.e., at the prototype steel guideway (beam 213, ①), the prototype concrete guideway (beam 215, ②), the reference concrete guideway (beam 229, ③), and the hybrid guideway (beam 267, ④). Four different arrays were used, each comprising 29 microphones: (i) the wayside vertical line array (WV array), (ii) the wayside horizontal line array (WH array), (iii) the wayside X-shaped two-dimensional array having 32 cm (1.05 ft) microphone spacing (WX32 array), and (iv) the wayside X-shaped two-dimensional array having 16 cm (0.52 ft) microphone spacing (WX16 array).

Whereas the WX arrays were positioned about in the middle of the height of the TR08, the WV array was placed in both a low and a high position at each measuring site. These two different heights are necessary to improve the resolution of the array if the sound-source distribution is to be determined over the total height of the vehicle. The variation of the array height could easily be achieved by mounting the array on a hydraulic tower which was supplied and operated by the IABG. On the other hand, due to this tower and the different conditions of the surrounding land, particularly due to the position of a drainage ditch running parallel to the eastern side of the track, it was not possible to keep the distance between the array and the guideway constant at the different measuring sites. Thus, a slight variation of this distance had to be accepted.

Data for the microphone arrays were recorded at vehicle speeds of about 150, 200, 300, and 400 km/h (93, 124, 186, and 249 mph). Three passbys of the TR08 were planned to be measured at each of these speeds. In the case of the array measurements, the actual vehicle speed during each individual passby was calculated from the signals of two light barriers that were a part of the array system.

The following photographs give an impression of the characteristics of the measuring sites and the microphone arrays. Figure 2-2 shows the measuring site with the prototype steel guideway (①), the associated WV array in its high position mounted on the hydraulic tower as well as the single microphone at a distance of 6.5 m (21.3 ft) from the track centerline in the foreground. Figure 2-3 has been taken in the region of the reference concrete guideway (③). It illustrates the tower placed between column 228 and 229 with the WX32 array attached to it. Furthermore, the van containing the electronic equipment can be seen in the background. The last photograph in this series, Figure 2-4, shows the WH array. The photo has also been taken in the region of the reference concrete guideway. For each measuring site, the details are given in the following lists, where height of the array means the position of the middle microphone (WV array), the central microphone (WX arrays) or the row of microphones (WH array):

Prototype steel guideway (①)
- Location: between column 212 and 214,
- beam length: 50 m (164 ft), so-called "double-span" beam,
- array configuration: WV array, low position (August 21, 2001),
  WV array, high position (August 22, 2001),
- measuring plane: at column 213,
- distance of array: 4.54 m (14.89 ft) from guideway centerline,
- height of array: 0.06 m (0.20 ft) above guideway (WV array, low position), 2.14 m (7.02 ft) above guideway (WV array, high position),
- vehicle speeds: from 149.3 to 399.7 km/h (92.8 to 248.4 mph), see Table 2-2,
- weather conditions: for August 21 and 22, 2001, see Table 2-2,
- remark: for noise-reduction reasons, the prototype steel beam is filled with light expanded clay aggregate.

Prototype concrete guideway (②)
- Location: between column 214 and 216,
- beam length: 50 m (164 ft), so-called "double-span" beam,
- array configuration: WV array, low position (August 23, 2001),
  WV array, high position (August 24, 2001),
- measuring plane: at column 215,
- distance of array: 4.54 m (14.89 ft) from guideway centerline,
- height of array: 0.06 m (0.20 ft) above guideway (WV array, low position), 2.14 m (7.02 ft) above guideway (WV array, high position),
- vehicle speeds: from 149.5 to 399.6 km/h (92.9 to 248.3 mph), see Table 2-2,
- weather conditions: for August 23 and 24, 2001, see Table 2-2.

Figure 2-2. The TR08 passing by the measuring site with the prototype steel guideway (column 213) and the WV array (in high position) mounted on the hydraulic tower; the 6.5 m-microphone (in low position) can be seen in the foreground.

Reference concrete guideway (③)
    Location: between column 216 and 266,

    beam length: 25 m (82 ft),

    array configuration: WX32 array (August 14, 2001),
                            WX16 array (August 15, 2001),
                            WV array, low and high position (August 16, 2001),
                            WH array (August 17, 2001),

    measuring plane: between column 228 and 229,

    distance of array: 4.98 m (16.34 ft) from guideway centerline,

    height of array: 0.80 m (2.62 ft) above guideway (WX32 and WX16 array), 0.06 m (0.20 ft) above guideway (WV array, low), 2.14 m (7.02 ft) above guideway (WV array, high), 0.80 m (2.62 ft) above guideway (WH array), vehicle speeds: from 149.6 to 401.3 km/h (93.0 to 249.4 mph), see Table 2-2,

weather conditions: for August 14 to 17, 2001, see Table 2-2.

Hybrid guideway (④)

Location: between columns 266 and 268,

beam length: 62 m (203 ft), so-called "double-span" beam,

array configuration: WV array, low position (May 16, 2002), WV array, high position (May 16, 2002),

measuring plane: at column 267,

distance of array: 4.95 m (16.24 ft) from guideway centerline,

height of array: 0.06 m (0.20 ft) above guideway (WV array, low position), 2.14 m (7.02 ft) above guideway (WV array, high position),

vehicle speeds: from 150.1 to 395.6 km/h (93.3 to 245.8 mph), see Table 2-2

weather conditions: for May 16, 2002, see Table 2-2

Figure 2-3. The TR08 passing by the measuring site with the reference concrete guideway and the hydraulic tower positioned between column 228 and 229 with the attached WX32 array.

In contrast to measurements with single microphones, the short beam length of the prototype steel and concrete guideway does not pose a problem in terms of relevance of results for array measurements. Due to the short distance between the arrays and the vehicle and guideway, side effects from adjacent guideway types can be ruled out, especially, since the WV array was always positioned in the center of the 50 m (164 ft) double-span beam.

Figure 2-4. The WH array mounted on the hydraulic tower in the area of the reference concrete guideway.

2.3 INSTRUMENTATION AND PROCEDURES

2.3.1 Measuring Equipment

Single-Microphone Measurements

Microphones.   Microphones #1 and #2: 1/2" cartridges, B&K type 4134, with preamplifiers, B&K type 2639, and windscreens, B&K type UA 0237; microphones #3 and #4: 1/2" cartridges, B&K type 4191, with preamplifiers, B&K type 2619, and windscreens, B&K type UA 0237; microphones #5 and #6 (part of sound-level meter, B&K type 2230).

Measuring amplifiers.   Microphones #1, #2, and #3: B&K type 2636; microphone #4: B&K type 2610; microphones #5 and #6: sound-level meter, B&K type 2230.

Data storage.   8-channel Digital Tape Recorder TASCAM DA-88, sampling frequency 48 kilohertz (kHz) (DAT).

The complete equipment from microphones to storage was calibrated each measuring day with an acoustical calibrator, B&K type 4230, having an accuracy of ±0.3 decibels (dB). Using this calibrator and the other measuring equipment, the requirements for class one instruments specified in DIN EN 60651 ("Sound level meters") and DIN EN 60942 ("Electroacoustics - sound calibrators") were met.

Microphone-Array Measurements

Microphones.   30 1/2" cartridges, B&K type 4191, with preamplifiers, B&K type 2619, and windscreens, B&K type UA 0237.

Amplifier.   32-channel microphone amplifier made by L&P.

Data storage.   PC with 486 processor, equipped with four A/D-boards type RTI 860 made by Analog Devices, sampling frequency 23.8 kHz.

While 29 channels of the aforementioned microphone amplifier were needed for the array microphones, channel #30 was used for the single microphone placed at a distance of 6.5 m (21.3 ft) from the track centerline. Correspondingly, the signal from this microphone was stored along with the data of the array microphones. The complete equipment from microphones to storage was calibrated each measuring day by use of an acoustical calibrator, B&K type 4230 (see above).

Vehicle detector.   To ascertain the relation between recorded sound pressure and vehicle coordinate, and to evaluate the exact vehicle speed, two light barriers (owned by the IABG) were used. When the front and the rear part of the

TR08 pass the light barriers, signals are generated, which were stored in channel #31 and #32 simultaneously with the microphone signals.

2.3.2 Principles and Application of Microphone-Array Technology

A precondition for the microphone arrays in the measurements described here was the use of the same array configurations as those already employed in 1999 during the sound-source localization on the ACELA train on the test track of the Transportation Technology Center (TTC) in Pueblo, Colorado (FRA 2000). The aim was to ensure comparability of the array results of both high-speed vehicles. Therefore, the array configurations for the measurements on the TR08 were not adjusted to any specific conditions.

However, due to the local site characteristics, in particular, to a drainage ditch running alongside the track, the arrays at the prototype steel and the prototype concrete guideway were situated approximately 0.5 m (1.6 ft) closer to the vehicle. Furthermore, since the width of the TR08 is larger than that of a normal railway train, the distance between the arrays and the outer surface of the TR08 is smaller than for the ACELA train. However, this has the positive effect that adjacent sound sources can be resolved more clearly and therefore does not constitute a disadvantage. The prescribed array geometries also define the analyzable frequency ranges for the array measurements. The following paragraphs give a short description of the characteristics of the arrays used.

Basic Information and Description of the Nested Line Arrays.
Principle. By positioning a number of microphones in a line, sound sources can be located in one dimension. The analysis is carried out by selecting a point on the measured object (focal point). The signals measured by the microphones at time t are summed by accounting for the propagation time t_{pn} from the focal point to the n-th microphone; i.e.,

$$p(t) = \frac{1}{N}\sum_{n=1}^{N} p_n(t + t_{pn}) \tag{2-1}$$

where p(t) is the output of the microphone array and t corresponds to the time at which the source is focused (time of emission). For a source located at the focal point, the phase differences between the signals $p_n(t+t_{pn})$ of the different microphones are zero. Thus, the summed pressure p(t) is the sound pressure of the source that would be measured by a single microphone. When the source is not at the position of the focal point, the phase differences between the signals $p_n(t+t_{pn})$ are not zero which leads to partial cancellation and a smaller summed pressure. From the output of the array, a beam pattern characterized by a main lobe and several lower-level side lobes is formed. More details concerning the principles of microphone-array technology can be found in Barsikow et al. (1987), Barsikow et al. (1988), and Barsikow (1996).

Nesting the arrays. In order to make array measurements in a wide frequency range, it is necessary to use several sub-arrays with different microphone spacings. In the present study, the line arrays consisted of three sub-arrays comprising 15 microphones each. By nesting the microphones, 29 of them were sufficient for the total array. The chosen microphone spacings were 32, 16, and 8 cm (1.05, 0.52, and 0.26 ft) which result in the notations "WV32", "WV16", and "WV08" for the vertical sub-arrays, "WV" for the total vertical array, "WH32", "WH16", and "WH08" for the horizontal sub-arrays, and "WH" for the total horizontal array.

Frequency range (WV and WH array). Evaluating the data measured with microphone arrays is restricted by a lower and an upper frequency limit. With respect to these limitations, the three sub-arrays of both the vertical and horizontal line arrays had the following frequency ranges:

- 32-array: six third-octave bands (315, 400, 500, 630, 800, and 1000 Hz), frequency range 280 to 1120 Hz;
- 16-array: three third-octave bands (1250, 1600, 2000 Hz), frequency range 1120 to 2240 Hz;
- 08-array: three third-octave bands (2500, 3150, and 4000 Hz), frequency range 2280 to 4500 Hz.

Position of WV and WH array. The positions of the WV and WH array have already been given in the lists of details in Section 2.2.2. Note that the WV array was oriented parallel to the side contour of the TR08, whereas the WH array was mounted parallel to the centerline of the track.

Position of focal points (WV array). The test vehicle is divided into 71 vertical focal positions lying between heights of -3.0 (-9.8) and +4.0 m (+13.1 ft) related to the upper surface of the guideway and having a spacing of 10 cm (0.33 ft). In the horizontal direction where the resolution equals that of a single omnidirectional microphone, a 20 cm (0.66 ft) spacing for each measuring point was chosen. The focal points are positioned in a plane whose lateral distance from the row of microphones is 3.15 m (10.33 ft) in the cases of the reference concrete and hybrid guideways, and 2.70 m (8.86 ft) in the cases of the prototype steel and prototype concrete guideways (i.e., in a plane that always corresponds to the side contour of the TR08).

Position of focal points (WH array). In the horizontal direction where the resolution is high, the spacing of the focal points is 10 cm (0.33 ft). Results will only be presented in this one dimension. The focal plane does not differ from that selected for the WV array, so that the lateral distance between the focal points and the row of microphones is 3.15 m (10.33 ft), see above.

Evaluation of data (WV and WH array). For each focal point, each recorded vehicle passby, and each sub-array, narrow-band spectra were calculated from

512 data samples, resulting in a bandwidth of 46.5 Hz. Subsequently, they were averaged over a time corresponding to the vehicle travelling 3.5 m (11.5 ft). Thus, integration time depends on speed. Every sampling point in the related FFT input signals represents an addition of sound pressures measured by the 15 microphones of one sub-array (see Eq. (2-1)). The signals include time shifts to account for the sound propagation from the focal point to each microphone. Respecting the frequency range for each sub-array, unweighted and A-weighted sound-pressure levels for the total frequency range were obtained by summing the sound levels measured with the three sub-arrays.

Shading. In order to be able to distinguish between several sources of different intensities, it is necessary to reduce the side-lobe levels of the array directivity pattern. This can be done by using Dolph-Chebychev shading (Skudrzyk 1971) whereby all shading slightly increases the width of the main lobe, and the width of the main lobe determines the ability of the array to separate adjacent sources. The Dolph coefficients were chosen in such a way that all side lobes were 25 dB down relative to the main lobe instead of 13 dB for the unshaded array. This shading was found to be a good compromise between main-lobe width and side-lobe suppression.

Swept focus. When measuring moving sound sources with an array, it is necessary to satisfy the criterion that the product of frequency bandwidth, B, and integration time, T, (BT product) must be great enough. An elimination of convective effects is also needed if narrow-band results are calculated. These problems can be solved by incorporating into the analysis the changing distance of the moving source during its passby. This procedure is usually applied when sound-source distributions are measured with a horizontal array, but can also be used for data recorded by a vertical array or even a single microphone. Although with a vertical array, beam steering to scan the distribution of sources is only possible in the vertical direction, the signals from each microphone can also be de-Dopplerized by accounting for the position of the source under consideration.

Resolution. Figure 2-5 depicts the width of the main lobe in the direction of the row of microphones as a function of frequency for the VV and WH array, viz., for the three sub-arrays whose related working frequency ranges are also indicated in the Figure. These results are based upon a numerical simulation of data that would be measured by the array if a point monopole sound source moved along the track at 200 km/h (124 mph). The geometric dimensions are those of the real measuring situation, the height of the source equals the middle of the array, and the source is positioned at a lateral distance of 3.15 m (10.33 ft) from the array, which corresponds to the side contour of the TR08 at the measuring sites with the reference concrete and the hybrid guideways.

As can be seen in Figure 2-5, the main lobe in the working frequency ranges of the 16- and 08-array has an average width of about 0.34 m (1.12 ft), so that these sub-arrays provide a very good separation of sound sources. When the frequency range of the 32-array is divided in two, the same average resolution power will be obtained for the upper range between 560 and 1120 Hz, but for lower frequencies between 280 and 560 Hz only a main lobe with an average width of about 0.68 m (2.23 ft) can be achieved.

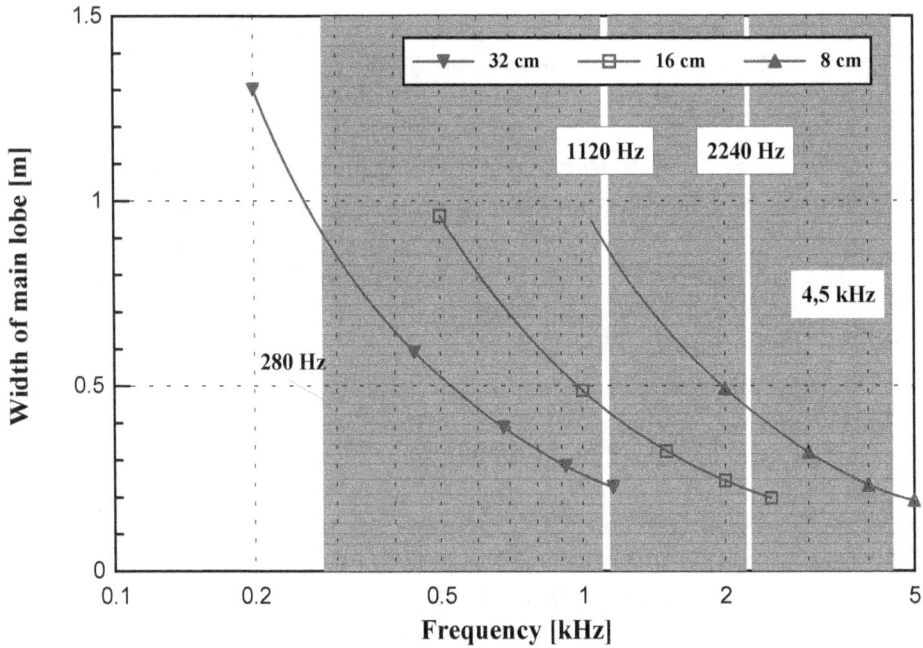

Figure 2-5. Width of main lobe as a function of frequency for the WV and WH array, viz., for the three sub-arrays whose working frequency ranges are indicated; results are with 25 dB Dolph-shading.

Description of the X-Shaped Arrays

Principle. A localization of sound sources with a line array can only be made in one dimension; in the case of the WV array, in the vertical one, and in the case of the WH array, in the horizontal direction. If a two-dimensional separation of sound sources is needed, other array configurations have to be used (e.g., an X-shaped array). In contrast to a cross-shaped array consisting of a horizontal and a vertical line array, sound-source localization with an X-array, which is also a cross array but turned by 45°, often yields a higher precision.

Since many sound sources on a moving train or maglev vehicle occur along either a horizontal or vertical line, these sources are difficult to separate in results measured with a cross array. The reason is that the side lobes of this array extend along the horizontal and vertical directions. Because these side lobes lie at most 12 dB below the level of the main beam, the results of such a measurement can include side-lobe leakage and thus might not be a true representation of the actual source strengths under investigation.

In the present study, however, measurements were made with 2 different wayside X-shaped arrays comprising 29 microphones. These arrays were formed by 2 line arrays having 15 microphones each, with common use of the middle microphone. Because of the diagonal structure of the WX array's side lobes, single sound sources can often be detected more easily compared with the results of cross arrays.

Frequency range. As the microphones of the WX arrays could not be nested due to the limited number of microphones, a relatively small working frequency range had to be chosen. In the present study, two X-shaped microphone arrays with different microphone spacings were used. The frequency range for the array with a microphone spacing of 32 cm (1.05 ft) is from 280 to 1120 Hz, while the other X-shaped array with 16 cm (0.52 ft) microphone spacing was used in the frequency range from 1120 to 2240 Hz. The microphone spacings of 32 and 16 cm result in the notation "WX32" and "WX16" for the arrays comprising the following third-octave bands:

> WX32 array: six third-octave bands (315, 400, 500, 630, 800, and 1000 Hz), frequency range 280 to 1120 Hz;
> WX16 array: three third-octave bands (1250, 1600, 2000 Hz), frequency range 1120 to 2240 Hz.

Position of WX32 and WX16 array. The positions of the WX32 and WX16 array have already been given in the lists of details in Section 2.2.2. Note that the plane formed by the array microphones was oriented parallel to the guideway as well as to the side contour of the TR08.

Evaluation of data. The TR08 is divided into both horizontal and vertical focal points whose positions and spacings are equal to those used for the processing of the WV and WH array's data. Also, the evaluation of data measured with the WX arrays does not differ from the procedure described above, with the exception for a crossed array that the spectra measured by the two partial line arrays are energetically multiplied.

Shading. By analogy with the WV array, a 25 dB Dolph-shading was applied. This difference in peak amplitudes between main and side lobes, however, is only valid for the two partial line arrays. For the entire array, side lobes are suppressed by at most 12.5 dB.

Swept focus. In the case of the WX arrays, it is also useful to sweep the focus with the source in the direction of motion. As was already described before, this provides a significantly improved resolution and a higher quality in spectral decomposition. In the present case, the focus is moved through a passby span of 3.5 m (11.5 ft).

Resolution. Figure 2-6 shows the width of the main lobe in the vertical direction as a function of frequency for the WX32 array. The calculations for this

Figure are based on the same procedure as was described in connection with Figure 2-5. In the frequency range from 280 to 1120 Hz, values for the width of the main lobe are obtained between 1.4 and 0.35 m (4.6 and 1.15 ft). Due to the geometrical similarity, the widths of the main lobe for the WX16 array in its working frequency range from 1120 to 2240 Hz are the same as the main-lobe widths of the WX32 array in the frequency range from 560 to 1120 Hz, viz., between 0.7 and 0.35 m (2.3 and 1.15 ft).

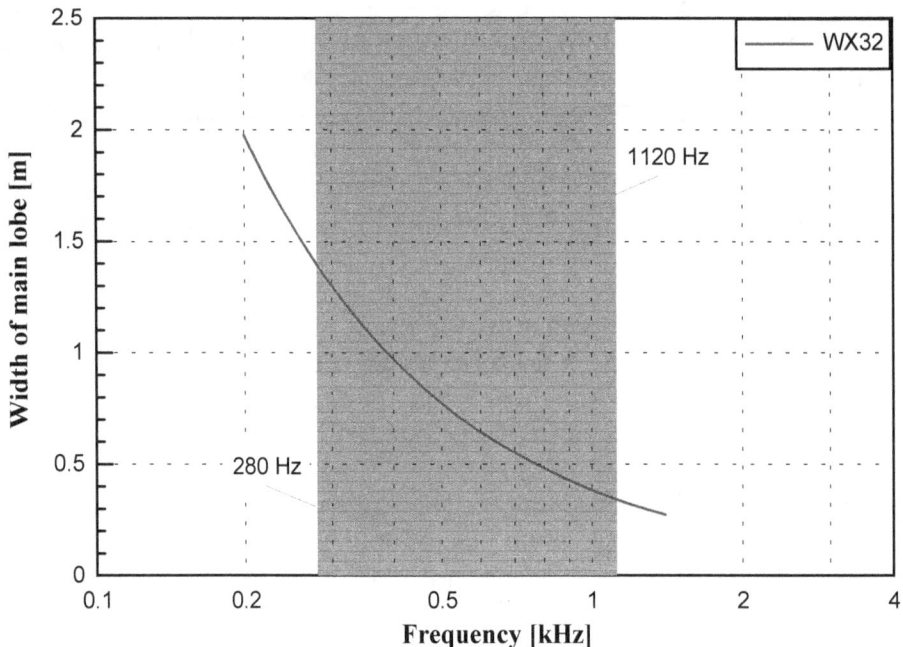

Figure 2-6. Width of main lobe in the vertical direction as a function of frequency for the WX32 array; results are with 25 dB Dolph-shading.

In practice, relatively strong side lobes arranged in a diagonal pattern around a localized sound source are frequently visible when results of measurements with an X-array are shown. This is a consequence of the less than ideal resolution of an X-array in the diagonal directions where the suppression of side lobes is quite low. Such diagonal patterns do not represent real sound sources except, of course, for the original source in the middle.

2.4 MEASUREMENT PROGRAM

2.4.1 Single-Microphone Measurements

As mentioned above, measurements using up to six single microphones were made at measuring sites with various guideway configurations at vehicle speeds between about 100 and 400 km/h (62 and 249 mph). A number of tables are given in Appendix B that contain the relevant information concerning the passbys of the TR08 measured at each site and each single-microphone position (in order not to be beyond the scope of this main report, the 34 tables are not given in the main body of the report). Tables B-1 to B-34 list the date and time of the measurement, the vehicle speed, and the single-number values calculated from the corresponding time histories of the A-weighted sound-pressure level. Each Table also identifies the measuring series to which the measurements belong (series A, B, G, H, I, J, K, L, Y, and Z as they are listed in the test matrix, Appendix A).

2.4.2 Microphone-Array Measurements

Sound measurements using various microphone-array configurations were made during passbys of the TR08 within the speed range from about 150 to 400 km/h (93 to 249 mph). In Table 2-2, all relevant information concerning each individual passby of the vehicle is gathered. The list of runs of the TR08 is sorted according to the passby ID numbers as they were introduced in the test matrix. The Table contains only those runs that have been analyzed.

It is documented in the test matrix (Appendix A) that three measurements were planned at each speed within measuring series A, B, C, D, E, G, H, I, J, K, and L in order to increase the statistical reliability of the results. As can be seen in Table 2-2, this was done successfully in measuring series B, C, G, H, I, J, and L. In series A, E, and K, the goal was achieved in all but one passby, which could either not be carried out at 300 km/h (186 mph) due to technical problems (series A and E) or could only be carried out with less than 400 km/h (249 mph); i.e., at 385 km/h (239 mph), and was therefore not processed (series K). In series D, a data error occurred at a passby at 400 km/h (249 mph), and another passby with a scheduled velocity of 400 km/h (249 mph) only reached 370 km/h (230 mph).

Table 2-2. Number, date, and time of vehicle passby, array type, vehicle speed, temperature, and wind speed.

Passby ID No.	Date	Time	Array Type	Speed [km/h (mph)]	Temperature [°C (°F)]	Wind Speed [m/s (kts)]
\multicolumn{7}{c}{Reference Concrete Guideway}						
A-150-1	2001-08-14	11:18	WX32	149.8 (93.1)	22 (72)	2.2 (4.3)
A-150-2	2001-08-14	12:49	WX32	149.7 (93.0)	24 (75)	3.0 (5.8)
A-150-3	2001-08-14	14:25	WX32	149.8 (93.1)	24 (75)	3.2 (6.2)
A-200-1	2001-08-14	11:29	WX32	199.5 (124.0)	23 (73)	2.5 (4.9)
A-200-2	2001-08-14	13:35	WX32	199.9 (124.2)	24 (75)	3.1 (6.0)
A-200-3	2001-08-14	13:46	WX32	199.5 (124.0)	24 (75)	2.8 (5.4)
A-300-1	2001-08-14	10:46	WX32	299.6 (186.2)	22 (72)	1.8 (3.5)
A-300-2	2001-08-14	11:34	WX32	299.8 (186.3)	23 (73)	2.7 (5.2)
A-400-1	2001-08-14	11:22	WX32	400.7 (249.0)	22 (72)	2.4 (4.7)
A-400-2	2001-08-14	12:54	WX32	400.8 (249.0)	24 (75)	3.2 (6.2)
A-400-3	2001-08-14	13:28	WX32	400.7 (249.0)	25 (77)	<1.0 (<2.0)
Reference Concrete Guideway						
B-150-1	2001-08-15	09:22	WX16	149.8 (93.1)	21 (70)	2.0 (3.9)
B-150-2	2001-08-15	10:15	WX16	149.6 (93.0)	21 (70)	2.4 (4.7)
B-150-3	2001-08-15	11:13	WX16	149.6 (93.0)	22 (72)	2.2 (4.3)
B-200-1	2001-08-15	09:40	WX16	199.7 (124.1)	21 (70)	2.0 (3.9)
B-200-2	2001-08-15	10:26	WX16	199.6 (124.0)	21 (70)	1.8 (3.5)
B-200-3	2001-08-15	11:25	WX16	199.4 (123.9)	23 (73)	4.3 (8.4)
B-300-1	2001-08-15	09:27	WX16	299.7 (186.2)	21 (70)	2.0 (3.9)
B-300-2	2001-08-15	10:31	WX16	299.6 (186.2)	22 (72)	2.9 (5.6)
B-300-3	2001-08-15	11:30	WX16	299.6 (186.2)	23 (73)	4.6 (8.9)
B-400-1	2001-08-15	10:19	WX16	400.8 (249.0)	21 (70)	2.2 (4.3)

B-400-2	2001-08-15	13:01	WX16	400.7 (249.0)	24 (75)	4.5 (8.7)
B-400-3	2001-08-15	14:28	WX16	400.8 (249.0)	26 (79)	4.0 (7.8)
Reference Concrete Guideway						
C-150-1	2001-08-16	12:44	WV, high	150.3 (93.4)	21 (70)	6.0 (11.7)
C-150-2	2001-08-16	13:27	WV, high	149.6 (93.0)	21 (70)	8.7 (16.9)
C-150-3	2001-08-16	14:23	WV, high	149.8 (93.1)	24 (75)	6.7 (13.0)
C-200-1	2001-08-16	12:55	WV, high	199.6 (124.0)	21 (70)	3.7 (7.2)
C-200-2	2001-08-16	13:38	WV, high	199.8 (124.1)	21 (70)	8.8 (17.1)
C-200-3	2001-08-16	14:34	WV, high	199.9 (124.2)	23 (73)	6.8 (13.2)
C-300-1	2001-08-16	11:45	WV, high	299.6 (186.2)	21 (70)	6.5 (12.6)
C-300-2	2001-08-16	13:43	WV, high	299.6 (186.2)	21 (70)	5.5 (10.7)
C-300-3	2001-08-16	14:39	WV, high	299.8 (186.3)	23 (73)	8.4 (16.3)
C-400-1	2001-08-16	11:34	WV, high	400.6 (248.9)	21 (70)	5.5 (10.7)
C-400-2	2001-08-16	13:32	WV, high	400.8 (249.0)	21 (70)	5.3 (10.3)
C-400-3	2001-08-16	14:28	WV, high	400.8 (249.0)	23 (73)	8.3 (16.1)
Reference Concrete Guideway						
D-150-1	2001-08-16	08:48	WV, low	149.9 (93.1)	21 (70)	3.9 (7.6)
D-150-2	2001-08-16	09:38	WV, low	149.7 (93.0)	21 (70)	4.0 (7.8)
D-150-3	2001-08-16	10:09	WV, low	149.7 (93.0)	20 (68)	2.5 (4.9)
D-200-1	2001-08-16	09:06	WV, low	199.5 (124.0)	22 (72)	3.7 (7.2)
D-200-2	2001-08-16	09:49	WV, low	199.8 (124.1)	20 (68)	2.8 (5.4)
D-200-3	2001-08-16	10:49	WV, low	199.7 (124.1)	21 (70)	3.1 (6.0)
D-300-1	2001-08-16	08:53	WV, low	299.6 (186.2)	21 (70)	2.1 (4.1)
D-300-2	2001-08-16	09:54	WV, low	299.5 (186.1)	20 (68)	2.8 (5.4)
D-300-3	2001-08-16	10:54	WV, low	299.6 (186.2)	21 (70)	3.2 (6.2)
D-400-1	2001-08-16	09:42	WV, low	369.9 (229.8)	21 (70)	4.2 (8.2)

D-400-2	2001-08-16	10:14	WV, low	400.9 (249.1)	21 (70)	3.2 (6.2)
Reference Concrete Guideway							
E-150-1	2001-08-17	09:46	WH	149.7 (93.0)	16 (61)	3.0 (5.8)
E-150-2	2001-08-17	10:45	WH	149.7 (93.0)	18 (64)	3.5 (6.8)
E-150-3	2001-08-17	11:17	WH	149.9 (93.1)	18 (64)	4.7 (9.1)
E-200-1	2001-08-17	09:58	WH	199.8 (124.1)	16 (61)	2.6 (5.1)
E-200-2	2001-08-17	13:13	WH	199.6 (124.0)	19 (66)	4.6 (8.9)
E-200-3	2001-08-17	13:59	WH	199.6 (124.0)	19 (66)	2.9 (5.6)
E-300-1	2001-08-17	09:51	WH	299.6 (186.2)	16 (61)	3.0 (5.8)
E-300-2	2001-08-17	14:15	WH	300.3 (186.6)	19 (66)	2.8 (5.4)
E-400-1	2001-08-17	10:02	WH	400.8 (249.0)	16 (61)	2.7 (5.2)
E-400-2	2001-08-17	11:22	WH	401.1 (249.2)	18 (64)	3.7 (7.2)
E-400-3	2001-08-17	12:29	WH	401.3 (249.4)	19 (66)	4.0 (7.8)
Prototype Steel Guideway							
G-150-1	2001-08-22	11:08	WV, high	149.3 (92.8)	23 (73)	5.9 (11.5)
G-150-2	2001-08-22	11:42	WV, high	149.7 (93.0)	24 (75)	5.9 (11.5)
G-150-3	2001-08-22	14:11	WV, high	149.8 (93.1)	26 (79)	4.9 (9.5)
G-200-1	2001-08-22	10:33	WV, high	199.5 (124.0)	22 (72)	4.6 (8.9)
G-200-2	2001-08-22	12:34	WV, high	199.6 (124.0)	24 (75)	4.3 (8.4)
G-200-3	2001-08-22	12:46	WV, high	199.5 (124.0)	25 (77)	4.3 (8.4)
G-300-1	2001-08-22	10:16	WV, high	299.5 (186.1)	21 (70)	4.4 (8.6)
G-300-2	2001-08-22	10:38	WV, high	299.6 (186.2)	22 (72)	5.9 (11.5)
G-300-3	2001-08-22	11:13	WV, high	299.7 (186.2)	23 (73)	5.9 (11.5)
G-400-1	2001-08-22	11:47	WV, high	399.3 (248.1)	24 (75)	5.9 (11.5)
G-400-2	2001-08-22	12:39	WV, high	398.6 (247.7)	25 (77)	5.0 (9.7)
G-400-3	2001-08-22	12:50	WV, high	397.8 (247.2)	25 (77)	3.6 (7.0)

Prototype Steel Guideway						
H-150-1	2001-08-21	09:19	WV, low	149.8 (93.1)	16 (61)	<1.0 (<2.0)
H-150-2	2001-08-21	09:55	WV, low	149.7 (93.0)	17 (63)	1.5 (2.9)
H-150-3	2001-08-21	10:34	WV, low	149.5 (92.9)	18 (64)	3.0 (5.8)
H-200-1	2001-08-21	08:56	WV, low	199.5 (124.0)	16 (61)	<1.0 (<2.0)
H-200-2	2001-08-21	11:14	WV, low	199.6 (124.0)	20 (68)	4.0 (7.8)
H-200-3	2001-08-21	11:26	WV, low	199.4 (123.9)	20 (68)	4.0 (7.8)
H-300-1	2001-08-21	08:43	WV, low	299.2 (185.9)	16 (61)	<1.0 (<2.0)
H-300-2	2001-08-21	09:24	WV, low	299.6 (186.2)	16 (61)	<1.0 (<2.0)
H-300-3	2001-08-21	11:19	WV, low	299.7 (186.2)	20 (68)	4.0 (7.8)
H-400-1	2001-08-21	10:39	WV, low	399.7 (248.4)	19 (66)	3.0 (5.8)
H-400-2	2001-08-21	12:27	WV, low	398.9 (247.9)	20 (68)	4.0 (7.8)
H-400-3	2001-08-21	13:12	WV, low	398.6 (247.7)	22 (72)	2.0 (3.9)
Hybrid Guideway						
I-150-1	2002-05-16	10:00	WV, high	150.1 (93.3)	19 (66)	<1.0 (<2.0)
I-150-2	2002-05-16	10:22	WV, high	150.4 (93.5)	19 (66)	<1.0 (<2.0)
I-150-3	2002-05-16	13:37	WV, high	150.6 (93.6)	20 (68)	3.2 (6.2)
I-200-1	2002-05-16	09:30	WV, high	200.2 (124.4)	18 (64)	<1.0 (<2.0)
I-200-2	2002-05-16	10:02	WV, high	200.3 (124.5)	19 (66)	<1.0 (<2.0)
I-200-3	2002-05-16	10:59	WV, high	200.3 (124.5)	19 (66)	2.1 (4.1)
I-300-1	2002-05-16	09:35	WV, high	300.4 (186.7)	18 (64)	<1.0 (<2.0)
I-300-2	2002-05-16	10:18	WV, high	300.3 (186.6)	19 (66)	<1.0 (<2.0)
I-300-3	2002-05-16	11:04	WV, high	300.3 (186.6)	19 (66)	2.1 (4.1)
I-400-1	2002-05-16	09:25	WV, high	393.6 (244.6)	18 (64)	<1.0 (<2.0)
I-400-2	2002-05-16	10:13	WV, high	395.6 (245.8)	19 (66)	<1.0 (<2.0)
I-400-3	2002-05-16	10:53	WV, high	392.0 (243.6)	19 (66)	1.5 (2.9)

			Hybrid Guideway			
J-150-1	2002-05-16	11:10	WV, low	150.4 (93.5)	19 (66)	1.8 (3.5)
J-150-2	2002-05-16	11:39	WV, low	150.6 (93.6)	19 (66)	2.2 (4.3)
J-150-3	2002-05-16	12:08	WV, low	150.3 (93.4)	19 (66)	3.0 (5.8)
J-200-1	2002-05-16	11:49	WV, low	200.2 (124.4)	19 (66)	3.1 (6.0)
J-200-2	2002-05-16	12:18	WV, low	200.2 (124.4)	19 (66)	<1.0 (<2.0)
J-200-3	2002-05-16	13:27	WV, low	200.3 (124.5)	20 (68)	2.8 (5.4)
J-300-1	2002-05-16	11:53	WV, low	300.2 (186.5)	19 (66)	1.7 (3.3)
J-300-2	2002-05-16	12:23	WV, low	300.4 (186.7)	19 (66)	<1.0 (<2.0)
J-300-3	2002-05-16	13:30	WV, low	300.4 (186.7)	20 (68)	5.0 (9.7)
J-400-1	2002-05-16	11:43	WV, low	392.4 (243.8)	19 (66)	2.6 (5.1)
J-400-2	2002-05-16	12:12	WV, low	392.1 (243.6)	19 (66)	1.9 (3.7)
J-400-3	2002-05-16	13:14	WV, low	392.0 (243.6)	20 (68)	3.2 (6.2)
			Prototype Concrete Guideway			
K-150-1	2001-08-24	08:42	WV, high	149.8 (93.1)	22 (72)	<1.0 (<2.0)
K-150-2	2001-08-24	09:26	WV, high	149.8 (93.1)	22 (72)	<1.0 (<2.0)
K-150-3	2001-08-24	10:13	WV, high	149.7 (93.0)	24 (75)	<1.0 (<2.0)
K-200-1	2001-08-24	09:00	WV, high	199.6 (124.0)	22 (72)	<1.0 (<2.0)
K-200-2	2001-08-24	10:25	WV, high	199.7 (124.1)	24 (75)	<1.0 (<2.0)
K-200-3	2001-08-24	11:30	WV, high	199.6 (124.0)	26 (79)	2.8 (5.4)
K-300-1	2001-08-24	08:47	WV, high	299.4 (186.0)	22 (72)	<1.0 (<2.0)
K-300-2	2001-08-24	10:30	WV, high	299.5 (186.1)	24 (75)	<1.0 (<2.0)
K-300-3	2001-08-24	12:49	WV, high	299.6 (186.2)	27 (81)	2.1 (4.1)
K-400-1	2001-08-24	09:31	WV, high	399.1 (248.0)	23 (73)	<1.0 (<2.0)
K-400-2	2001-08-24	11:35	WV, high	399.5 (248.2)	25 (77)	2.8 (5.4)
			Prototype Concrete Guideway			

L-150-1	2001-08-23	10:35	WV, low	149.5 (92.9)	24 (75)	4.8 (9.3)
L-150-2	2001-08-23	11:20	WV, low	149.9 (93.1)	25 (77)	3.8 (7.4)
L-150-3	2001-08-23	12:17	WV, low	149.7 (93.0)	26 (79)	3.0 (5.8)
L-200-1	2001-08-23	10:53	WV, low	199.2 (123.8)	24 (75)	4.0 (7.8)
L-200-2	2001-08-23	11:31	WV, low	199.5 (124.0)	25 (77)	3.4 (6.6)
L-200-3	2001-08-23	14:09	WV, low	199.7 (124.1)	24 (75)	4.2 (8.2)
L-300-1	2001-08-23	10:40	WV, low	299.3 (186.0)	24 (75)	4.0 (7.8)
L-300-2	2001-08-23	12:22	WV, low	299.8 (186.3)	26 (79)	5.0 (9.7)
L-300-3	2001-08-23	13:23	WV, low	299.4 (186.0)	25 (77)	4.4 (8.6)
L-400-1	2001-08-23	11:25	WV, low	399.0 (247.9)	25 (77)	4.2 (8.2)
L-400-2	2001-08-23	11:36	WV, low	399.4 (248.2)	25 (77)	4.7 (9.1)
L-400-3	2001-08-23	13:10	WV, low	399.6 (248.3)	26 (79)	4.0 (7.8)

It should be noted once again that the test runs of the TR08 listed in Table 2-2 only show those passbys that have been processed. Additional passbys have been recorded for each measuring series (i.e., between 3 and 11 runs); however, these data have not been processed due to data errors, external disturbances, or because they exceeded the number of runs to be processed at each vehicle speed (three).

2.5 ANALYSES AND RESULTS

In the main part of this report, only the main results of the single-microphone and array measurements are presented. More detailed results can be found in Appendices B-D.

2.5.1 Single-Microphone Measurements

Due to the compact test schedule and in order to complete the array and single-microphone measurements simultaneously, the data of the single microphones could not be processed immediately during the measuring campaign but only be stored digitally on tape for later analyses. The data processing was subsequently carried out in the akustik-data laboratory using a B&K measuring amplifier, type 2610, which was fed with the previously recorded raw microphone signals. These signals were then A-weighted and exponentially averaged using the "fast" averaging mode (time constant 125 ms) of the amplifier. Results of the data processing are time histories of the sound-pressure level measured at a maximum of six different microphone positions during passbys of the TR08 travelling on the various guideway types. These measured time histories are the basis for calculating single-number levels that describe the noise emission of a moving object (e.g., a railway train or a maglev vehicle).

Definition of Noise Descriptors
The following noise descriptors are used in this report:

- the **maximum** (sound-pressure) **level**, $L_{Amax,fast}$,
- the **event** (time-equivalent sound-pressure) **level**, $L_{Aeq,E}$,
- the **sound-exposure level** (SEL), and
- the **one-hour** (time-equivalent sound-pressure) **level**, $L_{Aeq,1h}$.

The maximum level refers to the peak level that occurs during the passby of the vehicle. Since the peak level is strongly dependent upon the averaging mode of the measuring instrument, the index "fast" indicates that an integration time according to the "fast" time constant has been used while processing the sound-pressure data. For information purposes, the maximum levels of all TR08 passbys are listed in the Tables in Appendix B.

In contrast to the relatively short time constant of the "fast" averaging, the other three single-number levels are measured by integrating the acoustical energy over time intervals that are in the order of the passage time of the vehicle or even longer. For the event level $L_{Aeq,E}$, the sound-pressure is integrated over a time interval that extends from t_1, the time point when the sound-pressure level of the oncoming vehicle lies ΔL below the maximum level, up to t_2, the time point when the sound-pressure level produced by the departing vehicle lies ΔL below the maximum level, where ΔL is a particular level difference. With the event time $t_E = t_2 - t_1$, the event level is defined by

$$L_{Aeq,E} = 10 \log \left\{ \frac{1}{t_E} \int_{t_1}^{t_2} 10^{L_A(t)/10} dt \right\} \text{ dB(A)}, \qquad (2\text{-}2)$$

where $L_A(t)$ is the A-weighted time-dependent sound-pressure level measured during the passby of the vehicle. For the sound-level difference ΔL, several values are used in practice, viz., $\Delta L = 10$ dB and $\Delta L = 20$ dB, or even higher values, if $L_A(t)$ is to be integrated over all "audible" sound. In order to make calculations with a clear definition, however, values of 10 or 20 dB for ΔL have been employed here. Whereas $\Delta L = 20$ dB could be used for all passbys measured with the microphone beneath the guideway and the 15.2, 25.0, and 30.5 m-microphones (50.0, 82.0, and 100.0 ft-microphones), the ΔL of only 10 dB was required if the recorded signal was so short that a decrease of 20 dB below the maximum level did not occur within the stored data file.

This latter case happened when the signal of the 6.5 m (21.3 ft)-microphone was stored along with the array-microphone data, viz., at the measuring sites with the reference concrete, the prototype steel, the prototype concrete, and the hybrid guideway. The event levels that were determined during the passages of the TR08 are also given in Appendix B. It may be mentioned here that the event level is called the "equivalent continuous sound-pressure level" in ISO 3095, with the measuring time $\underline{t}$ or any level difference ΔL being undefined.

By using a simple equation, the SEL can be derived from the event level:

$$SEL = L_{Aeq,E} + 10 \log (t_E / t_0) \; dB(A), \tag{2-3}$$

where $t_0 = 1$ s. Another simple equation links the event level to the one-hour level. The only difference from Equation (2-3) is that the reference time t_0 is one hour or 3600 s, respectively, instead of one second:

$$L_{Aeq,1h} = L_{Aeq,E} + 10 \log (t_E / 3600 \; s) \; dB(A). \tag{2-4}$$

The one-hour level is the decisive noise descriptor in the German regulations for railway and maglev noise. As a result, values for $L_{Aeq,1h}$ are given in Tables B-1 to B-34 in the appendix in order to fulfill the German requirements in this matter. On the other hand, the one-hour level is not applicable in the U.S. Therefore, the information presented in the next section is restricted to the sound-exposure level.

Summarized Presentations of the Measured Sound-Exposure Levels
Whereas the measured values for the maximum level, the event level, and the one-hour level are only listed in Tables B-1 to B-34 in Appendix B, the representations in Figure 2-7 to Figure 2-12 allow a good survey of the sound-exposure levels measured at the various measuring sites. The aim of the Figures is two fold: on the one hand, they show the SEL as a function of vehicle speed, and, on the other hand, differences in noise emission between the various measuring sites become obvious by putting together all values measured at the same microphone position in one Figure. Accordingly, Figure 2-7 contains the SELs measured at 30.5 m (100.0 ft) distance; Figure 2-8 shows the values at 25.0 m (82.0 ft); the levels at 15.2 m (50.0 ft) can be found in Figure 2-9; the SELs measured with the 6.5 m (21.3 ft)-microphone are in Figure 2-10 (high position) and in Figure 2-11 (low position), and finally, the levels of the microphone beneath the guideway are depicted in Figure 2-12.

In addition to the measured levels, regression curves are also shown in Figures 2-7 through 2-12. These curves were calculated from second-degree polynomials that were fitted to the SELs by minimizing the square root of the variances; i.e., of the sum of the squares of the deviations between measured and computed values. These polynomials are valid only within the speed ranges represented by the measurements. An extrapolation beyond these ranges would not account for the changing character of the dominant sound sources at very low and very high speeds (i.e., there is no physics in the fitted polynomials). Consequently, the coefficients of the polynomials are not given in the report. The regression curves are intended for a better survey of the results only. Nevertheless it is obvious that the SELs are increasing strongly with increasing vehicle speed. AS can also be seen in the Figures, at speeds below about 150 km/h (93 mph), the SELs begin to increase as well. This effect is due to the long event times (t_E) at low vehicle speeds and the mor or less constant emission of the dominant sound sources at low speeds (see Equation 2-3).

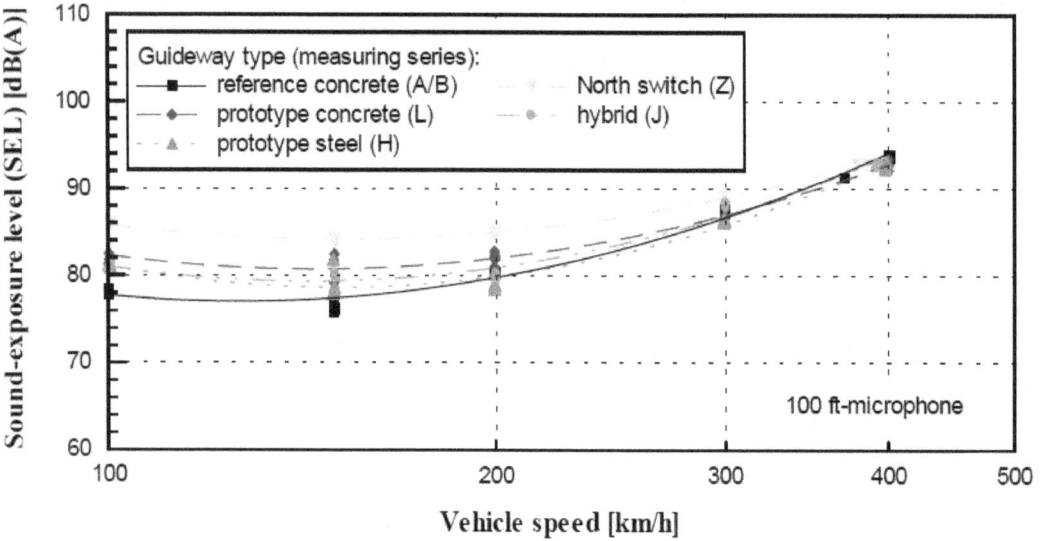

Figure 2-7. Sound-exposure level as a function of vehicle speed measured at 30.5 m (100.0 ft) distance from track centerline and 1.2 m (4.0 ft) above the ground for various guideway types.

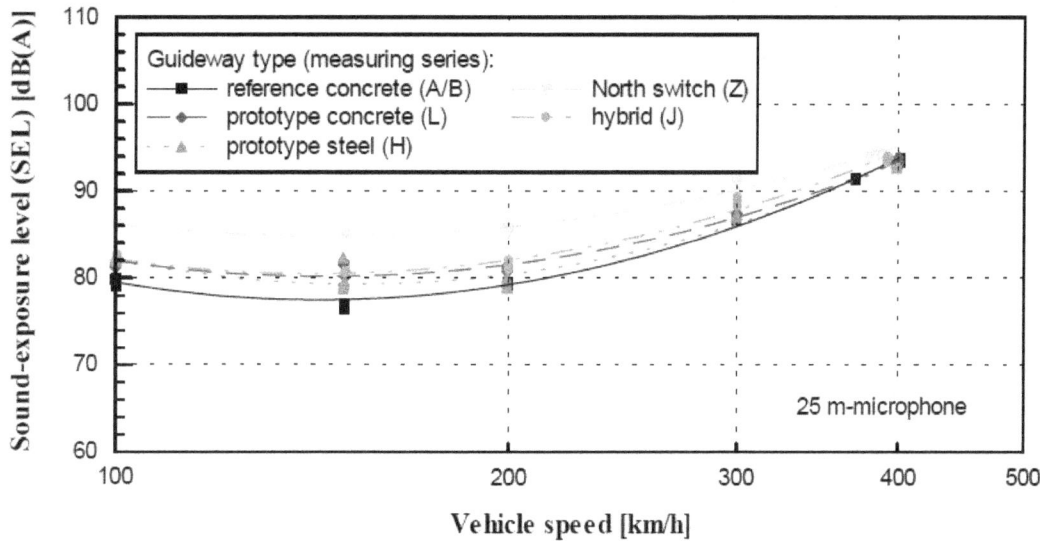

Figure 2-8. Sound-exposure level as a function of vehicle speed measured at 25.0 m (82.0 ft) distance from track centerline and 3.5 m (11.5 ft) above the ground for various guideway types.

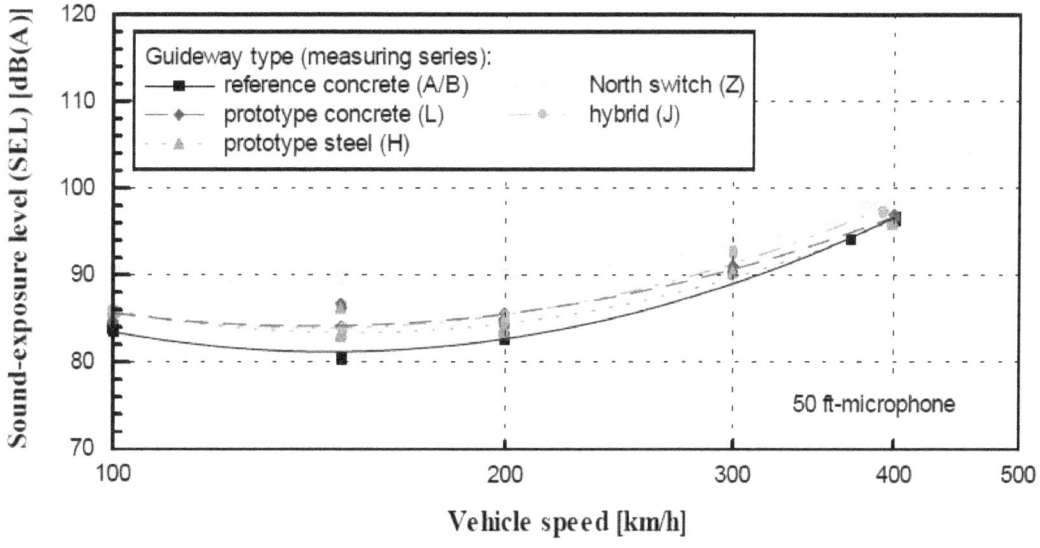

Figure 2-9. Sound-exposure level as a function of vehicle speed measured at 15.2 m (50.0 ft) distance from track centerline and 1.5 m (5.0 ft) above the ground for various guideway types.

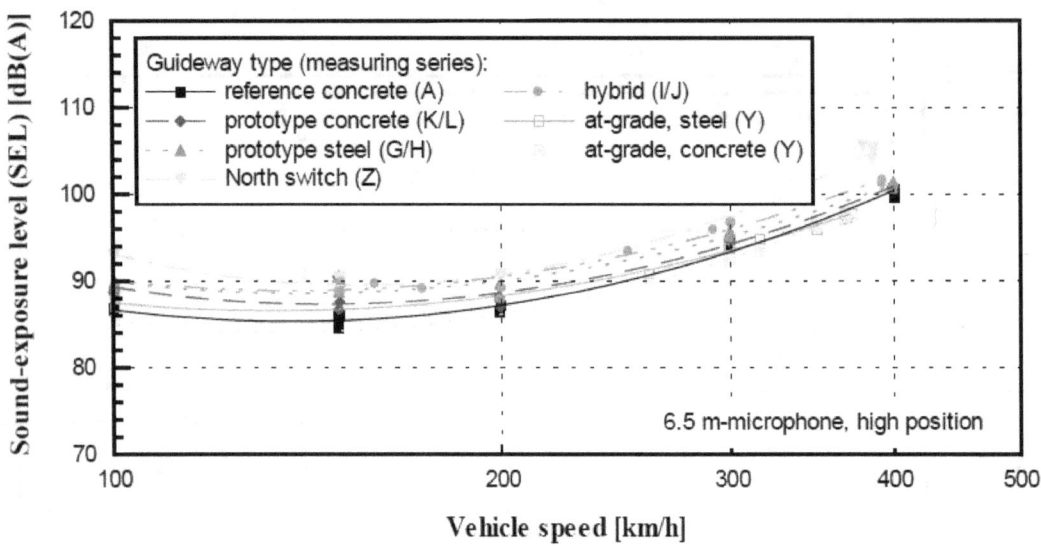

Figure 2-10. Sound-exposure level as a function of vehicle speed measured at 6.5 m (21.3 ft) distance from track centerline and the height of the upper surface of the guideway for various guideway types.

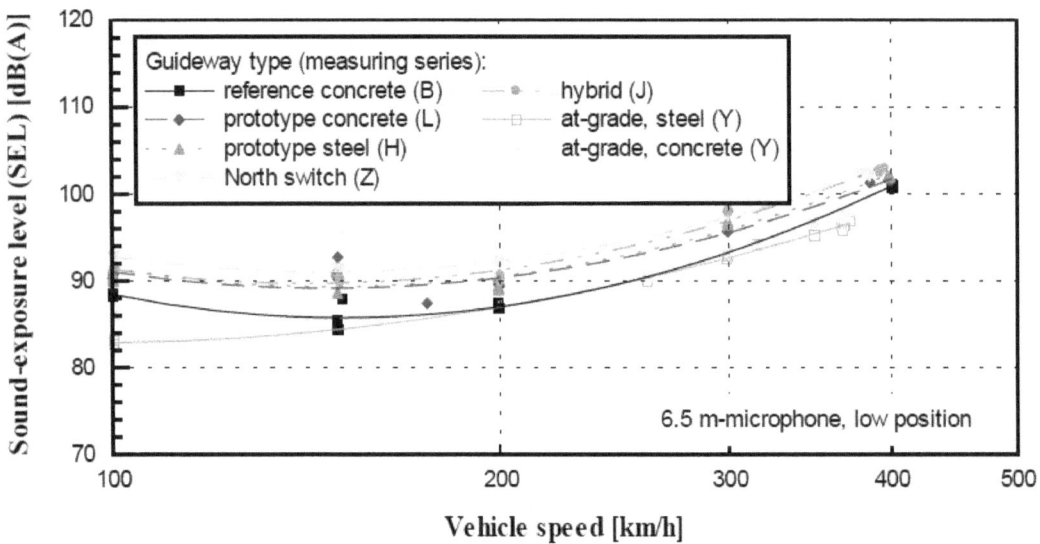

Figure 2-11. Sound-exposure level as a function of vehicle speed measured at 6.5 m (21.3 ft) distance from track centerline and 1.5 m (5.0 ft) below the upper surface of the guideway for various guideway types.

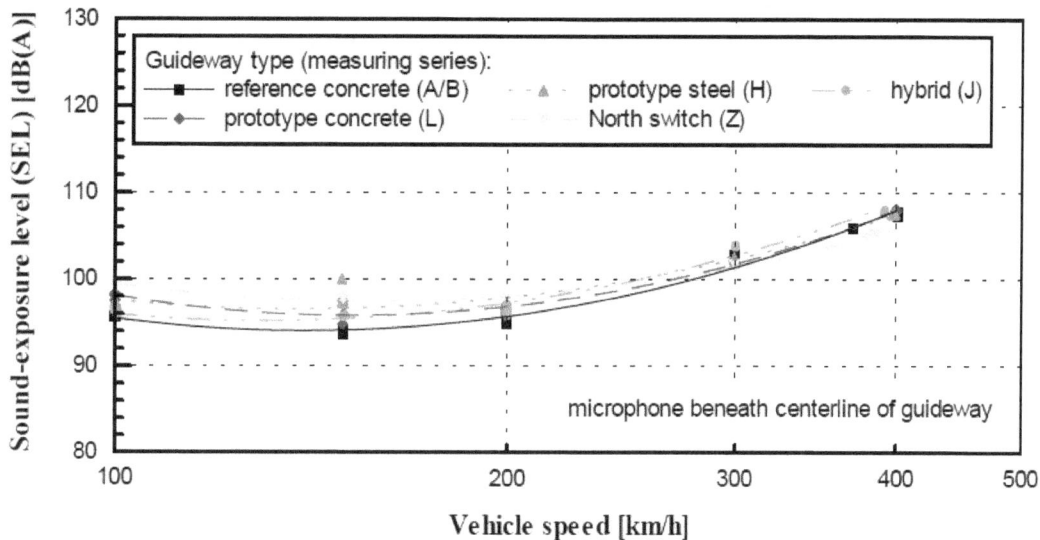

Figure 2-12. Sound-exposure level as a function of vehicle speed measured beneath guideway centerline at a height of 1.5 m (5.0 ft) above the ground for various guideway types.

For each lateral position of the observer, a clear ranking concerning the radiated noise at the various measuring sites can be drawn from Figures 2-7 through 2-12. Generally, passbys of the TR08 travelling on the reference concrete guideway are the quietest, whereas passages on the North switch produce the highest SELs. The other guideway types can be found somewhere in between the reference concrete guideway and the North switch. At a distance of 6.5 m (21.3 ft) from the simulated at-grade guideways however, the measured sound-pressure levels at higher vehicle speeds seem to be as low or even lower than the corresponding levels measured close to the reference concrete guideway. Since investigations at the at-grade guideways have only been made at 6.5 m (21.3 ft) distance, it is not clear if this result is also valid for other measuring distances.

A very promising result concerning the prototype steel guideway can be drawn from Figure 2-7 to Figure 2-9. At 15.2, 25.0, and 30.5 m (50.0, 82.0, and 100.0 ft) measuring distance and at speeds above about 200 km/h (124 mph), the steel beam radiates essentially the same SELs as does the reference concrete guideway. Only at lower speeds between 100 and 150 km/h (62 and 93 mph), the steel beam is 3 to 4 dB(A) louder. As compared with the acoustically untreated steel beam of the North switch, the filling of the prototype steel guideway with light expanded clay aggregate obviously resulted in a substantial reduction of noise.

Another, albeit not unexpected result is a significant diminution of the level differences between the various guideway types with increasing speed. The reason is that at low speeds the sound generated by the guideway determines the total noise emission, whereas at high speeds the aerodynamic sound sources on the vehicle dominate the total noise, and the latter sources having a much higher speed exponent than the mechanically induced sources of the guideway.

As an exception, the measurements beneath the centerline of the guideway do not show the strong dependence upon the guideway type at speeds above about 200 km/h (124 mph). Obviously, the sound radiation from the guideway in the direction of the microphone is so weak that the (aerodynamic) sound sources situated in the levitation frame or guidance chassis of the vehicle play the decisive role.

2.5.2 Microphone-Array Measurements

While the results of the single-microphone measurements shown in the previous section describe the total noise emission during passbys of the TR08, the investigations with the various microphone arrays enable a detailed identification of individual sound sources on both the vehicle and guideway. In this context, determining the acoustical center of the sound radiation during passbys of the TR08 is of great importance, since these results are to be used in the planning phase of a new maglev connection to assess the need for passive sound-abatement measures. Therefore, the microphone-array measurements described here emphasize data from the vertical line array (the WV array) employed at all four measuring sites. However, this array is only able to separate sound sources in the vertical direction (see Section 2.3.2), and not in the horizontal direction. Therefore, two different versions of an X-shaped microphone array (the WX32 and the WX16 array) as well as a horizontal line array (the WH array) were used at the reference concrete guideway in order to separate horizontally adjacent sound sources.

The microphone-array data were processed and analyzed using proprietary software developed by akustik-data Engineering. Version 7.0 of program AD-AUS was employed in the present project. Time-domain beamforming is the fundamental algorithm in AD-AUS. This software cannot be purchased. On the other hand, diagrams and sound-source distributions were plotted using Stanford Graphics 2.1, commercially available software.

Results of the WV Array Positioned at the Reference Concrete, the Prototype Steel, the Prototype Concrete, and the Hybrid Guideway
Although a vertical line array of microphones is only able to separate sound sources in the vertical direction, a quasi two-dimensional presentation of the results can be produced. This is achievable by lining up the one-dimensional distributions measured at different vehicle coordinates during the passby of the vehicle (here, every 20 cm (0.66 ft) along the vehicle, see Section 2.3.2) in one diagram, whereby the sound-pressure levels are plotted by the use of colour-coding. There are two reasons that give these diagrams a two-dimensional character: firstly, the sound-pressure level due to a sound source decreases

progressively with increasing distance from it, and secondly, the localized sources very likely have directional radiation patterns with their maxima directed more or less towards the wayside. Nevertheless, the localized sound sources cause horizontal "stripes" in the presentations due to the fundamental lack of the WV array's resolution in the horizontal direction.

Figure 2-13 through 2-16 show such two-dimensional sound-source distributions measured at the four sites using the WV array. In order to obtain these distributions, the decisive and time consuming step after processing the data of all individual passbys was to average the results measured at the same vehicle speed, the same array height and the same measuring site. The two resulting average sound-source distributions taken from the high and the low measuring position of the array were then merged to form a single graphical representation at each speed and each measuring site. Consequently, the data of a total of 94 passages of the TR08 (see Table 2-2) have been processed for Figure 2-13 to Figure 2-15. In these Figures, the sound-source distributions for heights between -2.5 and +1.5 m (-8.2 and +4.9 ft) relative to the upper surface of the guideway stem from the results of the WV array in its low position, whereas for heights between +1.5 and +4.0 m (+4.9 and +13.1 ft), the distributions have been taken from the high position of the WV array. Any possible unsteadiness in sound-pressure levels that might appear at the line of intersection of both distributions (i.e., at 1.5 m (4.9 ft) height above the guideway) has no physical meaning.

The data in Figure 2-13 through Figure 2-16 are also presented in Figures C-1 through C-4 (Appendix C). While Figures 2-13 through 2-16 present a clear trend in the location of sound sources as a function of speed, Figures C-1 through C-4 present each speed's sound sources in more detail.

Figure 2-13. Averaged sound-source distributions measured with the WV array during passbys of the TR08 on the reference concrete guideway at speeds between 150 and 400 km/h (93 and 249 mph).

Figure 2-14. Averaged sound-source distributions measured with the WV array during passbys of the TR08 on the prototype steel guideway at speeds between 150 and 400 km/h (93 and 249 mph).

Figure 2-15. Averaged sound-source distributions measured with the WV array during passbys of the TR08 on the prototype concrete guideway at speeds between 150 and 400 km/h (93 and 249 mph).

Figure 2-16. Averaged sound source distributions measured with the WV array during passbys of the TR08 on the hybrid guideway at speeds between 150 and 400 km/h (93 and 249 mph)

As can be seen in Figures 2-13 through 2-16, the regions of maximum sound radiation depend mainly on the vehicle speed, with the guideway type having an influence on the ranking of the individual sound sources. These Figures and, particularly, Figures C-1 to C-4 in Appendix C with their individually adjusted dynamic ranges of the sound-pressure level, show that at speeds of 150 and 200 km/h (93 and 124 mph), the emission of the guideway is by far the most important one. In this context, the sound level produced by the reference concrete guideway is the lowest, for the prototype concrete guideway it is approximately 2 dB(A) higher and for the prototype steel guideway, the level is about 4 dB(A) higher relative to the reference concrete guideway. The strength of the hybrid guideway's radiation ranks approximately between the prototype concrete and the prototype steel guideway.

For the reference concrete guideway, at 200 km/h (124 mph), individual sound sources can already be detected on the vehicle (e.g., near the front of the vehicle at the gap between vehicle floor and the upper surface of the guideway, at the beginning of the cover of the levitation frame as well as along the entire lower edge of this cover). At a height of about 2 m (6.6 ft), a single sound source or sources detected in the area of transition between vehicle sections 1 and 2 (and, to a minor degree, between sections 2 and 3) also occur with the same intensity in Figure 2-14 and 2-16, and Figures C-1 to C-4, respectively. Furthermore, at 100 and 200 km/h (93 and 124 mph), individual points of emission to be found during all passbys of the TR08 at a height of approximately 0.6 m (2.0 ft) above the guideway are obviously related to a series of inlets or outlets of cooling fans for the levitation and guidance magnets in that area. These openings are covered with louvers, which are also known to be substantial sources of aerodynamic noise.

With increasing speed, compact local sound sources on the vehicle become more important in terms of intensity and significance, whereas the guideway becomes a less important factor. This can be observed particularly well for those sources located at a height of 2 m (6.6 ft) in the two areas of transition between vehicle sections, whose source strength is only exceeded by the two sources at the head of the vehicle described above. Sound sources at such large heights of 2 m (6.6 ft) are of particular importance for the localization, since they have a decisive influence on the dimensioning of passive sound-abatement measures.

The differences in the maximum levels in the graphical representations measured during passbys at the same speed, which seem to contradict the above explanations, can basically be attributed to the different measuring distances required for the reference concrete and the hybrid guideway on the one hand and the two prototype guideways on the other. This means that compared with the reference concrete and the hybrid guideway, for each localized sound source an approximately 1.5 dB(A) higher level can be expected at the prototype steel and the prototype concrete guideway.

The quantification of the sound radiation from the individual sound sources already localized will be done in a subsequent section, where the results of the WH and WX arrays can also be included. In the following section, vertical sound-source distributions averaged along the vehicle will be presented.

Vertical Distributions of the Sound-Exposure Level
The average quasi two-dimensional sound-source distributions presented in the preceding section, form the basis of the vertical distributions that will be shown in this section. These vertical distributions will primarily be produced in order to give a good visual impression of the regions that contribute the most acoustical energy to the total sound radiation during passbys of the TR08. Imagine now that horizontal cuts can be made through the two-dimensional representations in Figure 2-13 to 2-16. Data for such cuts exist in the corresponding data files for height coordinates from -2.5 to +4.0 m (-8.2 to 13.1 ft) in steps of 10 cm (0.33 ft), which is the spacing of the focal points in the vertical direction (see Section 2.3.2).

The data of each cut (which are also available for a certain distance in front of the vehicle and behind it) represent a time history of the sound-pressure level at the selected height coordinate. Each of these time histories can then be averaged, resulting in a single-number level. Here, an integration over the total time of the existing data was carried out. By referring this time (the event time) to one second, the sound-exposure level is calculated. When these SEL values are lined up and plotted at each vehicle speed and for each guideway type, vertical sound-source distributions emerge and are shown in Figure 2-, Figure 2-, Figure 2-, and 2-20 for speeds of 150, 200, 300, and 400 km/h (93, 124, 186, and 249 mph), respectively.

Firstly, these vertical distributions confirm that at low speeds the dominating sound originates from the guideway, whereas the higher the speed the more the sound sources on the vehicle also contribute essentially to the total noise, until, at about 400 km/h (249 mph), the sources on the vehicle determine the overall sound radiation (at least during passbys on the reference concrete and the prototype concrete guideway). At all speeds, a significant difference of the sound radiation in the region of the guideway between about -2.0 and -1.0 m (-6.6 and -3.3 ft) can be seen in the Figures. By taking into account the different measuring distances (see above), the prototype concrete, the prototype steel, and the hybrid guideway generate levels that are at an average approximately 1 to 3 dB(A) higher than those from the reference concrete guideway.

At the height coordinate of about -0.9 m (-3.0 ft), in all vertical distributions a local maximum appears that is most pronounced in the case of the hybrid guideway. This maximum obviously corresponds to the sources that have already been recognized in Figure 2-13 to 2-16Figure 2-15 at the lower edge of the cover of the levitation frame. A further local maximum can be found in all vertical distributions at 300 and 400 km/h (186 and 249 mph) at a height of about +0.5 m

(1.6 ft). Again, the sound sources responsible for this maximum have been mentioned in connection with the two-dimensional distributions, viz., the row of openings for the cooling of the levitation and guidance magnets. Moreover and particularly at 400 km/h (249 mph), a third maximum at about 2.0 m (6.6 ft) is visible in the vertical distributions. This maximum stems from the strong sound sources located at that height in the regions of the transition from vehicle sections 1 to 2 and from sections 2 to 3. These sources could also be seen clearly before in the two-dimensional distributions.

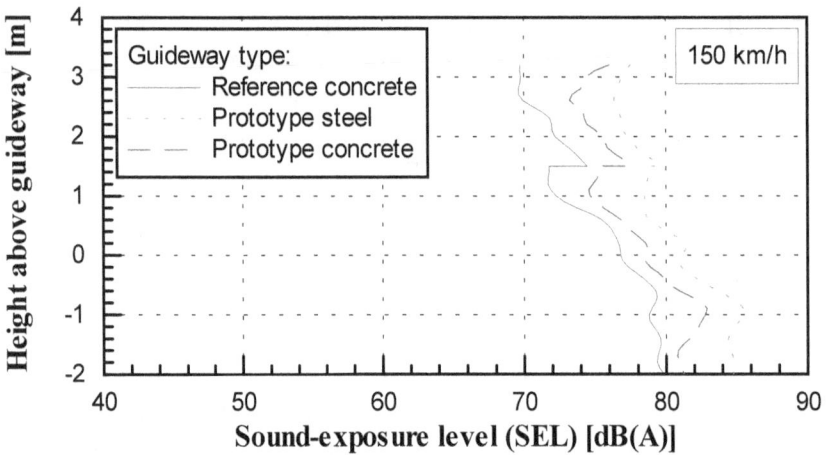

Figure 2-17. Vertical sound-source distributions measured with the WV array during passbys of the TR08 on the different guideway types at a speed of 150 km/h (93 mph).

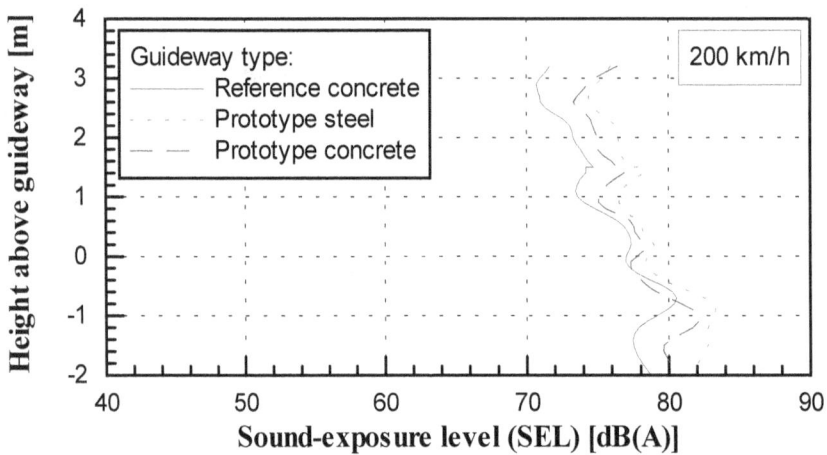

Figure 2-18. Vertical sound-source distributions measured with the WV array during passbys of the TR08 on the different guideway types at a speed of 200 km/h (124 mph).

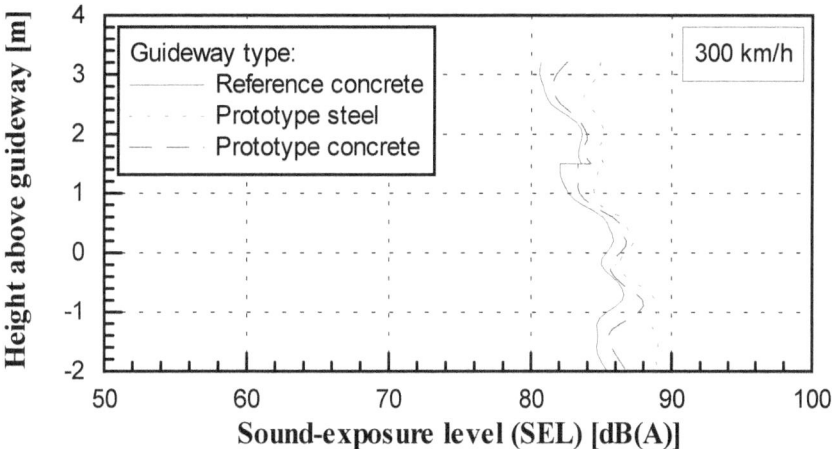

Figure 2-19. Vertical sound-source distributions measured with the WV array during passbys of the TR08 on the different guideway types at a speed of 300 km/h (186 mph).

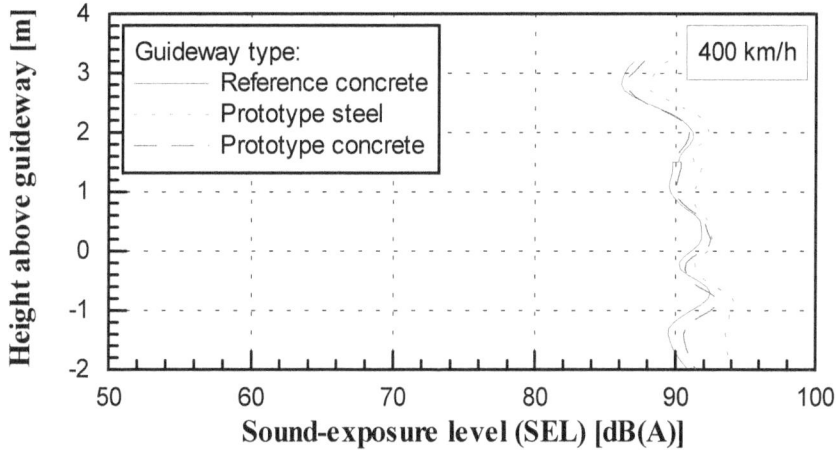

Figure 2-20. Vertical sound-source distributions measured with the WV array during passbys of the TR08 on the different guideway types at a speed of 400 km/h (249 mph).

The vertical distributions also show that the level differences in the region of the guideways are more or less independent of vehicle speed, whereas the level differences at higher vehicle coordinates vanish with increasing speed. Since different measuring distances had to be used at the various measuring sites, the remaining level differences at higher speeds must be attributed to these circumstances. In principle, it was expected that also at the lower speeds the sound originating from the upper region of the vehicle is more or less identical for passbys on different guideway types. In Figure 2- and Figure 2-, however, a considerable deviation of the measured sound-source distributions can be observed in this region. Possibly, this phenomenon is connected with different vibration levels of the vehicle's outer surface while travelling on different guideway types.

Results of the WH Array Positioned at the Reference Concrete Guideway
Figure 2- shows one-dimensional sound-source distributions measured with the WH array during passbys of the TR08 travelling on the reference concrete guideway at about 150, 200, 300, and 400 km/h (93, 124, 186, and 249 mph). Due to the ability of the WH array to separate sound sources in the direction of the vehicle only, sound produced by sources located at different heights cannot be resolved. Particularly, no differentiation is possible between noise coming from the guideway or vehicle. Therefore, the results of the WH array should be interpreted in combination with the results of the WV and the WX arrays.

Nevertheless, the sound-source distributions in Figure 2- contain some interesting results. First of all, at low speeds (150 and 200 km/h (93 and 124 mph)) a rather regular pattern of peaks occurs in the curves. Many of these peak sound levels can easily be attributed to the noise coming from the ventilation openings that are arranged in regular distances along the vehicle at about 0.5 m (1.6 ft) above the guideway. With increasing speed, however, this regularity in the curves disappears, since the acoustical emission of the ventilation openings is superimposed by other sound sources.

At the higher speeds of 300 and 400 km/h (186 and 249 mph), the peak level of an isolated single source in the region of the vehicle's nose is clearly visible in the curves. This source, which can also be detected at the same vehicle coordinate at the two lower speeds, is localized by the WH array exactly at the beginning of the cover of the levitation frame. This probably indicates that the flow separating from the contour of the cover is the origin of the sound source.

At 400 km/h (249 mph), the highest sound level appears at the vehicle coordinate of about 5.5 m (18.0 ft). It is not clear whether this radiated sound is caused by the antenna on the TR08's roof or some component of the front door (or in its close vicinity), which is also located at the same vehicle coordinate. This latter explanation should be considered, since distinct peaks also occur in the sound-source distributions close to the doors at the transition from vehicle section 1 to section 2. It should be mentioned here, too, that the average sound levels measured at vehicle coordinates, say, between 5 and 20 m (16 and 66 ft) are significantly higher compared with the average levels in the middle or rear part of the TR08. Further results of the WH array (i.e., results of the WH08 sub-array) will be presented in the following section in connection with selected results of the WX16 array.

Figure 2-21. Horizontal sound-source distributions measured with the WH array during passbys of the TR08 on the reference concrete guideway at speeds between 150 and 400 km/h (93 and 249 mph).

Results of the WX Arrays Positioned at the Reference Concrete Guideway
In addition to determining the acoustical center of the vehicle's sound emission, it is particularly interesting to identify the relevant individual sound sources. Therefore, measurements have also been carried out with two X-shaped arrays which make it possible to produce truly two-dimensional sound-source distributions.

Figure 2- shows such distributions measured with the WX32 array in the frequency range from 280 to 1120 Hz during passbys of the TR08 travelling on the reference concrete guideway at speeds from about 150 to 400 km/h (93 to 249 mph). In the restricted frequency range of the WX32 array and at low speeds, the most important sound sources can be found at one particular ventilation opening of each vehicle section slightly above the guideway. These sources are located at the vehicle coordinates of about 8.5, 33.5, and 58.5 m (27.9, 109.9, and 191.9 ft) with respect to the beginning of the vehicle. At the higher speeds, only one sound source seems to dominate the sound emission of the vehicle: it is localized in the lower region of the front shape of the vehicle and is obviously generated by flow interactions with the vehicle floor, the upper surface of the guideway, and the beginning of the cover of the levitation frame.

In Figure 2-, the results of the WX16 array in its working frequency range from 1120 to 2240 Hz show an even more precise localization of the aforementioned sound sources at one particular ventilation opening of each section of the TR08 as compared to the results of the WX32 array in Figure 2-. Moreover, in the entire speed range from 150 to 400 km/h (93 to 249 mph), additional sound sources can be pinpointed at a height of about 2.0 m (6.6 ft) above guideway in the regions of the transition from section 1 to 2 and the transition from section 2 to 3. To be more exact, sources are found in this region between the doors and the adjacent window where the inlet or outlet openings of another cooling system, i.e., of the air-conditioning units, are situated. Depending possibly on the actual power of the individual units, these sources are clearly visible in Figure 2- or even disappear in the background noise.

In Figure 2-, a superposition of results of the WX16 array and the WH08 array in the frequency range from 2240 to 4500 Hz is presented. On the one hand, it can be stated that the agreement of the one- and the two-dimensional sound-source distributions measured with two completely independent array configurations is very good, even though different passbys of the TR08 have been investigated. For instance, the three peak levels in the one-dimensional distribution from the WH08 array appear exactly at the locations where the ventilation openings are detected by the WX16 array, viz., at the above mentioned vehicle coordinates. This can be recognized for 150 and 200 km/h (93 and 124 mph).

Figure 2-22. Two-dimensional sound-source distributions measured with the WX32 array during passbys of the TR08 on the reference concrete guideway at speeds between 150 and 400 km/h (93 and 249 mph).

Figure 2-23. Two-dimensional sound-source distributions measured with the WX16 array during passbys of the TR08 on the reference concrete guideway at speeds between 150 and 400 km/h (93 and 249 mph).

Figure 2-24. Simultaneous presentation of one- and two-dimensional sound-source distributions measured with the WH08 and the WX16 array during passbys of the TR08 on the reference concrete guideway at speeds between 150 and 400 km/h (93 and 249 mph).

On the other hand, the results of the WH08 array reveal additional sound sources at the vertical contours of each door of the vehicle at the higher speeds. Although each of these sound sources is only of minor importance for the overall sound emission of the TR08, the elimination of all of these sources could both reduce the exterior noise of the vehicle and, by reducing the interior noise, also improve the travelling comfort of the passengers.

Localized Sound Sources of the TR08 Resulting from the Array Measurements

As a result of the measurements using the WV, WH, WX32, and WX16 array, a number of individual sound sources could be localized on the TR08 and the different guideway types. Each of these sources has already been described in the preceding sections and how it has been detected. For some of these sources, the results measured with the different arrays were combined in order to calculate the strength and speed exponent of the source under investigation. Both locally compact regions of sound radiation and sound emissions more or less distributed along a line were found in Figure 2-13 to 2-16 and Figure 2-1 to 2-24. For the first category, the sources can be represented with sufficient accuracy by point sources, whereas the radiation of line sources will be modeled by a row of several point sources having a certain spacing in the horizontal direction.

The emission of the individual point sources (index i) on the TR08 and the guideway can be described by equations of the kind

$$L_{A,i}(5\ m) = L_{A,200,i} + \alpha_i \cdot 10 \log (v / 200\ km/h)\ dB(A), \qquad (2\text{-}5)$$

where $L_{A,i}$ is the A-weighted sound-pressure level at 5.0 m (16.4 ft) distance from the source and v is the speed of the vehicle. For each sound source i, the value of the reference level $L_{A,200,i}$ at the reference speed of 200 km/h (124 mph) and the speed exponent α_i are listed in Table 2-3. Equation (2-5) is based on the so-called "power law", where sound pressure is proportional to vehicle speed raised to a certain power and the magnitude of the speed exponent depends on the sound generating mechanism.

For each localized, locally compact sound source, the reference level $L_{A,200}$ and the speed exponent α have been calculated by feeding the sound-pressure levels measured with the different arrays into a regression analysis and then transforming the reference level to represent its value at 5.0 m (16.4 ft) distance. This procedure was applied in the cases of the flow-induced source located close to the nose of the TR08, the four sources caused by the air-conditioning units at 2.0 m (6.6 ft) height near the transitions of the vehicle sections, and the three sources in connection with particular ventilation openings at about 0.5 m (1.6 ft) height.

Table 2-3. Parameters of localized sound sources

Sound Source	Vehicle Coordinate [m (ft)]	Height Above Guideway [m (ft)]	Reference Level $L_{A,200}$ [dB(A)]	Speed exponent α	Length of Line Source [m (ft)]	Spacing between Point Sources [m (ft)]
Separating flow at vehicle nose	2.0 (6.6)	0	76.4	6.0	-	-
Air-Conditioning units	25 (82)	2.0 (6.6)	69.7	7.8	-	-
	29 (95)					
	50 (164)					
	54 (177)					
Ventilation openings	8.5 (27.9)	0.5 (1.6)	72.7	1.4	-	-
	33.5 (109.9)					
	58.5 (191.9)					
Lower edge of levitation frame	-	-0.9 (-3.0)	70.9	4.3	75.0 (246.0)	5.0 (16.4)
Reference concrete guideway (150 km/h)	-	-1.5 (-4.9)	70.0	-	90.0 (295.3)	2.5 (8.2)
Reference concrete guideway (200 to 400 km/h)			72.5	4.5		
Prototype concrete guideway (150 km/h)	-	-1.5 (-4.9)	71.0	-		
Prototype concrete guideway (200 to 400 km/h)			73.5	4.5		
Hybrid guideway (150 km/h)		-1.5 (-4.9)	72.0	-		
Hybrid guideway (200 to 400 km/h)			74.5	4.5		

Prototype steel guideway (150 km/h)	-	-1.5 (-4.9)	73.0	-		
Prototype steel guideway (200 to 400 km/h)			75.5	4.5		

For the localized line sources, viz., the emission coming from the lower edge of the cover of the levitation frame and the guideway, another procedure was used to determine the reference level $L_{A,200}$ and the speed exponent α. First of all, the values for the single (identical) point sources that form the corresponding line source were estimated on the basis of the results of the WV array. Whereas the speed exponent found this way remained unchanged, the reference level $L_{A,200}$ was varied until the best agreement between measurement and prediction could be achieved. Moreover, it was assumed that a short length of the guideway in front of the vehicle and behind it is also contributing to the radiated sound. This assumption leads to an effective length of the guideway, i.e., the row of point sources, of 90 m (295 ft), whereas the row of point sources representing the radiation from the lower edge of the levitation frame has a length of 75 m (246 ft), which is the actual length of the levitation frame.

Probably due to resonance effects, the sound radiation from the guideways can only be described by equation (2-5) between 200 and 400 km/h (124 and 249 mph). Accordingly, the reference levels listed in Table 2-3 are only valid in this restricted speed range. Therefore, the emission level at 150 km/h (93 mph) is given separately in the Table.

Sound-Pressure Spectra Measured with the WV Array
The following unweighted narrow-band and one-third octave-band spectra of the sound pressure have been obtained by using the WV array. In each of the following figures, Figure 2- to Figure 2-, spectra measured at five different vehicle heights are shown. The selected height coordinates having an interval of 0.7 m (2.3 ft) range from -1.4 to +2.1 m (-4.6 to +6.9 ft) relative to the upper surface of the guideway. The spectra have been obtained during passages of the TR08 as it traveled on the reference concrete, prototype steel, prototype concrete, and hybrid guideway at speeds of approximately 150, 200, 300, and 400 km/h (93, 124, 186, and 249 mph). The spectra at heights between -1.4 and +0.7 m (-4.6 and +2.3 ft) stem from the WV array in its low position, whereas the spectra at +2.1 m (+6.9 ft) height have been measured using the high WV array.

Except for the height coordinate of 0 m, the other heights selected here correspond to the heights of the local maxima that could be noticed in the vertical sound-source distributions in Figure 2- to Figure 2-. Moreover, all spectra were averaged over the total time of the existing data as it was the case for the distributions shown in these Figures. This means that the integration of the data was not only carried out along the entire length of the vehicle, but also a certain length in front of it and behind it. By doing so, the spectra represent the sound-pressure levels of the above distributions at selected heights (-1.4, -0.7, 0, +0.7, and 2.1 m (-4.6, -2.3, 0, +2.3, and +6.9 ft)). Due to the long integration time, however, no detailed spectral information concerning individual sound sources can be expected from the spectra. Thus, only a general overview of the spectral contents at the different vehicle heights can be given.

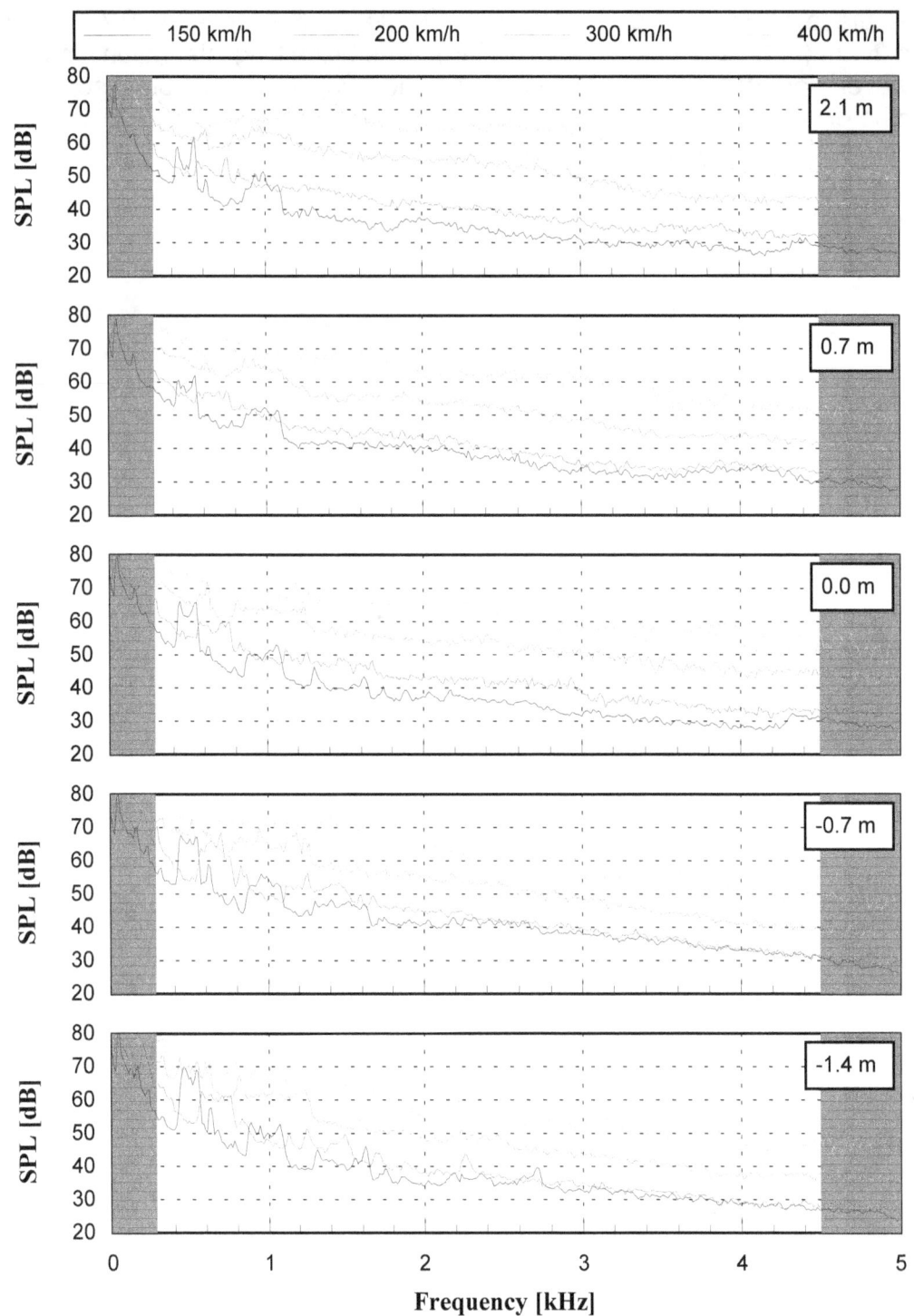

Figure 2-25. Unweighted narrow-band spectra of the sound-pressure level (SPL) measured with the WV array during passbys of the TR08 on the reference concrete guideway at speeds between 150 and 400 km/h (93 and 249 mph); Δf = 11.6 Hz.

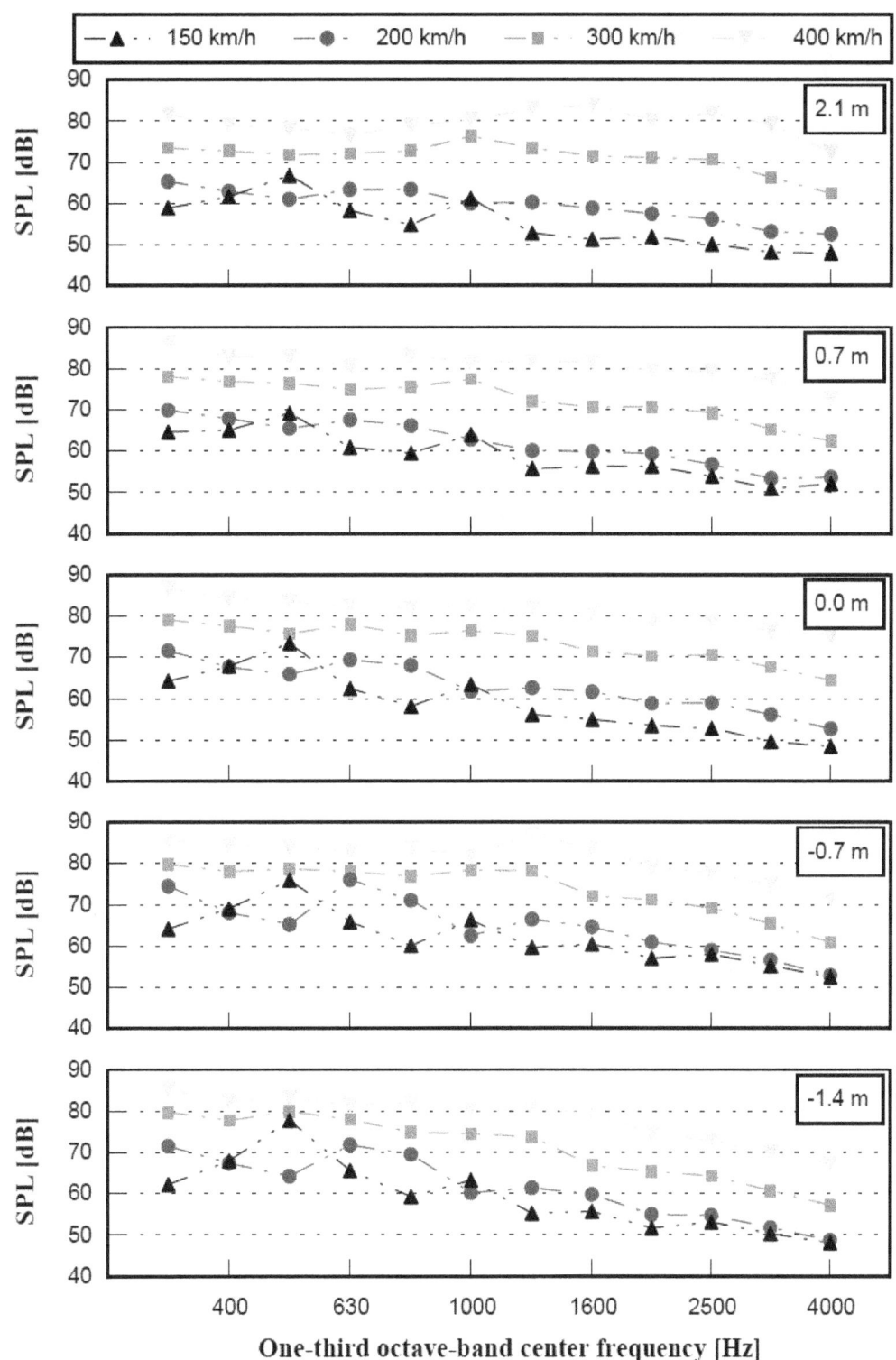

Figure 2-26. Unweighted one-third octave-band spectra of the sound-pressure level (SPL) measured with the WV array during passbys of the TR08 on the reference concrete guideway at speeds between 150 and 400 km/h (93 and 249 mph).

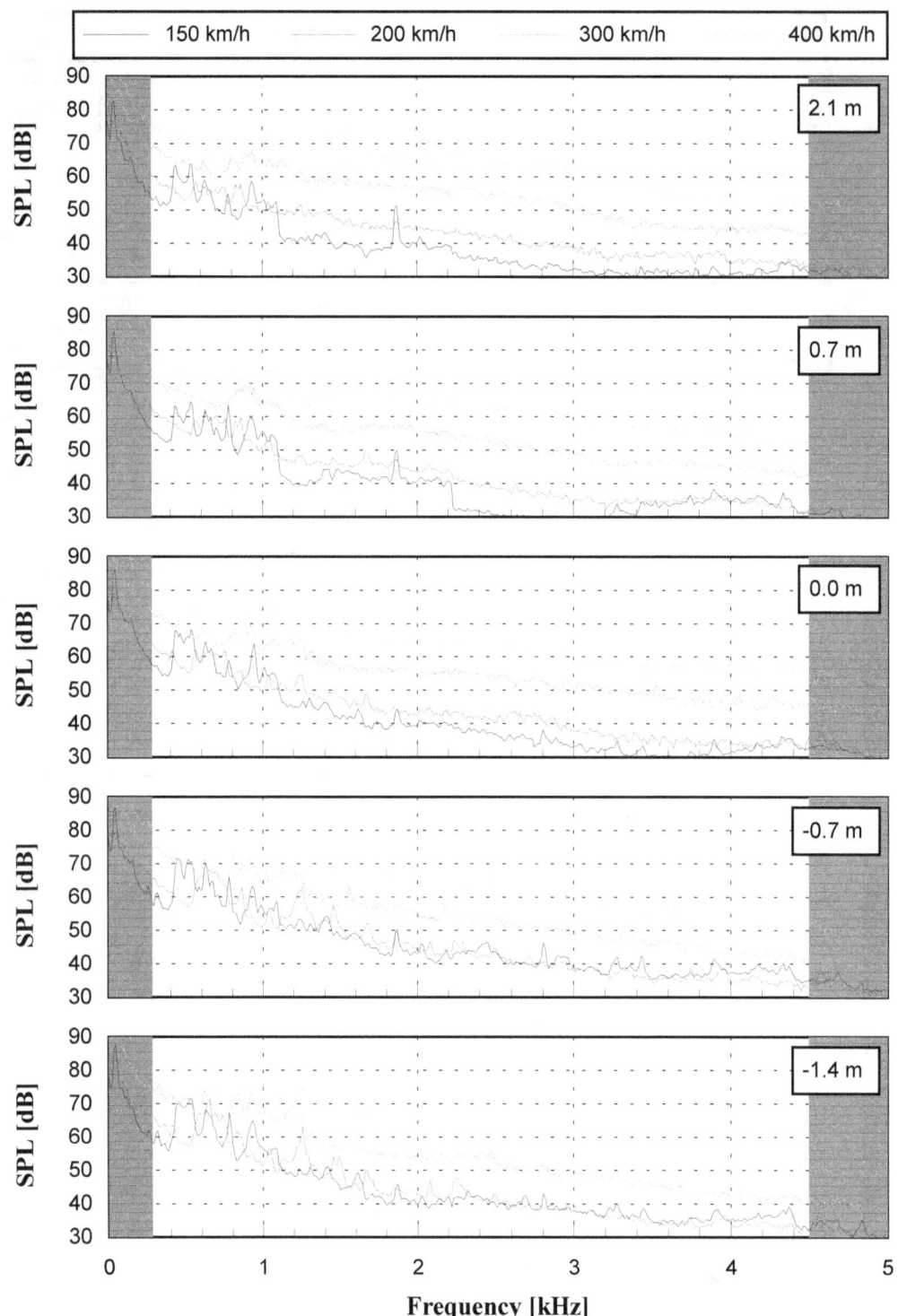

Figure 2-27. Unweighted narrow-band spectra of the sound-pressure level (SPL) measured with the WV array during passbys of the TR08 on the prototype steel guideway at speeds between 150 and 400 km/h (93 and 249 mph); Δf = 11.6 Hz.

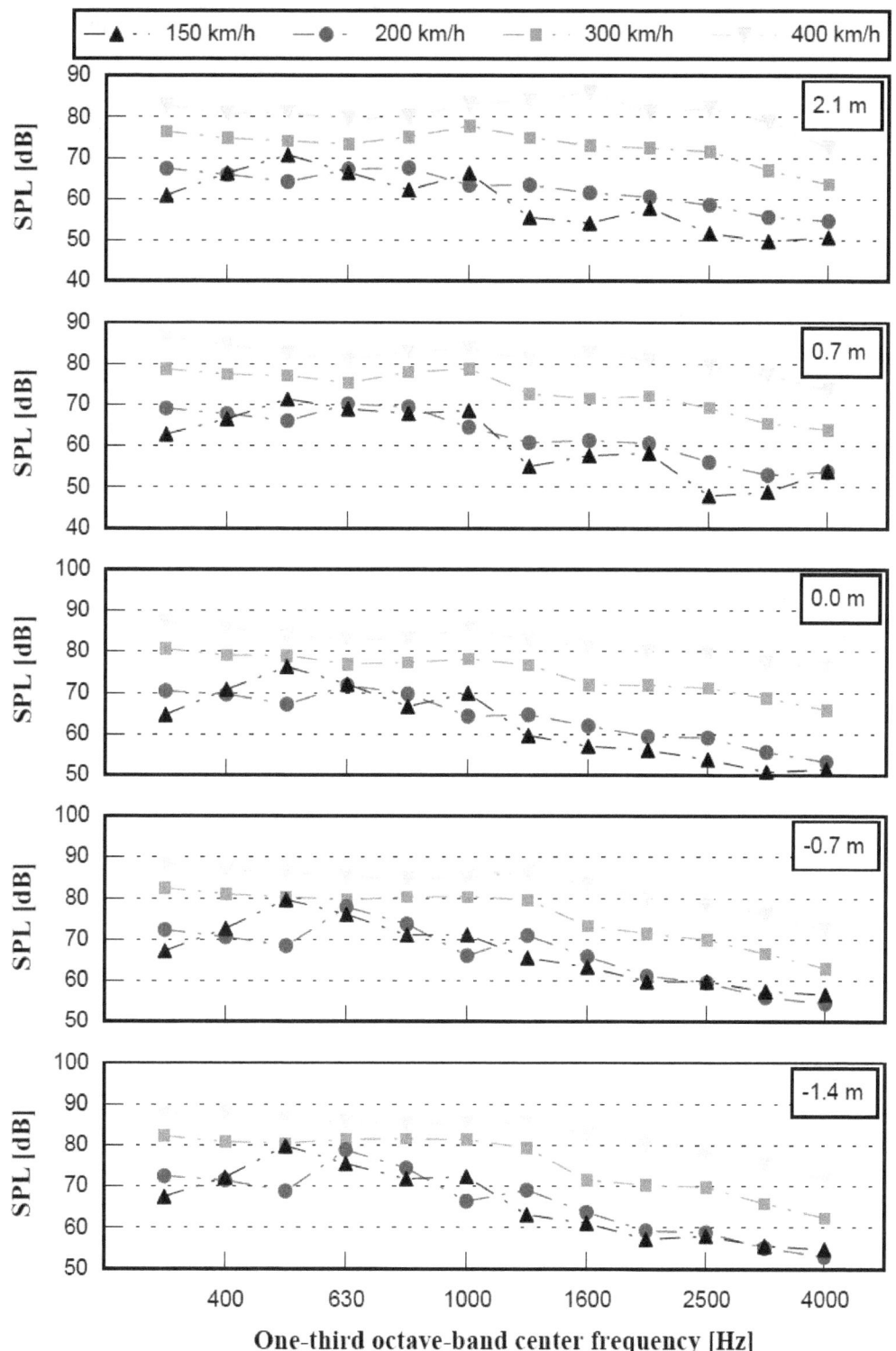

Figure 2-28. Unweighted one-third octave-band spectra of the sound-pressure level (SPL) measured with the WV array during passbys of the TR08 on the prototype steel guideway at speeds between 150 and 400 km/h (93 and 249 mph).

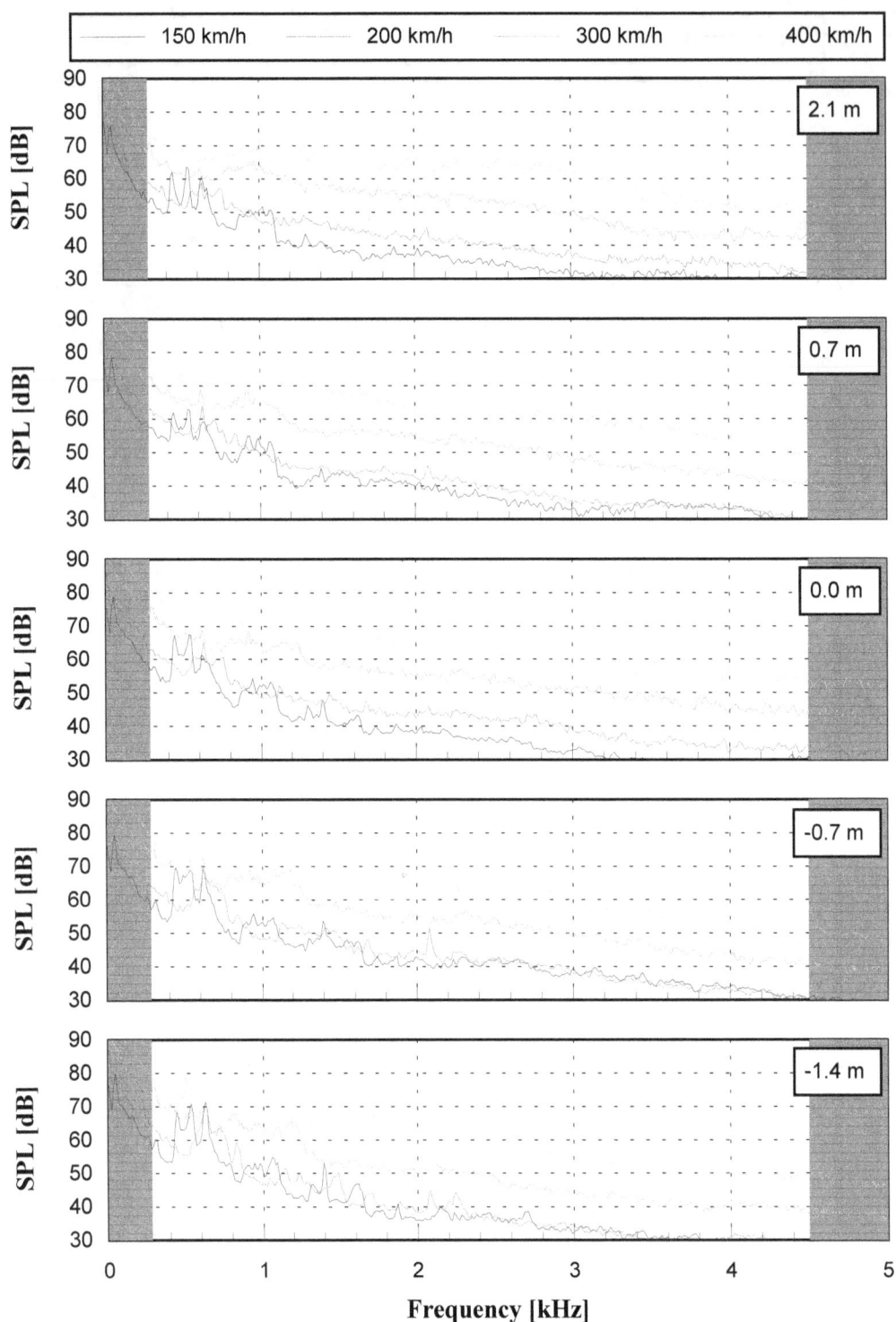

Figure 2-29. Unweighted narrow-band spectra of the sound-pressure level (SPL) measured with the WV array during passbys of the TR08 on the prototype concrete guideway at speeds between 150 and 400 km/h (93 and 249 mph); $\Delta f = 11.6$ Hz.

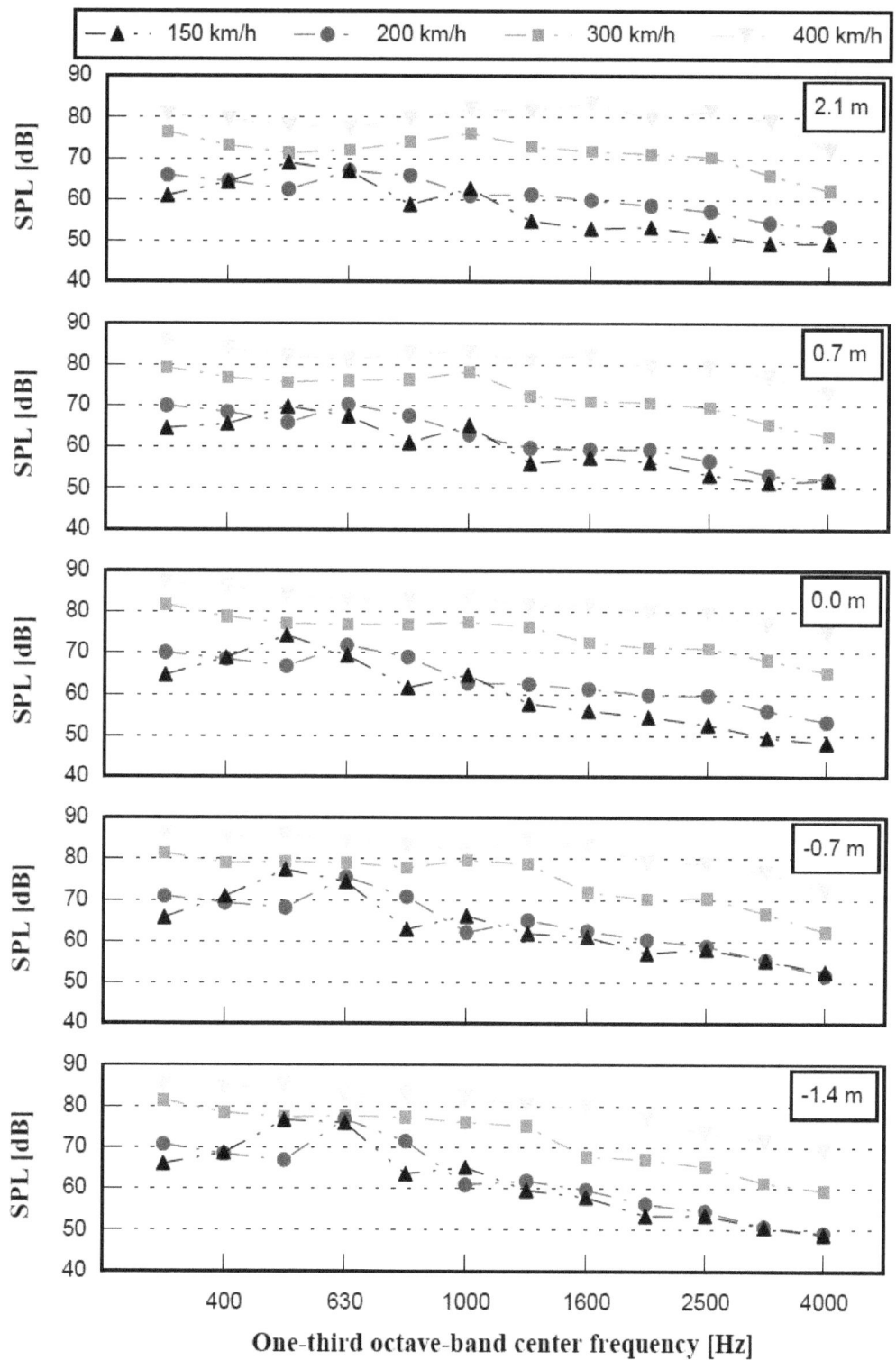

Figure 2-30. Unweighted one-third octave-band spectra of the sound-pressure level (SPL) measured with the WV array during passbys of the TR08 on the prototype concrete guideway at speeds between 150 and 400 km/h (93 and 249 mph).

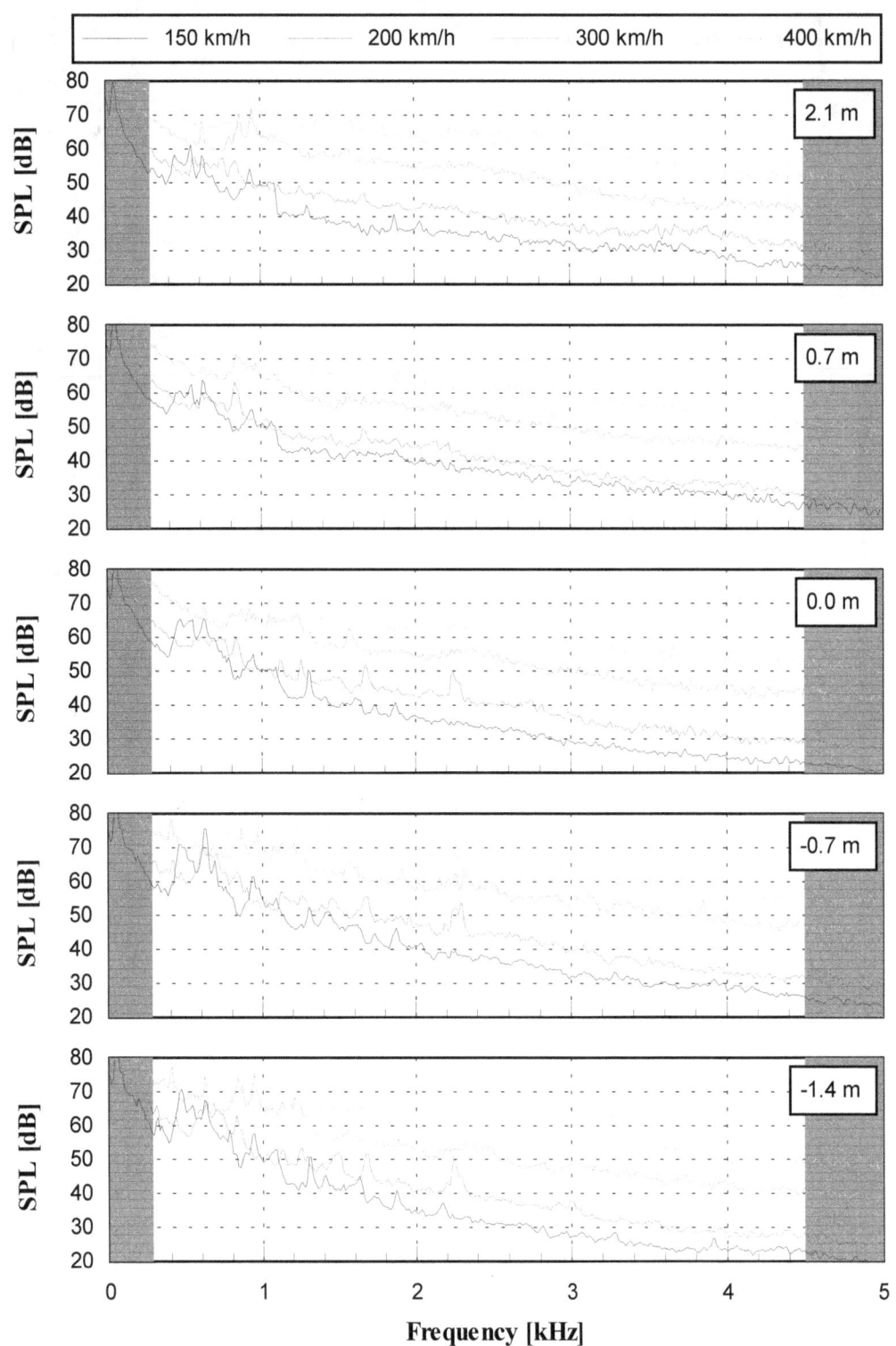

Figure 2-31. Unweighted narrow-band spectra of the sound-pressure level (SPL) measured with the WV array during the passbys of the TR08 on the hybrid guideway at speeds between 150 and 400 km/h (93 and 249 mph); Δf=11.6 Hz.

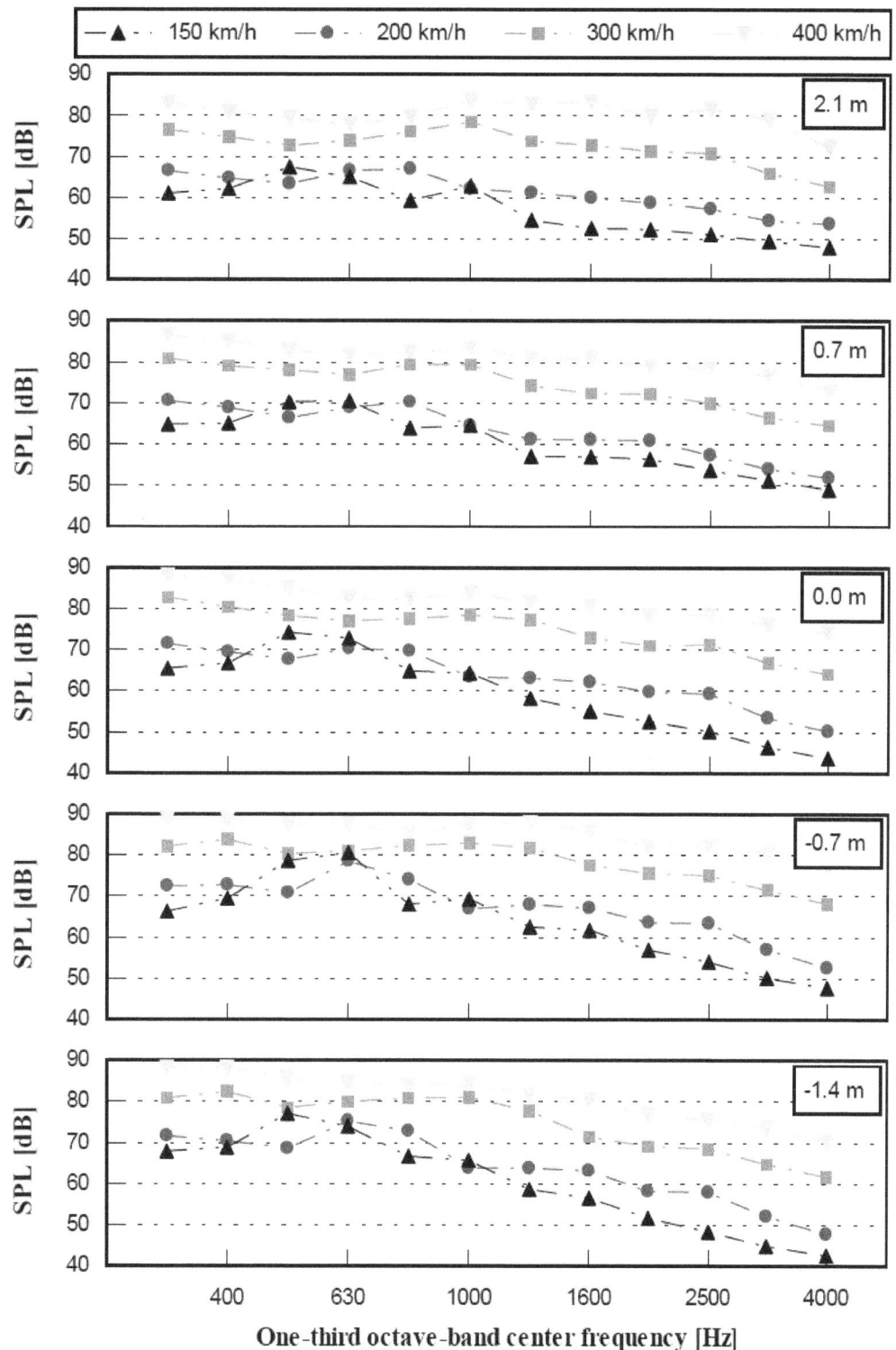

Figure 2-32. Unweighted one-third octave-band spectra of the sound-pressure level (SPL) measured with the WV array during passbys of the TR08 on the hybrid guideway at speeds between 150 and 400 km/h (93 and 249 mph)

Appendix D presents tabular one-third octave-band data corresponding to that presented in Figures 2-26, 2-28, 3-30 and 2-32.

Whereas the spectra are rather smooth at the higher speeds, some details can be observed in the narrow-band spectra at 150 and possibly 200 km/h (93 and 124 mph). Particularly at the height of the guideway, a distinct frequency peak is visible around 500 Hz (along with its first harmonic around 1000 Hz) for all guideway types at 150 km/h (93 mph). At 200 km/h (124 mph), the peak of the fundamental tone is shifted to frequencies around 650 Hz according to the higher vehicle speed. This frequency peak is caused by the grooved structure of the stators attached to the underside of the guideway's cover plate. In the cases of the steel and hybrid guideways, additional distinct frequency peaks can be found in the narrow-band spectra at 150 km/h (93 mph) between 600 and 1000 Hz, whose origins, however, are not clear.

The frequency peaks around 500 and 1000 Hz (at 150 km/h) and around 650 Hz (at 200 km/h) are of course reflected in the one-third octave-band spectra. The corresponding distinct one-third octave-band levels are clearly visible in the spectra at those vehicle speeds. At the higher speeds, the one-third octave-band levels decrease continuously with increasing frequency at the lower vehicle heights, whereas at the height of +2.1 m (+6.9 ft) it seems as though the peak frequencies are shifted to higher frequencies with increasing speed. Such behavior of the spectra measured in the upper part of the vehicle indicates that aerodynamic sound sources might dominate the total noise in this region of the TR08.

2.6 MODEL CALCULATIONS

2.6.1 The AD-PRO Prediction Model

In order to predict the noise radiated by a railway or maglev vehicle or the related impact noise at a selected location lateral to the track, akustik-data Engineering developed the AD-PRO computer model. Version 2.21 of the program was used for the present study. AD-PRO can predict sound-pressure levels generated by a train during unaccelerated passbys at any realistic speed for ground vehicles. In addition, the program can calculate noise levels for a stationary vehicle with some of its systems under power. A detailed description of AD-PRO can be found in Klemenz et al. (1997). As yet, the AD-PRO software is proprietary and not available commercially.

AD-PRO computes the total radiated noise of a train by summing up the contribution made by individual point sources, each of which represents the acoustical component generated by a relevant sound source. To account for the particular characteristics of each source, its strength, geometrical location on the vehicle, and horizontal directivity pattern are taken into account. The strength of a source is calculated so as to mirror its value at a reference speed of 200 km/h (124 mph), and its dependence on speed is given by a speed exponent. Depending upon the type of sound source, i.e., upon its generating mechanism, the speed exponent can have values ranging from 0, where the source strength is independent of vehicle speed, to about 7.0, where the source has a strong dependence on speed. Exponents having values approaching this latter limit are typical for flow-generated (aerodynamic) noise sources.

For the version of AD-PRO used here, the source strengths inputted to the program are expressed as A-weighted sound-pressure levels encompassing all relevant frequencies, i.e., over a wide range of frequencies. The values for the source strengths and speed exponents used for predicting the wayside noise of the TR08 are based on the values listed in Table 2-3. In accordance with the experience of akustik-data, it is assumed that each source comprises a particular mixture of monopole and dipole components that determine its directivity pattern.

At high vehicle speeds, the effects of motion on the sound sources (i.e., the convective effects), must be taken into account. The AD-PRO does take these effects into consideration. Other effects of sound propagation accounted for by AD-PRO are those in VDI 2714 guidelines entitled "Outdoor sound propagation". Some of the influences considered in the program are ground reflections, atmospheric absorption, and other damping by the ground and meteorological conditions. In addition, AD-PRO also calculates the sound-level reduction resulting from the presence of sound barriers of any height and distance from the track. This particular software feature was not, however, used for the present project because only predictions under unobstructed conditions were required.

2.6.2 Predictions of the Wayside Noise Generated by the TR08 Travelling on the Reference Concrete Guideway

The following figures show time histories that have been predicted using AD-PRO compared with time histories that have been measured during passbys of the TR08 travelling on the reference concrete guideway. For these comparisons, microphone positions 15.2, 25.0, and 30.5 m (50.0, 82.0, and 100.0 ft) have been chosen. Figure 2-33 through 2-36 present the results of the comparisons at vehicle speeds of about 150, 200, 300, and 400 km/h (93, 124, 186, and 249 mph). As mentioned before, the predictions take into consideration all sound sources that have been detected in the course of the investigations of the TR08 as listed in Table 2-3. The initial comparison of time histories showed that at 300 and 400 km/h (186 and 249 mph) no complete conformity could be achieved. This meant that an important sound source for high speeds was obviously not being considered. Only after the noise of the turbulent boundary layer (TBL noise) was included in the model calculations, do the results (illustrated in Figure 2-35 and 2-36) show agreement between prognosis and measurement. Since the TBL noise was not the subject of the measurements presented here, it was modeled based on results taken from an earlier publication (Barsikow 1990).

According to these results, the TBL noise is formulated as a standard feature in AD-PRO and can be used for any guided vehicle.

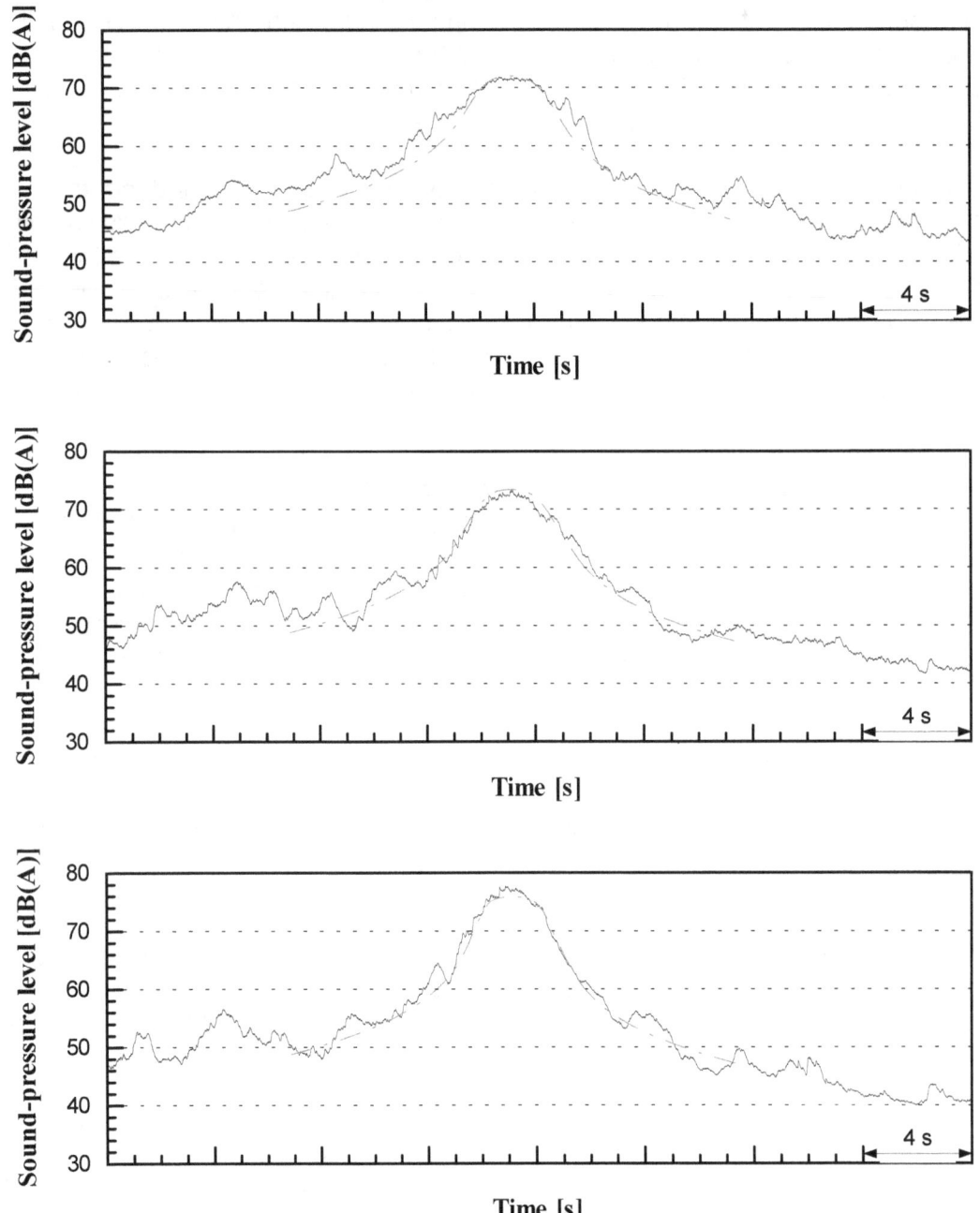

Figure 2-33. Measured (——) and predicted (- - - -) A-weighted time histories during passbys of the TR08 on the reference concrete guideway at speeds of 150 km/h (93 mph); measuring distances are 30.5 m (100.0 ft) (top), 25.0 m (82.0 ft) (mid), and 15.2 m (50.0 ft) (bottom).

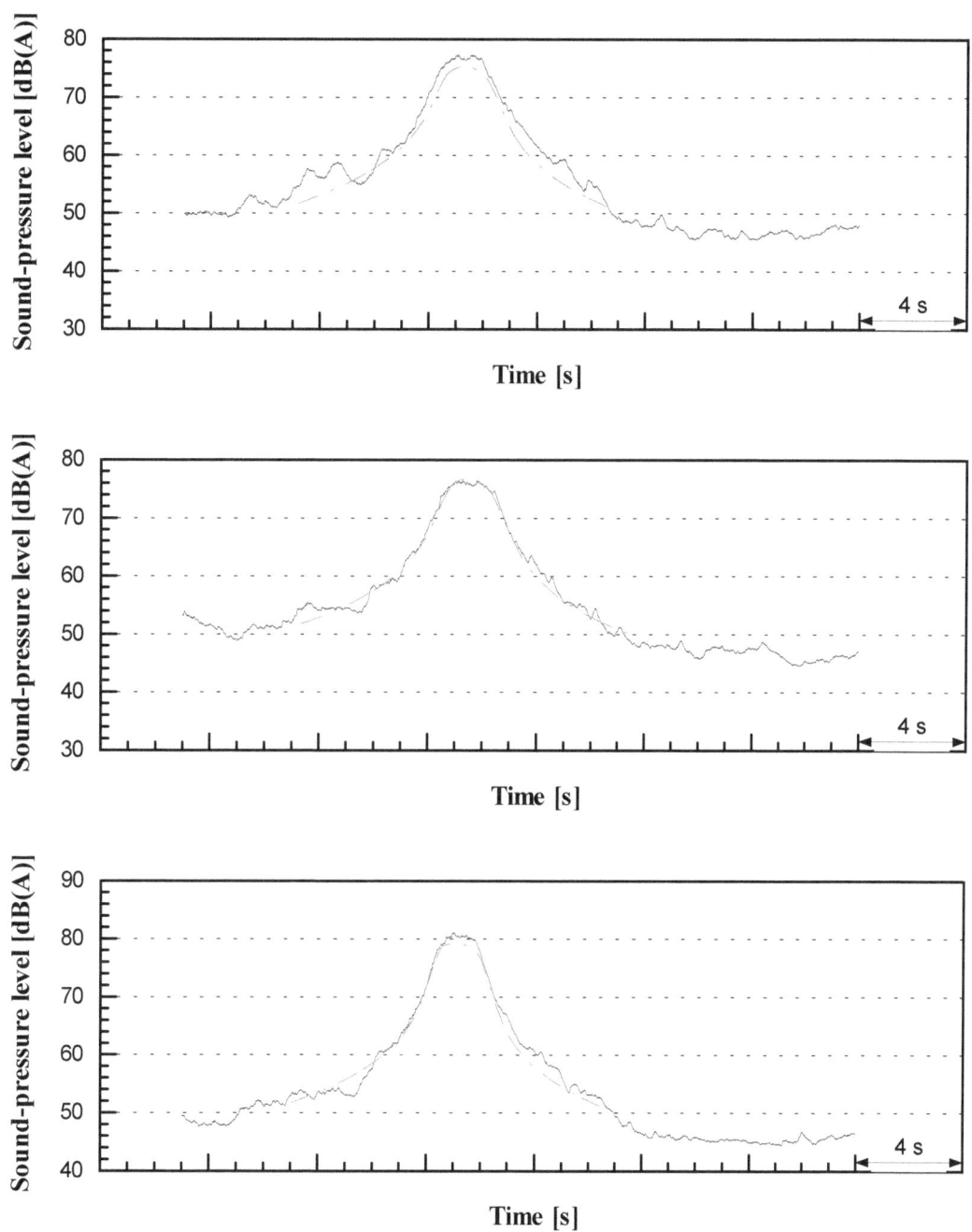

Figure 2-34. Measured (——) and predicted (- ·· - ·· -) A-weighted time histories during passbys of the TR08 on the reference concrete guideway at a speed of 200 km/h (124 mph); measuring distances are 30.5 m (100.0 ft) (top), 25.0 m (82.0 ft) (mid), and 15.2 m (50.0 ft) (bottom).

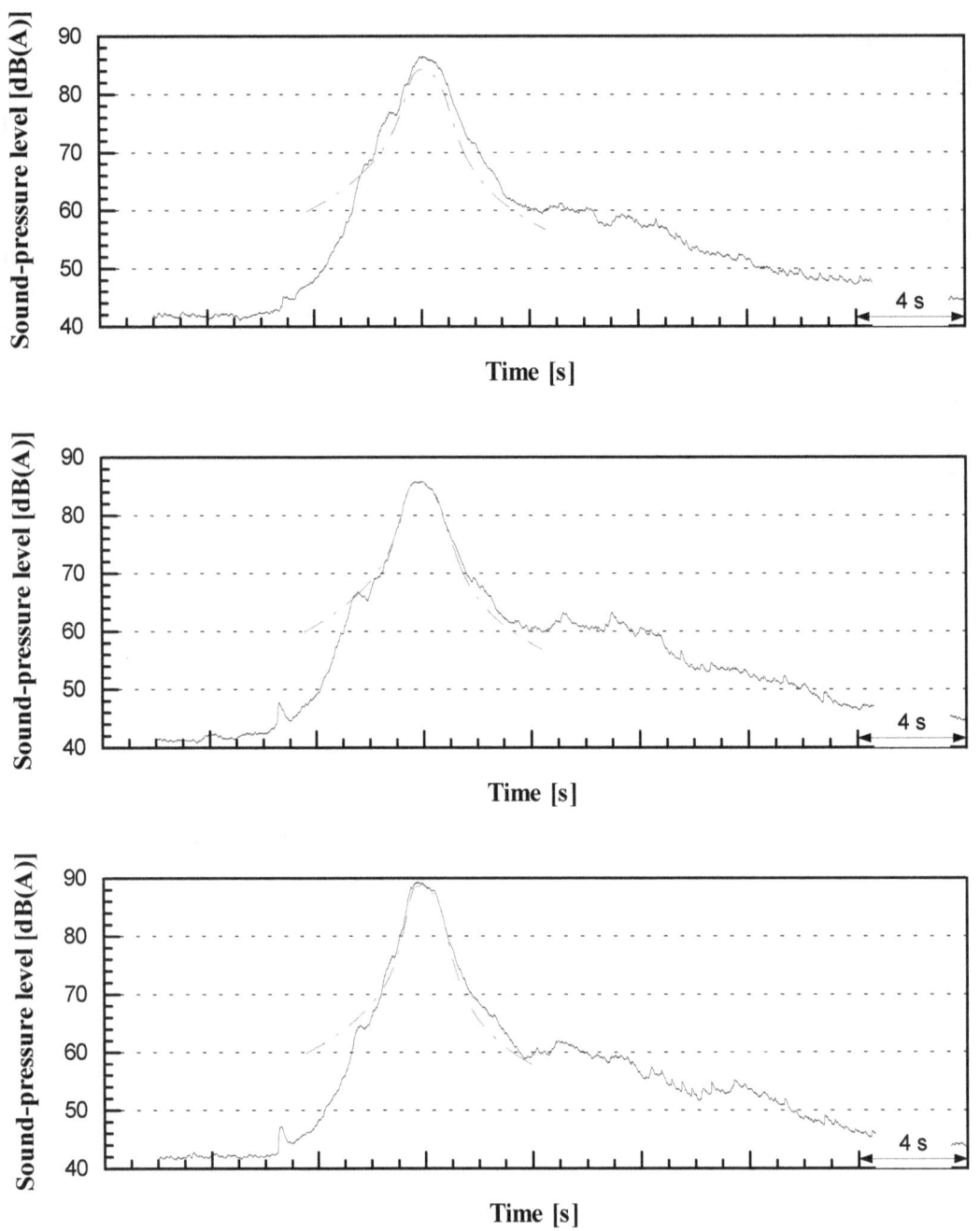

Figure 2-35. Measured (———) and predicted (-·-·-) A-weighted time histories during passbys of the TR08 on the reference concrete guideway at a speed of 300 km/h (186 mph); measuring distances are 30.5 m (100.0 ft) (top), 25.0 m (82.0 ft) (mid), and 15.2 m (50.0 ft) (bottom).

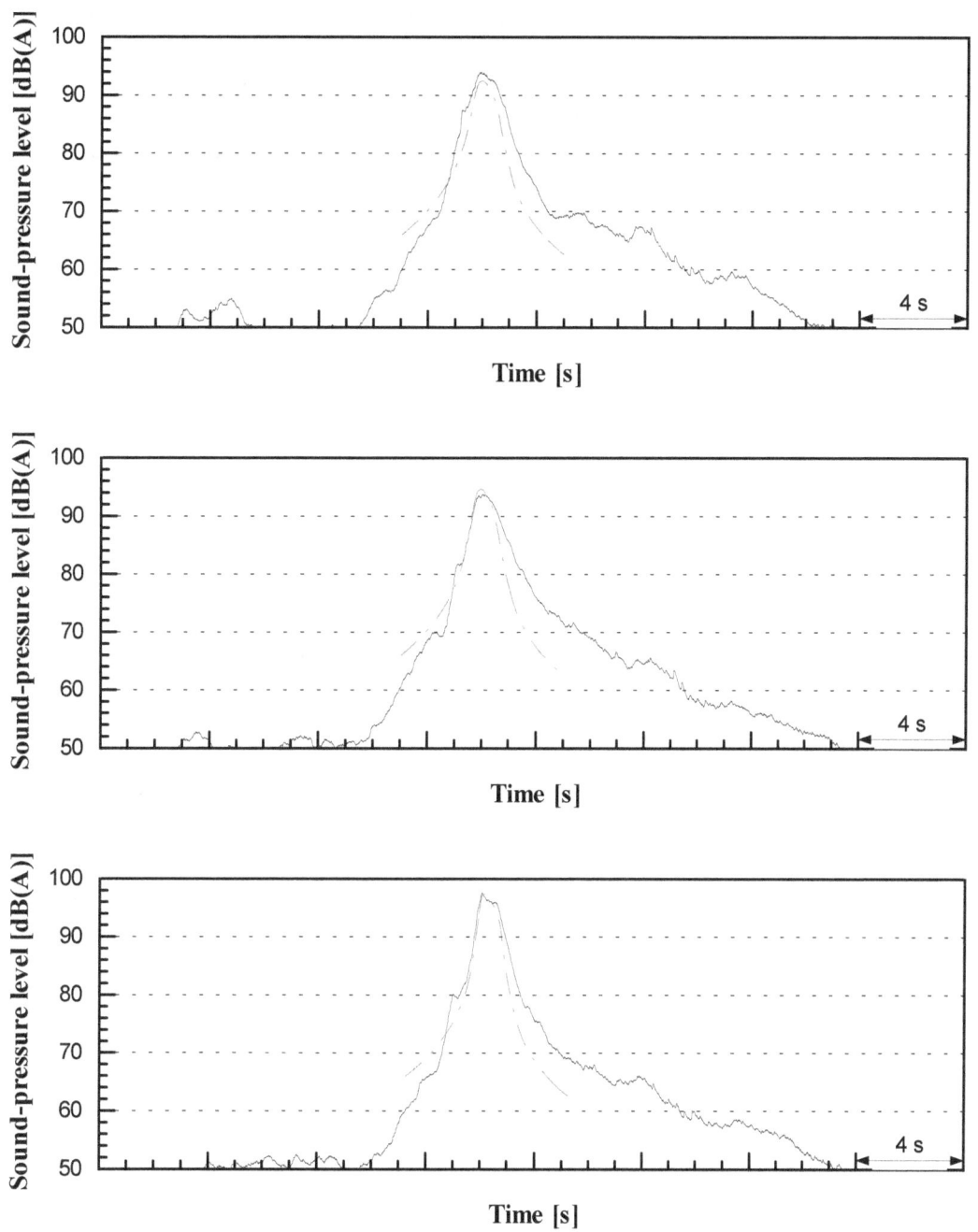

Figure 2-36. Measured (——) and predicted (- · - · -) A-weighted time histories during passbys of the TR08 on the reference concrete guideway at a speed of 400 km/h (249 mph); measuring distances are 30.5 m (100.0 ft) (top), 25.0 m (82.0 ft) (mid), and 15.2 m (50.0 ft) (bottom).

The overall good conformity between predicted and measured time histories does not only apply to the vicinity of the maximum level, but also to the curve characteristics when the vehicle is approaching or leaving. However, modeling yields better results for lower speeds than for higher speeds. At 400 km/h (249 mph), a clear difference between measurement and prediction becomes obvious when the TR08 is leaving. The difference can be attributed to the fact that none of any possible sound sources behind the vehicle is considered in AD-PRO. These sound sources, which have not been identified up to now, include the sound generated aerodynamically in the wake of the vehicle and a potential lingering sound from the guideway.

The good visual agreement between measurement and the predicted values can also be supported by the measured and predicted single-number levels. For this purpose, Table 2-4 shows a comparison of measured and calculated sound-exposure levels. In the case of the measured SELs, the average values determined at each of the four speeds are listed. According to Table 2-4, the prediction for the SEL at 15.2 m (50.0 ft) distance from the centerline of the guideway is a bit too low (up to a maximum of -1.0 dB(A)) at each speed. At 25.0 m (82.0 ft) distance, the prediction is very precise and shows only at 150 km/h (93 mph) a deviation of +1.0 dB(A). At 30.5 m (100.0 ft) distance, however, the predicted values are not as satisfactory, since deviations up to -2.4 dB(A) occur. These level differences are a result both of unprecise modeling in the vicinity of the maximum level and, in the case of high speeds, of the deviation in the time histories when the vehicle is leaving.

Table 2-4. Measured and predicted sound-exposure levels during passbys of the TR08 on the reference concrete guideway.

Vehicle Speed [km/h (mph)]	SEL [dB(A)] at 15.2 m (50.0 ft) Distance from Guideway Centerline		SEL [dB(A)] at 25.0 m (82.0 ft) Distance from Guideway Centerline		SEL [dB(A)] at 30.5 m (100.0 ft) Distance from Guideway Centerline	
	measured	predicted	measured	predicted	measured	predicted
150 (93)	80.5	79.8	76.7	77.7	76.3	76.4
200 (124)	82.7	81.8	79.4	79.6	80.4	78.3
300 (186)	89.8	88.9	86.8	86.7	87.6	85.2
400 (249)	96.4	95.7	93.6	93.3	93.9	91.5

Nevertheless, the predicted values for passbys of the TR08 on the reference concrete guideway show that the individual sound sources listed in Table 2-3 (except for the TBL noise) are sufficient for the acoustical modeling of the TR08. This is above all an important statement for the measuring distance of 15.2 m (50.0 ft), since in the following section, the source reference SELs for this distance will be calculated by means of AD-PRO.

2.6.3 Source Reference Sound-Exposure Levels at 50 ft Distance from Guideway

The relevant sound sources (except for TBL noise) and their reference sound-exposure levels (SEL_{ref}) to be used in computing the noise exposure of the TR08 are listed in Table 2-5. In this Table, SEL_{ref} for each source is given for the reference distance of 15.2 m (50.0 ft) from the centerline of the guideway. Also given in the Table are the reference value of the associated length of each source and the height of each source above the upper surface of the guideway.

Table 2-5. Reference sound exposure levels of individual sound sources of the TR08 system

Sound Source	Height Above Guide-way [m (ft)]	Reference Length [m (ft)]	SEL_{ref} [dB(A)]			
			150 km/h	200 km/h	300 km/h	400 km/h
Separating flow at vehicle nose	0	-	60.9	67.2	76.2	82.9
Air-conditioning units	2.0 (6.6)	29.0 (95.1)	54.5	62.9	75.2	84.1
Ventilation openings	0.5 (1.6)	50.0 (164.0)	68.3	69.0	70.2	71.3
Lower edge of levitation frame	-0.9 (-3.0)	75.0 (246.0)	69.7	73.9	80.1	84.7
Reference concrete guideway	-1.5 (-4.9)	90.0 (295.3)	78.9	80.3	86.8	91.7
Prototype concrete guideway	-1.5 (-4.9)	90.0 (295.3)	79.9	81.3	87.8	92.7
Hybrid guideways	-1.5 (-4.9)	90.0 (295.3)	80.9	82.3	88.8	93.7
Prototype steel guideway	-1.5 (-4.9)	90.0 (295.3)	81.9	83.3	89.8	94.7

The reference SELs in Table 2-5 are the results of model calculations using the AD-PRO software. For each of the localized sound sources or category of localized sound sources listed in Table 2-3, time histories have been calculated for passbys at 150, 200, 300, and 400 km/h (93, 124, 186, and 249 mph). On the basis of these time histories, the corresponding sound-exposure levels could then be determined by using the definition given in Section 2.5.1 and by setting the sound-level difference ΔL to 20 dB. Only in the case of the source caused by the separating flow at the vehicle's nose one single point source was considered in the model calculations. In all other cases, a number of point sources were taken into account for each passby, viz., four sources representing the noise produced by the air-conditioning units, and three sources representing the noise originating from the particular ventilation openings. Of course, the emissions from the lower edge of the levitation frame and the different guideways were modeled by line sources, whose lengths have already been given in Table 2-3.

2.7 SUMMARY OF RESULTS

The wayside noise of the maglev vehicle TR08 was investigated during tests in August, 2001, and May, 2002, at the TVE in the Emsland region of Germany. As specified in the test matrix (Appendix A), the noise data of a number of passbys of the TR08 were recorded at various measuring sites representing different guideway types, namely, the reference concrete guideway, the prototype steel and concrete guideways, the hybrid guideway, two versions of a simulated at-grade guideway, and the North switch. Measurements were made within the speed range from 100 to 400 km/h (62 to 249 mph) at all of these sites using up to six single microphones, whereas investigations have been carried out at the reference concrete, the prototype steel, the prototype concrete, and the hybrid guideway only employing different configurations of microphone arrays.

The aim of the single-microphone measurements was to document time histories of the A-weighted sound-pressure level at different microphone positions, both lateral to the track and beneath the guideway, during passages of the vehicle. From these time histories single-number levels have been calculated that describe the total noise emission of the TR08. In order to fulfill German and American interests, different noise descriptors had to be introduced (i.e., the maximum level, the SEL, and the one-hour level). Values of these levels are given in tabular form for all measured passbys. In addition, graphical representations of the SEL as a function of vehicle speed are shown for each microphone position. A ranking of the different guideway types concerning their noise emission can be drawn from these figures—passbys of the TR08 on the reference concrete guideway are the quietest whereas passbys on the North switch are the loudest. The results also show that at speeds above about 200 km/h (124 mph), the steel guideway is acoustically equivalent to the reference concrete guideway.

Using a vertical nested line array (WV array), the same array in the horizontal direction (WH array), as well as two X-shaped arrays (WX32 and WX16 array), one- and two-dimensional sound-source distributions of the vehicle and the guideway could be prepared. The results of these microphone-array measurements show that at low and intermediate speeds the sound emission during passages of the TR08 maglev vehicle is dominated by the radiation of the guideway and from mechanical sound sources in the lower region of the vehicle, whereas strong aerodynamically generated sound sources can be identified at high speeds. The aerodynamic sound sources are generated by flow interactions with the levitation frame as well as with the louvers covering the inlet or outlet openings of various ventilation systems. The strongest single sound source was localized at the front of the vehicle about at the height of the guideway's upper surface.

Including the above described sound sources into the prediction model AD-PRO, time histories during passbys of the TR08 could be calculated and compared with the corresponding measured time histories. In order to achieve a good agreement between predicted values and actual measurements, it was necessary to also include the TBL noise into the predictions at the higher vehicle speeds. After that, the agreement at all speeds was good to excellent, particularly for the 15.2 m (50.0 ft) and 25.0 m (82.0 ft) microphone positions. Again using the AD-PRO software, the reference sound-exposure levels of the localized sound sources of the TR08 at the reference distance of 15.2 m (50.0 ft) were calculated. These levels are given in tabular form for 150, 200, 300, and 400 km/h (93, 124, 186, and 249 mph) together with the associated reference lengths and source heights.

CHAPTER 3   VALIDATION AND VERIFICATION

3.1   DESCRIPTION OF VALIDATION AND VERIFICATION EFFORT

Volpe Center personnel provided validation and verification during all phases of the TR08 noise characterization program. Specifically, members of the Volpe Center Acoustics Facility reviewed, commented on and ultimately recommended to the FRA that the test proposal and test protocol be approved. Verification of contractor measurement techniques was provided during the measurement program at the Transrapid Test Facility (TVE) in Germany. Further, during field testing, independent acoustic data were collected to be used as a part of the eventual verification of contractor data. Finally, validation of contractor data, including analysis techniques and data presentation, was conducted. The following sections present the specifics of these individual aspects of the validation and verification efforts.

3.2   TEST PROTOCOL

3.2.1   Measurement Proposal

A preliminary measurement proposal was submitted by the contracting team to the Volpe Center in September 2000 (HMMH 2000a). Volpe Center staff reviewed the proposal for both reasonableness and completeness. The Transrapid (TRI), and magnetic levitation (maglev) in general, is a relatively new technology in terms of implementation. As such, there is not a wide body of knowledge on its noise characteristics. In response to the initial proposal, based on its comprehensive experience with noise measurement and analysis of other more traditional modes of transportation, the Volpe Center: (1) suggested more detail be provided in some sections; (2) requested specific issues be addressed in the proposal; and (3) provided guidance on final data presentation. A final measurement proposal (HMMH 2000b) reflecting the initial comments was submitted to the Volpe Center in October 2000 and was subsequently approved by the FRA upon the recommendation of the Volpe Center.

3.2.2   Test Plan

An initial Test Plan was submitted to the Volpe Center in October 2000 (HMMH 2000c). The Test Plan incorporated information from the previous measurement proposals, but also included specifics pertaining to schedule, microphone locations and orientation, actual measurement locations along the track, as well as data analysis and presentation details.

First, the Test Plan was evaluated to ensure that data collected would fully meet all requirements in appropriate guidance documentation. The primary document directly relevant to these measurements is the FRA's *High-Speed Ground Transportation Noise and Vibration Impact Assessment* (FRA 1998). This document focuses on the impact assessment phase of an environmental analysis for high-speed ground transport. While it does not cover in detail the requirements for source-noise data collection, it does detail analyses for which source-noise data are required. As such, the document served as a reference to ensure that data collected per the Test Plan would be adequate for any future environmental analyses. A second document, the FRA's *Handbook for the Measurement, Analysis and Abatement of Railroad Noise* (FRA 1982), was also used to evaluate the Test Plan. While this document is somewhat older, and many sections are outdated, it was considered an important resource. Among other things, the document states that measurements should be taken during "normal operating conditions." Given the relatively young state of maturity of maglev technology and the fact that precise maglev operating conditions within the U.S. are not yet known, it was considered important to, at a minimum, fully document all operating conditions during measurements. Therefore, once actual maglev operating conditions are better defined for the U.S., any potential limitations of the TR08 noise characterization and analysis may be understood.

The Test Plan was also evaluated to ensure that it was reasonable to accomplish in the time allotted for measurements at TVE. Although considered ambitious, the Test Plan schedule was deemed reasonable given: (1) favorable weather conditions allowing for completion of all related logistics, and (2) consistent and predictable TVE operations.

Additionally, the Test Plan was evaluated to ensure sufficient data were collected such that the modeling of potential noise impacts from the TR08 using state-of-the-art noise modeling computer software would be enabled. Modeling of these potential impacts might help planners to incorporate mitigation strategies into final design of maglev systems as appropriate.

The Volpe Center submitted comments to the contracting team requesting that more details be added in several sections and some additional tests be performed. As a result, a revised Test Plan was submitted to the Volpe Center at the end of November 2000. The revised Test Plan (Appendix A) was evaluated for completeness and accuracy, and ultimately approved by the FRA upon the recommendation of the Volpe Center.

3.3 FIELD VERIFICATION OF MEASUREMENTS AND PROTOCOL

During the period from August 12 through 18, 2001, Volpe Center staff participated in the field measurements at the TVE in the Emsland region of Germany. Several steps were taken during the measurements to verify contractor measurements and adherence to the approved protocol. Prior to the measurements, it was decided that the simplest means for verification of contractor data was for the Volpe Center to independently collect a limited amount of simultaneous data for comparison. Considering that the contractor was undertaking both single-microphone and array measurements, it was felt that setting up three microphone systems in one general area at the same time may cause unnecessary logistical problems. Fortunately, the TR08 is scheduled to run a standard speed profile during an entire loop around the test track. Therefore, it was decided that data collected during measurements at the same location as utilized by the contractors, but on a different day would both: (1) serve to verify contractor data, and (2) assuming positive verification of contractor data, confirm repeatability of TR08 acoustic data.

In addition to collecting independent acoustic data, Volpe Center staff witnessed the acoustic calibration process undertaken daily by the contractor. Placement and alignment of both array and single microphones were verified and documented using 35 mm photography. Additionally, the set-up and functionality of the measurement system, including signal processing and data storage devices, were demonstrated to Volpe Center personnel by the contractor. The general practices of the contractor, including field documentation and protocol were also evaluated.

3.4 VALIDATION OF CONTRACTOR DATA AND ANALYSES

Volpe Center staff collected acoustic data at several distances from, and multiple locations along, the guideway. Additionally, measurements were made beneath the guideway at several locations. In general, excellent agreement was observed between Volpe and contractor data. Comparisons of some of the Volpe and contractor data are presented below for the purpose of validation and as an example of the repeatability of TR08 acoustic data.

3.4.1 Measurements Alongside Guideway

Maximum Sound Level Data
Figure 3-1 presents a comparison of Maximum Sound Level (L_{AFmx}) data measured by the Volpe Center and contractor for the reference concrete beam guideway. Data collected at a distance of 50 feet from the guideway are presented as a function of speed. Though the data were collected on separate days, excellent agreement is illustrated. The largest range of sound level data for any given speed is 2.1 A-weighted decibels (dBA). Linear regressions for both sets of data, as well as the difference between these regressions, are presented in the figure. Similar data for the other guideway types suggest that sound levels are highly repeatable for a given guideway type, vehicle speed and distance away from the guideway.

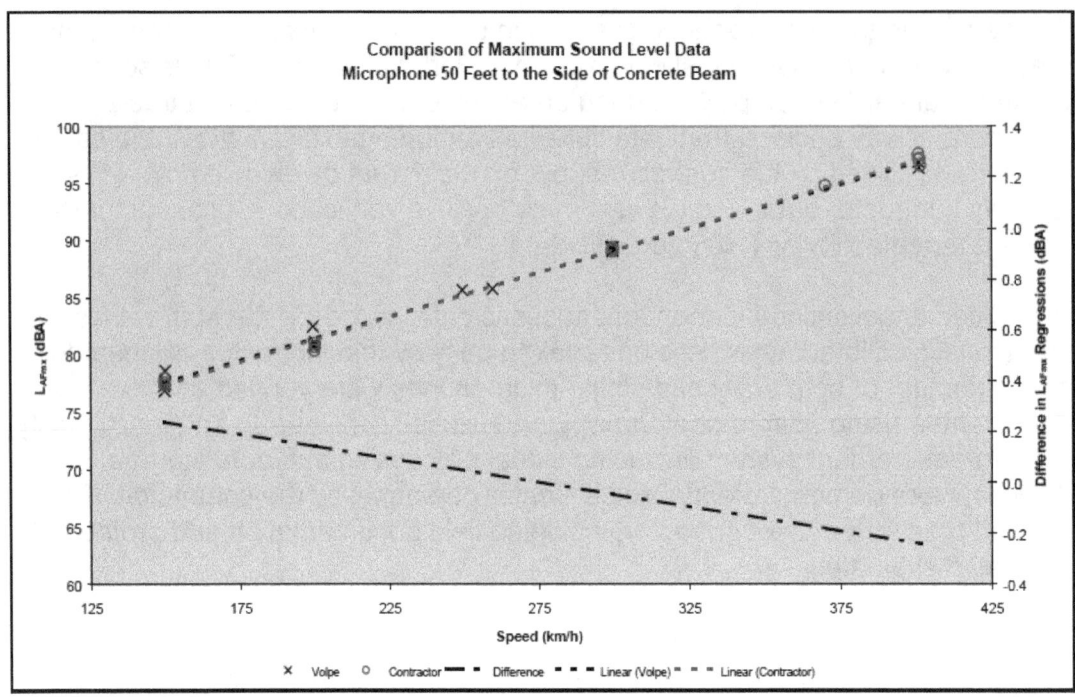

Figure 3-1 Comparison of Maximum Sound Level Data

Time History and Onset Rate Data

Figure 3-2 presents a representative sound level time history for a 300 kilometer per hour (km/h) pass-by of the TR08 measured at 50 feet. Superimposed on the figure is the associated onset rate of the sound level in dBA per second (dBA/sec). Onset rate is defined as the average rate-of-change of increasing sound pressure level, in dBA/sec, associated with the rapid approach of a high-speed vehicle. Onset rates, and the resultant increase in annoyance associated with high onset rates in particular, have been the subject of much research related to low-flying military aircraft. As discussed in Section 4.17 of the FRA's *Final Programmatic Environmental Impact Statement for the Maglev Deployment Program* (FRA 2001), onset rates of greater than 15 dBA/sec have been shown to cause annoyance in humans (Stusnick 1993).

For illustrative purposes, a 1–second moving average of the onset rate time history is presented in Figure 3-2. As can be seen in the figure, the onset rate is highest just prior to the maximum sound level for the passby. For this example event, a maximum onset rate of just under 17 dBA/sec is calculated. The average onset rate for the audible portion of the event is 7.5 dBA/sec. The onset rate is 15.7 dBA/sec for the portion of the event within 10 dBA of the maximum sound level (L_{AFmx}) and 15.3 dBA/sec within 15 dBA.

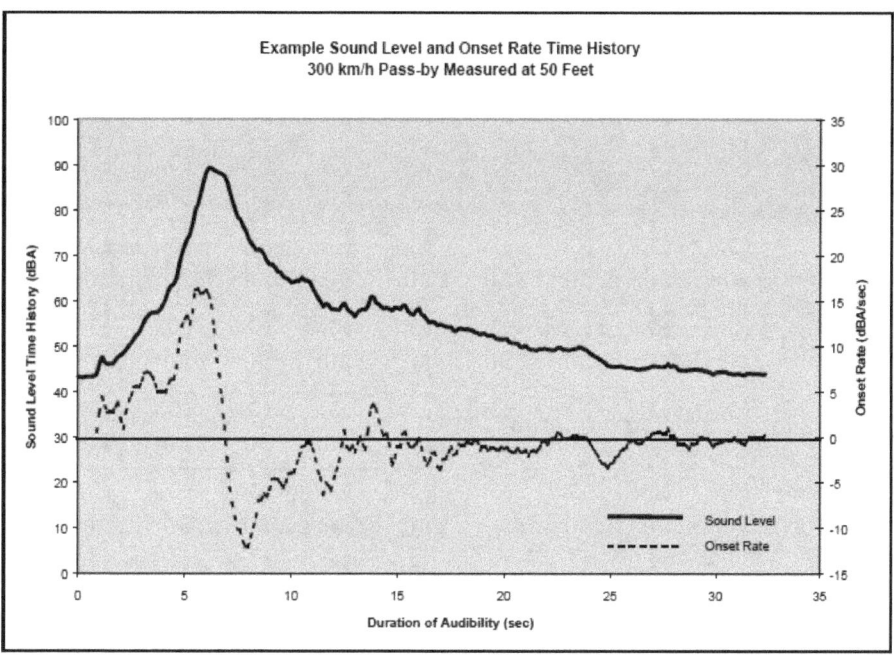

Figure 3-2 Example Sound Level and Onset Rate Time History Data

Figure 3-3 presents onset rate data for multiple maglev pass-bys, measured at distances of 50 and 100 feet from the centerline of the concrete guideway. Data are presented for speeds between 150 and 400 km/h and are derived using the average onset rate during the 1-second moving-averaged portion of the sound level history within 15 dBA of the maximum sound level. For reference, a straight line is presented in the figure representing a 15 dBA/sec onset rate, the value shown to be a threshold for human annoyance with low-flying military aircraft (Stusnick 1993). While the applicability of this threshold to maglev technology has not been demonstrated, intuition indicates the threshold may in fact be higher for maglev technology. This is based on the fact that the flight tracks and schedules of military aircraft training routes may basically appear random in nature to a particular individual on the ground. Conversely, maglev vehicles travel on a predictable route (the guideway) and are likely to follow regular schedules (e.g., every 15 minutes). It is hypothesized that given the consistent nature of maglev routes and schedules, the increased annoyance due to high onset rates may be lower than that experienced for low-flying military aircraft. Regardless, the 15 dBA/sec threshold is presented as the best currently available data. Further research would be necessary to establish a comparable threshold for maglev technology. For reference, the linear regressions of the 50- and 100-foot data cross the 15 dBA/sec threshold at approximately 335 km/h and 370 km/h, respectively.

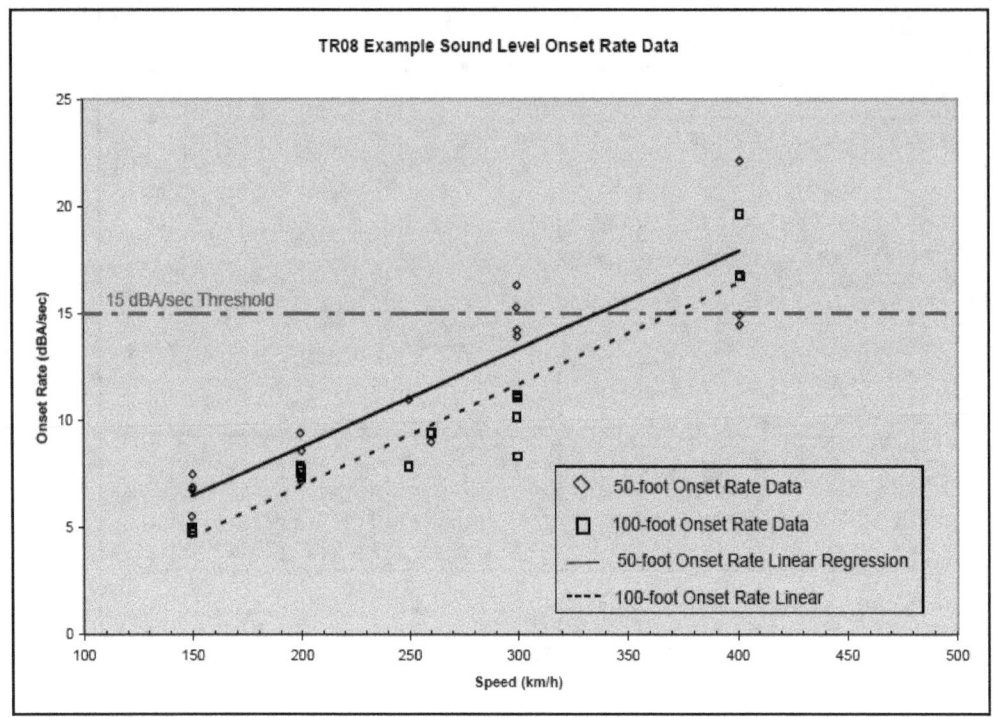

Figure 3-3 TR08 Sound Level Onset Rate Data

3.4.2 Measurements Beneath Guideway

Volpe staff collected acoustic data beneath both concrete and steel guideway sections. These data demonstrate higher variability than do the data alongside the guideway. Figure 3-4 presents a comparison of contractor acoustic data beneath the guideway (and linear regressions through the data) and Volpe data. The greater variability in both data sets notwithstanding, excellent agreement is seen between the Volpe and contractor data.

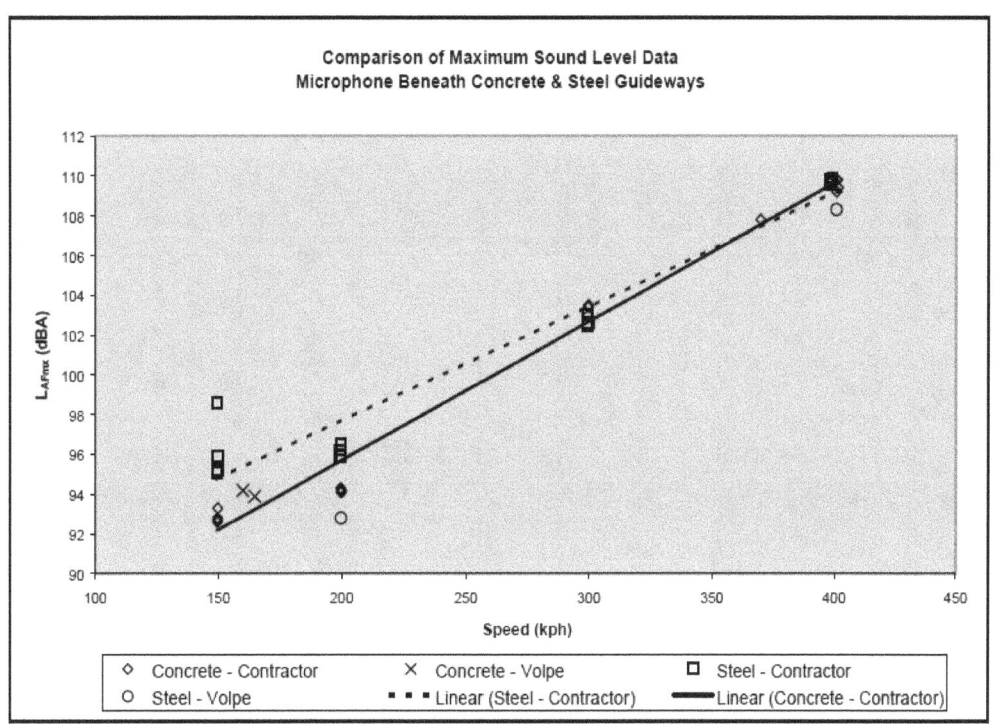

Figure 3-4. Comparison of Sound Level Data Beneath Guideway

3.5 FINDINGS AND CONCLUSIONS

TR08 acoustic data measured, analyzed, and presented by the contractor in the preceding sections of this report have been verified and validated. In general, excellent agreement has been illustrated between contractor single-microphone data and the limited acoustic data collected by the Volpe Center. Additionally, the contractor has demonstrated good technical practices throughout the project, including the design of the general approach and test protocol, measurements, and, finally, data analysis and presentation.

CHAPTER 4 SUPPLEMENTAL DATA AND DISCUSSION

4.1 COMPARISON OF TR08 NOISE WITH THAT FROM OTHER HIGH-SPEED TRANSPORTATION SYSTEMS

The results of the noise measurements of the TR08 can be compared with other high-speed ground transportation systems documented in the FRA's draft noise and vibration manual (FRA 1998). The results are shown in Table 4-1 in terms of sound exposure levels (SEL) at a distance of 30.5 m (100 ft) for representative speeds, with all trains normalized to the same length.

Table 4-1. Comparison of TR08 sound exposure levels with those of other high-speed ground transportation systems.

Speed [km/h (mph)]	SEL (dBA) at 30.5 m (100 ft)*						
	Maglev Technology					Wheel-on-Rail Technology	
	TR08				TR07		
	reference concrete guideway	prototype concrete guideway	prototype steel guideway	hybrid beam	reference concrete guideway	Acela	TGV
100 (62)	83	86	85	85	(NA)	(NA)	(NA)
150 (93)	81	82	84	85	80	87	88
200 (124)	86	87	88	85	83	92	92
240 (150)	(NA)	(NA)	(NA)	(NA)	85	94	93
300 (186)	93	94	95	92	90	(NA)	97
400 (249)	99	99	100	98	93	(NA)	(NA)

[1]Trains normalized to 225 m (740 ft) in length.

Note that the SELs from the TR08 in Table 4-1 are greater than the measured values reported in Table 2-4 due to the adjustment for a greater length than the actual vehicle measured.

4.2 TR08 REFERENCE DATA

The results of the measurement program can be directly applied in the detailed noise analysis method of FRA's guidance manual (FRA 1998). The "Source Reference SEL's at 50 feet" from Table 2-5 of this report can be used in the equations given in Table 5-2 of the guidance manual with the terms given in Table 4-2 of this report.

Table 4-2. Source Reference SELs of the TR08 system at 15 m (50 ft) for application in FRA Detailed Noise Analysis

Sound Source	Height Above Guideway [m (ft)]	Reference Length [m (ft)]	SEL_{ref} [dB(A)]	S_{ref} [km/hr (mph)]	K
Separating flow at vehicle nose	0	-	60.9	150 (93)	50
Air-conditioning units	2.0 (6.6)	29.0 (95.1)	54.5	150 (93)	68
Ventilation openings	0.5 (1.6)	50.0 (164.0)	68.3	150 (93)	8
Lower edge of levitation frame	-0.9 (-3.0)	75.0 (246.0)	69.7	150 (93)	30
Reference concrete guideway	-1.5 (-4.9)	90.0 (295.3)	78.9	150 (93)	30
Prototype concrete guideway	-1.5 (-4.9)	90.0 (295.3)	79.9	150 (93)	30
Hybrid guideway	-1.5 (-4.9)	90.0 (295.3)	80.9	150 (93)	30
Prototype steel guideway	-1.5 (-4.9)	90.0 (295.3)	81.9	150 (93)	30

4.3 REGULATORY FRAMEWORK

The purpose of this document is to measure and report the noise characteristics of the Transrapid TR08 Maglev System. The commercial transportation application of magnetic levitation system technology is new to the United States. As such, no maglev-specific noise regulations currently exist. Accordingly, either existing standards originally developed for conventional rail technologies will be applied to maglev applications within the U.S. or new regulations and/or exceptions for the unique technology will be developed. A brief review of high-speed rail standards in Japan and Germany helps to provide context for the U.S.

The Japanese regulations were developed for the Shinkansen, a traditional high-speed rail system. Conversely, the German regulations were developed specifically for maglev technology. In both cases, maximum, A-weighted sound levels are specified for specific land use categories. By defining the noise level limits in terms of land use, the regulations specify maximum sound levels received at *receptors*, as opposed to specific sound levels emitted by the vehicle noise source. In neither case are measurement distances specified for these levels.

Table 4-3 presents the Japanese noise limits specific to the Shinkansen high-speed rail system. While the Shinkansen was originally opened in 1964, recognition of its unique capabilities and potential prompted the specification that noise limits would be instituted in several steps through the year 2001 (Ono 2000).

Table 4-3. Japanese Shinkansen Noise Limits

Land Use Category	Maximum Sound Level, L_{ASmx} (dB)
I: Residential	70
II: Non-Residential	75

Source: Japanese Ministry of the Environment, 2002

Table 4-4 presents the German noise limits specific to maglev technology. Whereas the Japanese noise limits are based on maximum sound levels (L_{AFmx}), the German limits are based on the hourly equivalent sound level ($L_{Aeq,1h}$).

Table 4-4. German Magnetic Levitation Noise Standards

Land Use Category	Hourly Equivalent Sound Level, $L_{Aeq, 1h}$ (dBA)	
	Day	Night
I: Hospitals, Schools, Spas, Retirement Homes	57	47
II: Residential	59	49
III: Core, Village, Mixed-Use	64	54
IV: Industrial	69	59

Source: Bundesrat, 2001

While no maglev-specific noise regulations have been developed for the U.S., guidance materials for high-speed rail in general have been developed. The FRA presented guidance for the determination of potential noise impacts in the *High-Speed Ground Transportation Noise and Vibration Impact Assessment* document (FRA 1998). Based in part on Federal Highway Administration (FHWA) methodologies, and similar to the Japanese and German regulations, this document also recommends the application of impact criteria based on land use categories.

The introduction of a new transportation technology, specifically the supersonic commercial jet, may also serve as a precedent with respect to the application of current noise regulations to maglev technology. When the Concorde SST was first introduced in the U.S., it did not meet existing aircraft noise regulations.

Recognizing the unique nature of the technology, operation of the Concorde was allowed given that its noise emissions were "reduced to the lowest levels that are economically reasonable, technologically practicable, and appropriate for the Concorde type design." (FAA 1978)

Throughout the TR08 noise characterization process, every effort has been made by the Volpe Center to ensure that the project adheres to appropriate guidance materials and regulatory documentation. In particular, the Environmental Protection Agency (EPA) regulations at Title 40, Part 201 of the Code of Federal Regulations (CFR), provide a regulatory framework for noise emissions from the operation of rail equipment. In part, these regulations preclude the operation of rail cars moving at speeds greater than 72 km/h (45 mph) which produce sounds levels in excess of 93 dBA measured 100 feet from the centerline of the track (see 40 CFR 201.13). Contractor data, validated and verified by the Volpe Center, illustrate that TR08 sound levels, measured at a distance of 100 feet are below 93 dBA at speeds of approximately 300 km/h (186 mph), whereas these sound levels measured for speeds of approximately 380 km/h (236 mph) exceed 93 dBA. Even if it is determined that the EPA standards do not apply to maglev technology, it is likely that maglev noise emissions will be addressed either in a regulatory process or as part of the environmental compliance process.

CHAPTER 5 REFERENCES

40 CFR Ch. 1 (7-1-01 Edition) Subchapter G- Noise Abatement Programs, Part 201- Noise Emission Standards for Transportation Equipment, Interstate Rail Carriers.

Barsikow et al 1987. Barsikow, B., W.F. King III, and E. Pfizenmaier. Wheel/Rail Noise Generated by a High-Speed Train Investigated with a Line Array of Microphones. *Journal of Sound and Vibration*, Vol. 118(1), 99-122. 1987.

Barsikow et al 1988. Barsikow, B., and W.F. King III. On Removing the Doppler Frequency Shift from Array Measurements of Railway Noise. *Journal of Sound and Vibration*, Vol. 120(1), 190-196. 1988.

Barsikow 1990. Barsikow, B. The Importance of Aerodynamic Noise for Tracked Vehicles at Speeds up to 500 km/h. Proc. Internoise 90, Gothenburg, 1437-1440. 1990.

Barsikow 1996. B. Barsikow. Experiences with Various Configurations of Microphone Arrays Used to Locate Sound Sources on Railway Trains Operated by the DB AG. *Journal of Sound and Vibration*, Vol. 193(1), 283-293. 1996.

Bundesrat 2001. Personal electronic communication to akustik-data regarding German Bundesrat Noise Standards, September 23, 1997. March 6, 2001.

FAA 1978. 36-10, 43 FR 28420, June 29, 1978.

FRA 1982. Handbook for the Measurement, Analysis and Abatement of Railroad Noise, Federal Railroad Administration, U.S. Department of Transportation. January 1982.

FRA 1992. Safety of High Speed Magnetic Levitation Transportation Systems, Magnetic Field Testing of the TR07 Maglev Vehicle and System, Vol. I-Analysis and Vol.II-Appendices. Federal Railroad Administration, U.S. Department of Transportation. April 1992.

FRA 1993. Noise from High Speed Maglev Systems. Federal Railroad Administration. January 1993.

FRA 1998. High-Speed Ground Transportation Noise and Vibration Impact Assessment, Final Draft. Federal Railroad Administration, U.S. Department of Transportation. December 1998.

FRA 2000. Noise Characteristics of Northeast Corridor High-Speed Trainsets. Federal Railroad Administration, U.S. Department of Transportation. September 2000.

FRA 2001. Final Programmatic Environmental Impact Statement and Record of Decision; Maglev Deployment Program. Federal Railroad Administration, U.S. Department of Transportation. April 2001.

HMMH 2000a. Personal electronic communication from Carl Hanson, Harris Miller Miller & Hanson, to Christopher Roof, Volpe Center, regarding "Noise and Vibration Measurements of TransRapid Maglev Vehicle TR08 HMMH Proposal No. P00-20115." September 19, 2000.

HMMH 2000b. Personal electronic communication from Carl Hanson, Harris Miller Miller & Hanson, to Christopher Roof, Volpe Center, regarding "Noise and Vibration Measurements of TransRapid Maglev Vehicle TR08 HMMH Proposal No. P00-20115B" October 6, 2000.

HMMH 2000c. Personal electronic communication from Carl Hanson, Harris Miller Miller & Hanson, to Christopher Roof, Volpe Center, regarding "Proposed Test Plan for Noise and Vibration Measurements of TransRapid TR08" October 23, 2000.

Japanese Ministry of the Environment 2002. Environmental Quality Standards for Shinkansen Superexpress Railway Noise: http://www.env.go.jp/en/lar/regulation/railway.html

Klemenz et al 1997. Klemenz, M., and M. Hellmig. Application of the AD-PRO Software for the Prediction of Sound Emission of Future High-Speed Trains. Proc. 2nd International Workshop on the AeroAcoustics of High-Speed Trains, Berlin. 1997.

NOISE-CON 93: The Effect of Onset Rate on Annoyance to Military Aircraft Noise, Stusnick, et. al., Wyle Laboratories. Arlington, VA: May 1993.

Ono 2000. Ono, Shigeaki, "Measurement and Analysis of Railway Noise in Japan." *Journal of the Acoustical Society of Japan*, Volume 21, No. 6. 2000.

RTRI 2001. Railway Technical Research Institute Home Page. http://www.rtri.or.jp/rd/maglev/html/english/maglev_frame_E.html December 2001.

Skudrzyk 1971. Skudrzyk, E. The Foundations of Acoustics, Chapter XXVI: Sound Radiation of Arrays and Membranes. Published by Springer-Verlag. 1971.

TRI 2001. Transrapid International Home Page.
    http://www.transrapid.de/en/index.html November 2001.

APPENDIX A. NOISE TEST PLANS

This appendix contains the text of the Final Test Plans for measurement of noise as well as vibration and electromagnetic fields (EMF) associated with the Transrapid TR08 Maglev System. Details of the EMF measurements have been omitted, as they are not relevant to this report; details about vibration measurements are included because they are integrated with the details about noise measurements.

The Final Test Plans were completed in June 2001. Harris Miller Miller & Hanson (HMMH) authored these plans, with input and additional material from technical staff of the Federal Railroad Administration (FRA), the John A. Volpe National Transportation Systems Center (Volpe Center), MAGLEV, Inc., Transrapid International (TRI), and IABG (Industrieanlagen Betriebsgesellschaft, a European scientific-technical services company). The Test Plans were also reviewed by representatives of the Baltimore-Washington Maglev Project, including an environmental planning staff person from the Maryland Transit Administration and a representative from Parsons-Engineering Science, a noise and vibration consultant for the Baltimore-Washington Maglev Project.

Magnetic Levitation Transportation Technology Deployment Program

Section XIV, Pre-Environmental Impact Statement (EIS) Activities

Field Measurement of the EMF, Noise and Vibration Characteristics of the TR08

Final Test Plans

SECTION XIV
Pre-Environmental Impact Statement (EIS) Activities
Part B and C Final Test Plans

Field Measurement of the EMF, Noise and Vibration Characteristics of the TR08

Disclaimer

This report was prepared as an account of work sponsored in part by the Federal Railroad Administration and in part by the Commonwealth of Pennsylvania, under the Federal Cooperative Agreement number DTFRDV-99-H-60009 Agreement 62N082 and performed by MAGLEV, Inc. Neither the Commonwealth of Pennsylvania, The Federal Railroad Administration, MAGLEV, Inc. or any party acting on behalf of the aforementioned parties (hereinafter referred to as "the Parties") makes any warranty or representation, express or implied, with respect to the accuracy, completeness, or usefulness of the information contained in this report. The Parties also assume no liability with respect to the unauthorized use of any information, apparatus, method or process disclosed in this report which may infringe privately owned rights. Nor do the Parties assume any liability with respect to the use of, or for damages resulting from the use of any information, apparatus, method or process disclosed in this report

Section 1: Introduction

This report documents the final test plans for conducting the field measurements of the electromagnetic fields (EMF), noise and vibration characteristics of the TR08. This document is being submitted to the FRA for approval of the test plans prior to implementation.

The purpose of these tests in part is to repeat previous EMF and noise measurements that were made on the TR07 in the early 1990's. The test plans are also supplemented to include vibration measurements that were not part of the original 1990 FRA test program. The completion of the test plans, data evaluation, and data analysis will result in detailed documentation of the electrical emissions, noise emissions and vibration characteristics of the Transrapid Maglev System and TR08 vehicle. This measurement data will be utilized to support the required environmental planning and deployment activities for any Transrapid Maglev Project in the United States. This work is being accomplished as part of the Federal Railroad Administration's (FRA) Scope of Work Amendment No. 2, Section XIV Pre-Environmental Impact Statement (EIS) Activities, part B and Part C.

Section XIV part B and C is restated here for reference.

B.   Field Measurement EMI, EMF and EMR Characteristics of TR08

In cooperation with assigned staff from the Volpe National Transportation Systems Center (VNTSC), the Grantee shall develop a plan to measure and analyze the electromagnetic interference (EMI), electromagnetic fields (EMF), and the electromagnetic radiation (EMR) caused by the operation of the TR08 vehicle in service at the Transrapid Maglev Test Track in Emsland, Germany. With the approval of the FRA, the plan will be implemented.

C.   Field Measurement of Noise and Vibration Characteristics of TR08

In cooperation with assigned staff from the Volpe National Transportation Systems Center (VNTSC), the Grantee shall develop a plan to measure and analyze the noise and vibration caused by the operation of the TR08 vehicle in service at the Transrapid Maglev Test Track in Emsland, Germany. With the approval of the FRA, the plan will be implemented.

MAGLEV, Inc. is performing the role of coordinating the diverse technical requirements and logistics associated with the planned test campaign as it relates to US maglev projects.

Section 2: Test Plan Development and Finalization

Under the direction of the FRA, MAGLEV, Inc. contracted the vendor Electric Research Management (ERM) for the electrical emissions testing and the vendor Harris, Miller, Miller and Hanson (HMM&H) for the noise and vibration testing of the TR08. The test plans contained in appendix A and B of this report are the collective efforts of the vendors, staff at Volpe and MAGLEV, Inc. These plans were developed in cooperation with and with the approval of Transrapid International and the TVE test center staff.

Numerous technical interchanges and discussions were held with Volpe personnel, Electric Research Management (EMF test vendor), Harris, Miller, Miller and Hanson (noise and vibration test vendor) and Transrapid in finalizing the plans for the environmental testing of the TR08. As part of this effort, Transrapid recommended and supported a pre-meeting with the vendors and MAGLEV, Inc at the test facility. The purpose of this meeting was to discuss and finalize the details of the draft test plans, familiarize the vendors with the test facility and to select the actual guideway and vehicle locations for measurements.

On February 15-16, 2001, the vendors and MAGLEV, Inc. traveled to the test facility in Emsland, Germany for a meeting with Transrapid and the TVE test facility personnel. As a result of this meeting, agreement was reached on a consolidated test matrix for the noise, vibration and EMF testing. The consolidated test matrix and the measurement locations are documented in appendix C of this report. The noise testing requires the most preparation and time to conduct and therefore drives the test matrix activities. The EMF and vibration measurement activities have been developed to permit all work to be done concurrently with the noise measurements.

Pending approval of the test plans by the FRA, the actual field measurements are scheduled to begin during the week of April 2, 2001 at the test facility.

Section 3: TVE Test Facility Layout and Operational Considerations

The TVE test facility layout and daily operating parameters were considered in developing the final plans. The TVE test facility consists of a 31.5 km track as shown in Illustration 1 below. Various guideway types and configurations (steel, concrete, hybrid, at grade and elevated) exist at the test track. The final plan includes measurements at each representative guideway type.

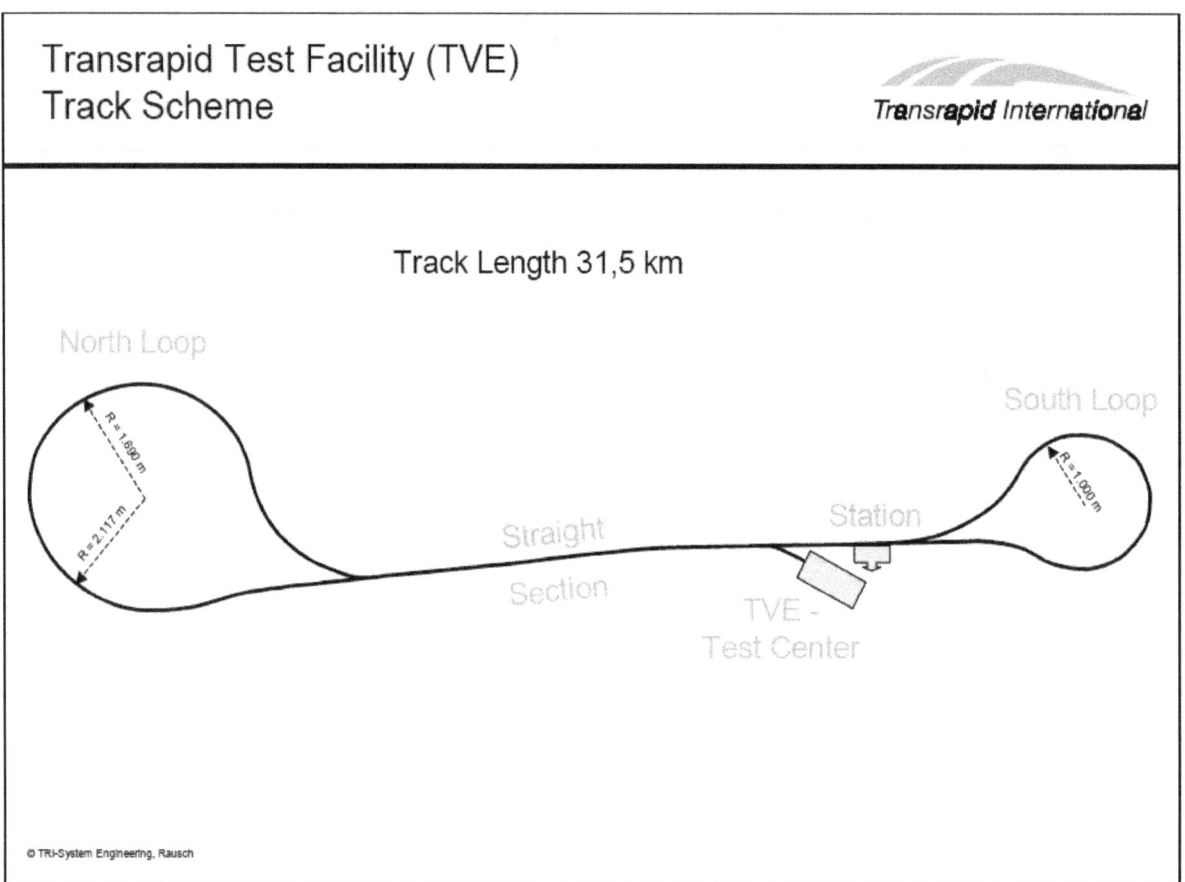

Illustration 1 Transrapid TVE Test Facility Track Scheme

The test facility normally operates from approximately 8:30 AM until 2:30 PM on Tuesday through Friday of each week. The train operates approximately every 30 minutes in the morning hours and every 45 minutes during the afternoon hours. Monday is reserved for facility maintenance and other test activities therefore no measurements requiring vehicle movements have been scheduled on Mondays. Based on the required measurements to fully characterize the noise, vibration and EMF of the system, a period of six days is required to conduct the field measurements.

Table 1: TVE Normal Operating Schedule

Trip Number [1]	Time of Departure [2]	Speed Profile [3]
N1	08:30 am	First Trip of the day, guideway inspection with low speed
N2	09:30 am	Demonstration speed profile
N3	10:00 am	Demonstration speed profile
N4	10:30 am	Demonstration speed profile
N5	11:00 am	Demonstration speed profile
N6	00:15 pm	Demonstration speed profile
N7	01:00 pm	Demonstration speed profile
N8	01:45 pm	Demonstration speed profile
	02:30 pm	Transfer into the Test Center

1) Each trip encompasses two rounds with a total trip length of 80 km
2) Departing at the TVE-Station
3) Speed profile can be adjusted according to the test requirements

Excluding the first inspection trip of the day, each trip consists of two complete runs of the entire test track length. This provides for a total of twenty-eight pass-bys (2 rounds x 2 pass-bys x 7 trips per day = 28 pass-bys) at each of the guideway locations in the straight sections of the track. Table 2 provides a summary of the trips and number of pass-bys.

Table 2, Pass-by Events along the Straight Section of the Guideway

Trip Number	Time of Departure At Station	No. Of Event Pass-by	Direction Towards	Vehicle Speed [1]
N2	09:30 am	1.	North, 1. Round	up to local max. Speed [2]
		2.	South, 1. Round	up to local max. Speed [2]
		3.	North, 2. Round	up to local max. Speed [2]
		4.	South, 2. Round	up to local max. Speed [2]
N3	10:00 am	5.	North, 1. Round	up to local max. Speed [2]
		6.	South, 1. Round	up to local max. Speed [2]
		7.	North, 2. Round	up to local max. Speed [2]
		8.	South, 2. Round	up to local max. Speed [2]
...	...	.		
...	...	.		
N8	01:45 pm	25.	North, 1. Round	up to local max. Speed [2]
		26.	South, 1. Round	up to local max. Speed [2]
		27.	North, 2. Round	up to local max. Speed [2]
		28.	South, 2. Round	up to local max. Speed [2]

1) Speed can be adjusted according to the test requirements
2) Local maximum speed depends on location

The speed profile for the normal operating sequence permits runs of up to 400 km/h in the straight section of the test track. Referring to Illustration 2, the normal operating sequence will provide a run of 150 km/h, 400 km/h, 200 km/h and 300 km/h in the straight section of the test track for each complete trip (two rounds). Trip time to complete the two rounds is approximately 22.5 minutes. Accounting for time necessary to download measurement data, approximately two trips will be made each hour. Additional trips can be added to the daily schedule in the event of delays in completing the measurements. The normal daily operating speed profile is shown in Illustration 2.

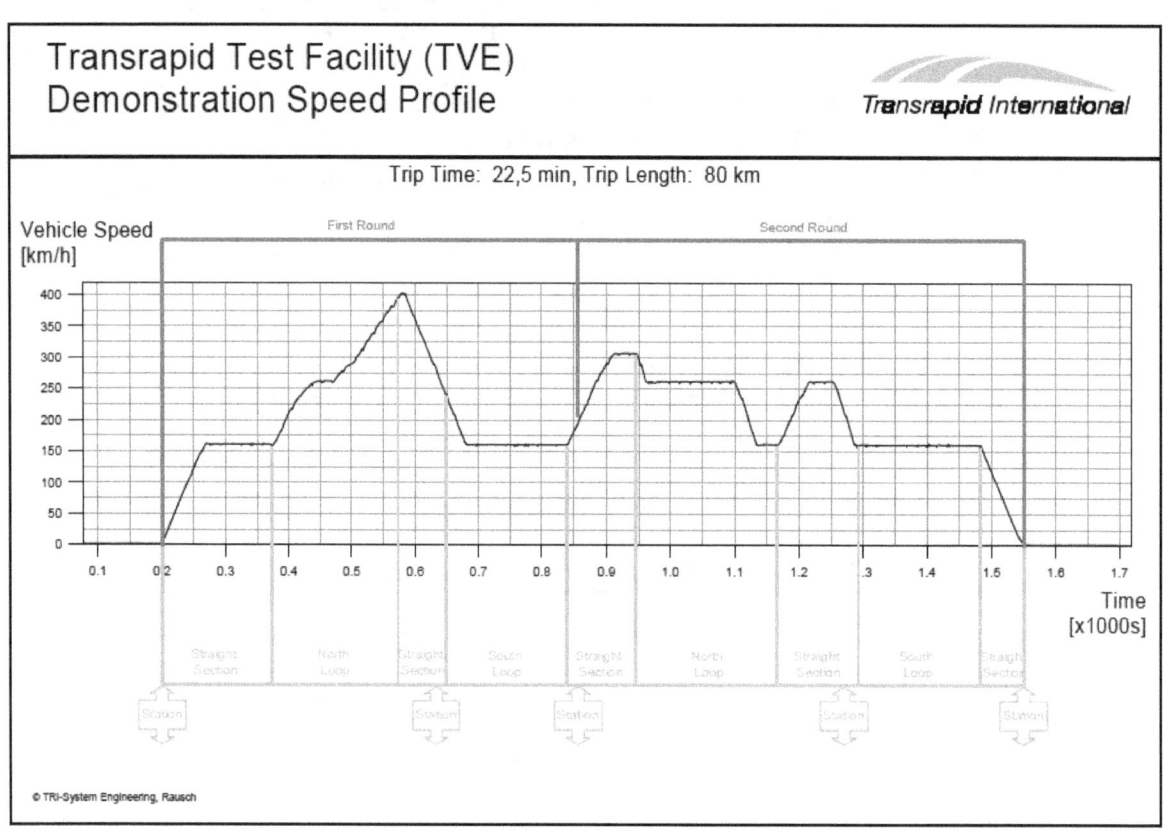

Illustration 2, Transrapid Test Facility (TVE) Demonstration Speed Profile

Guideway Measurement Locations and Local Speeds.
As part of the pre-meeting with Transrapid and the TVE facility personnel, the vendors conducted a survey of the facility and guideway locations. The purpose of this survey was to finalize the test locations for measuring the various guideway types (steel, concrete, hybrid, at grade and elevated) for guideway design and installations that represent the configurations that could be utilized for a US application. The survey also included a review of the local terrain conditions to determine the suitability of the site for measurement purposes. Based on the survey and Transrapid's description of the various guideway types and configurations, specific measurement sites were selected as shown in Illustration 3. The associated maximum speeds at each location are also identified in this Illustration.

Noise Characteristics of the Transrapid TR08 Maglev System

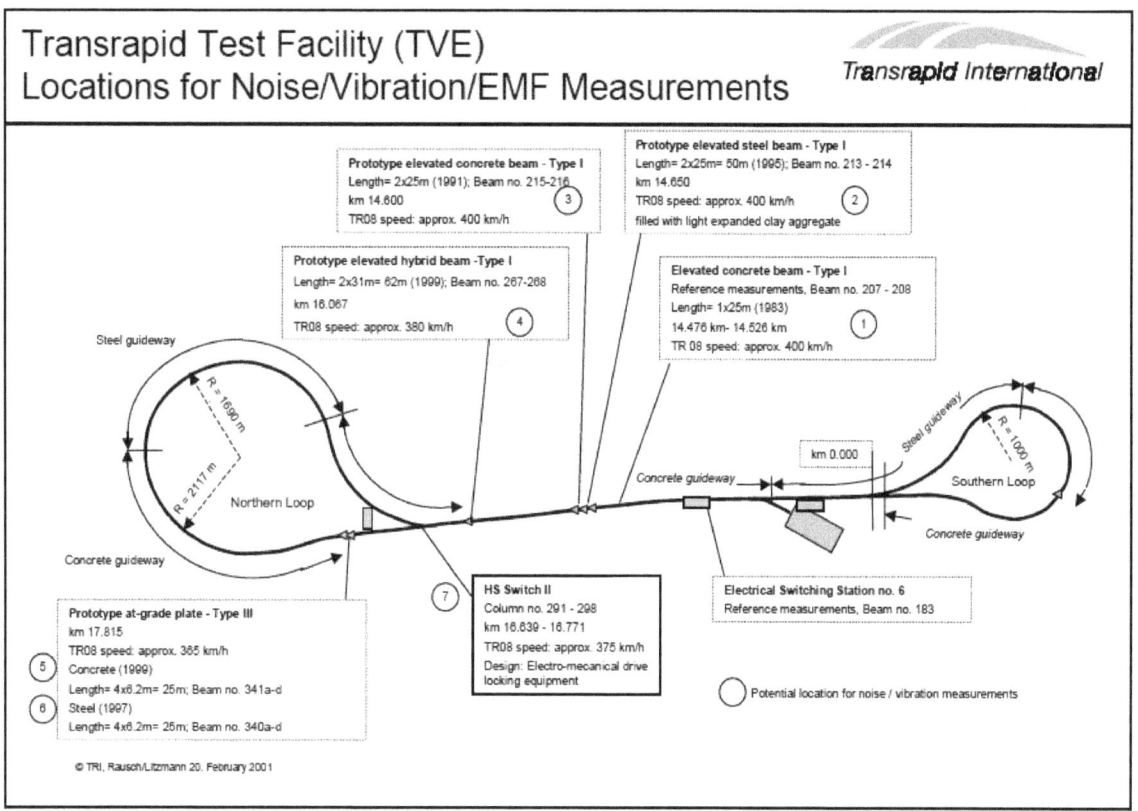

Illustration 3, Location of Measurement Sites and Associated Maximum Speeds

The numbered circles on Illustration 3 identify the basic order that the noise measurements will be conducted. As previously identified, the vibration and wayside EMF measurement teams will coordinate their activities with the noise measurement team. This way each vendor will have unrestricted access to each measurement location during the testing time.

Vehicle Measurement Locations
EMF measurements begin at several locations on the vehicles. These vehicle measurements will be made concurrently with the runs for the noise and vibration testing. Specific locations for measurements on the vehicle will be identified based on the location of propulsion, HVAC, communications and other onboard equipment. Measurements will be made at sixteen locations within the vehicle plus the attendant's compartment. This will result in a complete mapping of the spatial distribution of the electrical fields onboard the vehicle.

Section 4: Summary
MAGLEV, Inc. has worked with Electric Research Management, Harris, Miller, Miller and Hanson, Volpe, Transrapid International and the TVE test facility

personnel to develop the final test plan presented in this document. The plan incorporates all comments from the participants and represents a well-structured methodology to obtain, analyze and document the EMF, noise and vibration characteristics of the Transrapid Maglev System and TR08 vehicle.

The measurement location diagram and corresponding consolidated test matrix in Appendix C of this report summarizes each of the daily testing activities. The type of measurement (EMF, noise or vibration), trip number (N2 – N8), guideway measurement location, measurement equipment configurations and pass-by speeds for each series of measurements is identified in the matrix.

The testing is scheduled to commence during the week of April 2, 2001 and will be completed the following week. An additional day for making measurements has also been built into the plan to account for potential equipment failures or weather related delays.

Following completion of the testing in April, all data will be reduced, cataloged and evaluated by the vendors. The vendors are scheduled to complete the data evaluation and analysis and document the testing in a final report approximately two months following completion of the field measurements.

Section 5: Reporting of Results

Subsequent to preparation of Appendices A and B of this document, it was agreed that a collaborative effort as outlined in this Section will be used to prepare and publish final reports that provide documentation and results of the environmental tests.

Background. It is anticipated that the measurement results will be used in the development of maglev project environmental impact statements required under the National Environmental Policy Act (NEPA). The results should be presented in a manner that is acceptable from a technical and objective standpoint, and will minimize questioning of their validity. This information will also become a part of the permanent public record and will be posted by the FRA on a website for general access.

Report Format. The information will be documented in two separate FRA/Volpe Center official reports, one for EMF/EMR and the other for noise and vibration. These reports will be produced as a collaborative effort among the contractors, MAGLEV, Inc. and the federal staff of the Volpe Center. The "Authors" Box # 6 of the Report Documentation Page will list contributing staff members from contractors, MAGLEV, Inc. and the Volpe Center. The "Performing Organization" Box # 7 will identify their respective organizations. Box # 9 will show the FRA and the Port Authority as "Sponsors".

The Volpe Center staff will write an introductory chapter that describes the maglev deployment program, addresses the administrative process under which the tests were done, describes the MAGLEV, Inc. and FRA/Volpe Center validation, verification and quality assurance efforts, and gives a general summary, interpretation, and perspective of the results. The contractors will prepare the remainder of each report including appendices. Descriptive summary statistics of the EMF/EMR, noise (including spectral time histories), and vibration levels measured on or near the TR08 will be reported in tabular form and summarized graphically. These data will be event-based, i.e., provided for the specific locations and speeds of the vehicle at the time of the measurements. All documentation will be provided in hardcopy and electronic form. Electronic information will use Microsoft office products (WORD for text/ tables, Excel for spreadsheet and graphs) as well as portable document format (.pdf).

Test Plan for EMF Testing of the TR08

(Appendix A in original Test Plans)

APPENDIX A
NOT INCLUDED IN THIS REPORT

Test Plan for Noise and Vibration Testing of the TR08

(Appendix B in original Test Plans)

TEST PLAN FOR NOISE AND VIBRATION MEASUREMENTS OF TRANSRAPID TR08

INTRODUCTION

Noise and vibration measurements are proposed for the TR08 to be conducted at Transrapid Versuchanlage Emsland (TVE) in Germany. The noise tests are intended to replicate, where possible, previous testing programs on the TR07 vehicle. The testing program is requested by Federal Railroad Administration (FRA) as part of pre-environmental impact statement activities.[2]

The test plan calls for measurement of the TR 08 under specific operating conditions and on a limited number of sites. The plan is proposed to be performed over a three-week period with preparation and installation of microphone array equipment the week before the beginning of measurements and a testing period of up to two weeks. The noise measurement program is planned for six days, but it may have to be extended for additional days under conditions of unfavorable weather, test track operational problems, or last minute requests for additional information other than proposed. Of the foregoing causes for delay, weather is likely to be an important variable. Vibration measurements can, however, be performed with minimal limitations imposed by weather conditions.

The noise test plan has been prepared by Bernd Barsikow of akustik-data Engineering Office in Berlin, Germany. He is familiar with the test track having made similar measurements at that facility in the past. Akustik-data will be responsible for all the noise measurements.

Harris Miller Miller & Hanson Inc. (HMMH) will conduct the vibration measurements. The method will be performed according to the testing procedure described in FRA's guidance manual, High-speed Ground Transportation Noise and Vibration Impact Assessment.

Following sections describe the two test plans in general. Details are provided in the consolidated test plan matrix. Other tasks, including data analysis and reporting, are described in the Attachment : Scope of Work.

---

[2] **FRA Scope of Work Amendment No. 2, Section XIV. Pre-Environmental Impact Statement (EIS) Activities, Part C. Field Measurement of Noise and Vibration Characteristics of TR 08 (12/31/00).** In cooperation with assigned staff from the Volpe National Transportation Systems Center (VNTSC), the Grantee shall develop a plan to measure and analyze the noise and vibration caused by the operation of the TR 08 vehicle in service at the Transrapid Maglev Test Track in Emsland, Germany. With the approval of the FRA, the plan will be implemented.

1. SUMMARY OF TEST PLAN FOR NOISE MEASUREMENTS

The basis for the following test plan for measurements of airborne noise are the specifications for the various microphone arrays and their measuring stations along the guideway, the single microphones and their locations, and the number of passbys of the TR 08, as given in the consolidated test plan matrix. This measurement program has been cleared with the people who operate TVE and is based on experience gained by the akustik-data Engineering Office during several similar measuring projects carried out at TVE.

In particular, the scheduling of the various passbys on six days is contingent upon the test vehicle being fully operational for this period of time. Another consideration is the weather, which has to be suitable for acoustical measurements during the entire week[3]. Weather during the proposed measurement period is typically problematic in Northern Germany. Although the plan includes some flexibility in the work schedule for the operating crew, it may be necessary to add measurements in the following week as a contingency.

A test-run plan with four passbys of the TR 08 in 30 minutes is the basic assumption according to the standardized speed profile provided by TransRapid International (TRI). The speed ranges will cover 100 km/h to 400 km/h. At slower vehicle speeds (100 to 200 km/h), a complete circuit of the guideway will not be necessary because the vehicle only has to achieve the prescribed speed in the vicinity of the particular measuring station, and, in these cases, the acceleration phase will be relatively short. During the TR 07 measurements at low speed, the vehicle was operated in both directions to save time and energy: we propose to do the same for the TR 08, pending operational constraints imposed by TVE. We have chosen the range of 100 to 400 km/h for all tests (except one switch configuration) so that all guideway configurations will have the same speed ranges. Extrapolation to higher speeds can be performed, but with reduced accuracy. Operating condition of the vehicle will be monitored during test runs. One requirement is that all power sources be switched on during the pass-bys. We do not plan to monitor the passenger load in the vehicles specifically for the noise tests.

At the measuring stations, we have to make sure there will be good access to the elevated arrays (particularly for the IABG tower) as well as to the single microphones positioned at distances from the guideway of 50 and 100 feet and 6.5 and 25 meters. The boundary conditions (i.e., the topological conditions of the land) at these latter stations should also be suitable for acoustical measurements. These conditions should not, however, be a problem when the concrete guideway is mounted on pylons. There are many sections of concrete

---

[3] For acoustical measurements, wind velocity should be 6 m/s (12 mph) or less, and wind gusts should be no greater than about 9 m/s (20 mph). Tests can be conducted in light rain or drizzle if necessary, but not in heavy rain, sleet or accumulating snow.

guideway where the terrain is satisfactory for acoustical measurements[4]. The reference measuring station will be located at beam No. 208, the site of previous measurements along the high-speed straight section. There are fewer sections of steel guideway. One section of elevated steel guideway will be selected for measurement; beam No. 213 is a section similar to the length and type being proposed for U.S. applications. The hybrid beam No. 267 and the prototype concrete beam No. 215 will also be measured. Final selection was made during a field visit during 15 – 16 February.

In addition to the microphone array sites, locations will be selected for single-microphone measurements near the high-speed switch and along the at-grade guideway where it more or less lies at ground level. In these cases at the TVE, the geometry of the guideway and its immediate surrounds can set a limit to the maximum passby speed on these sections. For the switch, we propose to measure under both conditions straight ahead and turn-out. Speed restrictions apply to the turn-out condition, however, so we will not have as great a range of speeds as is possible for the straight configuration.

Noise test program details are shown in the consolidated test plan matrix.

2. VIBRATION TEST PLAN

The goal of the vibration tests is to obtain the "force density" of the TR08 for use in predictions of ground vibrations near proposed maglev systems at other locations than the test track in Emsland. The key parameters are expressed in the following equation:

$$LF = Lv - TM_{line}$$

where, assuming all quantities are in decibels with consistent reference quantities:

L_F = force density level for a line source (i.e. maglev on guideway)
L_v = rms vibration velocity level of a maglev passby
TM_{line} = line source transfer mobility from the guideway to the measurement site evaluated at a specific distance.

The line source transfer mobility (TM_{line}) represents an empirical relationship between a known linear vibration source and the resulting ground vibration. By measuring the TR08 vibration (L_v) at a site and using a special test to measure TM_{line} at the same site, the baseline force density (L_F) for the TR08 can be derived. The force density function is assumed to be independent of the ground characteristics, which means that it can be combined with a measured transfer mobility to predict ground-borne vibration at locations other than the test site.

---

[4] For example, site conditions satisfying ISO 3095-1975, Section 6 (Acoustical environment, meteorological Conditions, Background Level).

The force density is determined on the ground so the quantity includes the vehicle on its guideway; for example, the force density of the TR 08 on a concrete guideway may be different than that measured for the same vehicle on a steel beam. Analysis of the data will result in a catalog of force densities for TR08 on various guideway configurations.

HMMH test plan covers three days of measurement. Our proposed budget includes only one week of HMMH time at TVE. The test equipment will be shipped to Amsterdam ready for pick-up on Friday morning before the test week. After retrieving the equipment, HMMH will drive to the test track, unpack equipment and check all calibrations. Orientation and meeting with test personnel will take place on Monday morning. Vibration tests will be conducted Monday afternoon through Thursday. Repacking equipment for shipment will take place on Thursday night for return to Amsterdam for shipment back to the US on Friday. Any unforeseen delays in operations of the TR 08 on the test track will, of course, extend the testing period into the following week, as described above regarding the noise tests.

The details of the vibration test program are shown in the consolidated test plan matrix, organized to be almost parallel to the noise measurement program.

ATTACHMENT: SCOPE OF WORK

Noise and Vibration Measurements of TransRapid Maglev Vehicle TR08

Note: This Scope of Work is based on HMMH Proposal No. P00-20115 as modified by subsequent planning discussions. The Scope is included as an attachment to the Proposed Test Plan as an overview of the entire measurement program including data analysis and reporting.

Summary.
MAGLEV, Inc. has requested Harris Miller Miller & Hanson Inc. (HMMH) of Burlington, MA, to prepare a scope of work to measure and report the wayside noise and vibration characteristics of the TransRapid TR08. HMMH will team with akustik-data Engineering Office (a-d) of Berlin, Germany. Measurements will be performed at the TransRapid Test Facility (TVE) in Emsland, Germany. The team of HMMH and akustik-data has completed similar measurement and analyses of Amtrak's new high-speed train, Acela, and is in a good position to compare the noise and vibration characteristics of the different high-speed ground transportation platforms.

Task 1. Planning and Coordination
Dr. Carl Hanson of HMMH and Mr. Bernd Barsikow of a-d will develop a draft test plan for the measurements of noise and vibration of the TR08. The draft plan will be submitted to MAGLEV, Inc. for comments. After both MAGLEV, Inc. and the HMMH/a-d team agree, MAGLEV, Inc. will forward the plan to the Volpe National Transportation Systems Center (VNTSC) for comment and eventual approval. TVE will be consulted during the review process and must approve the test plan as it pertains to their operations. After agreements are reached, Barsikow will travel to the Transrapid Test Facility (TVE) in Emsland for a coordination meeting. Purpose of this meeting is to discuss details for implementing the test plan, select the measurement locations, check out the IABG telescopic tower and associated equipment, and make final plans for the field measurements.

Task 2. Field Testing
Akustik-data will conduct the noise tests and HMMH will conduct the vibration tests. IABG will provide technical support for the noise measurements and will provide information about the operating conditions of the vehicle during the tests.

Task 2A. Noise Measurements. Noise measurements, including preparation, set-up and data collection, will be performed by akustik-data engineering staff. Measurements will be made using akustik-data's microphone arrays placed close to the test track as well as single microphones at the European and FRA standard reference distances. Akustik-data's microphone arrays will be mounted on IABG's telescopic tower to position them appropriately for identifying noise sources on TR08.

Vertical arrays will be positioned to determine noise source heights on both vehicle and guideway. Horizontal and "X-arrays" will be used to pin-point noise source locations on the vehicle. All array configurations will be positioned 3.5 m from the outer wall of the TR 08, or about 5.0 m from the centerline of the track. This distance is close enough to focus on specific noise sources, but far enough away from the boundary layer of the vehicle to avoid gusts of moving air from the vehicle. An example of a microphone array and the TR07 is shown in the attached figure (Figure 1).

Measurements should be made for 3 passbys at each of four speeds ranging from 150 to 400 km/h for each array configuration. The test plan allows for a limited number of passbys at lower speeds of 100, km/h. This plan would result in a total of 178 passbys (11 series of array configurations and two series with single microphones) over a period of 6 days of measurement. Unfavorable weather conditions may extend the period of measurement. The final test plan matrix specifies the exact speeds required for each test, but in general the tests would be run as follows:

Three passbys at each of the following speeds – 150, 200, 300, and 400 km/h (two sets of speed doublings) for the following eleven array configurations (the concrete guideway beam No. 208 will be used as the baseline, as determined during the preliminary site visit):

1. Wayside vertical (WV) nested array in low and high position at the reference site with concrete guideway – for determining source height locations for vehicle;
2. Wayside horizontal (WH) nested array in low position at reference concrete guideway – for determining sources along length of vehicle at magnet locations;
3. Wayside vertical (WV) nested array in low and high position at a site with prototype concrete beam, a site with hybrid beam, and a site with steel beam – for determining source height locations for vehicle on those guideway configurations;
4. X-array with 32 cm microphone spacing (WX32) in mid-position at the reference concrete guideway – for locating low frequency sources; and
5. X-array with 16 cm microphone spacing (WX16) in mid-position at the reference concrete guideway – for locating mid-frequency sources.

Note that four sites would be selected for the array measurements: reference concrete beam, steel beam, prototype concrete beam and hybrid beam guideway locations. The speeds would have to be those on the standard profile at those sites. It is not practical to move the microphone arrays to different sites along the test track to obtain different speeds.

During all these tests, single microphones will be used to measure at three reference locations, 1.5 m above ground level at 50 feet (required by TNM and

FRA noise models), 1.2 meters above ground at 100 feet (required for EPA/FRA noise emission standards), 1.2 meters above ground level at 25 meters and two heights above ground at 6.5 meters (required for German standards). The single microphone set-up will be moved to cover six different guideway configurations: concrete beam, steel beam, prototype concrete beam, hybrid beam, at-grade beam and switch. The single microphone measurements at the concrete, steel and hybrid beams will take place during the array measurements, but a less comprehensive set of measurements (only a few speeds) at the switch will be carried out separately. The single microphone measurements will be used for four purposes: (1) to serve as reference points to sum up and model he sources, (2) to compare with other high-speed trains, (3) to compare with FRA and European standards, and (4) to compare noise characteristics among various guideway configurations.

Data will in all cases be recorded in digital form on magneto-optical disks for later analysis in akustik-data's laboratories. Final storage of data and results will be on CD media.

Task 2B. Vibration Measurements. A full vibration propagation test program will be performed to determine force density of the TR08 vehicle on four guideway configurations. The vibration propagation test procedure is described in Chapter 9 of FRA's guidance manual, *High-speed Ground Transportation Noise and Vibration Impact Assessment*[5]. The test consists of impacting the ground with a measured force and recording the resulting vibration pulses at various distances from the impact point. The relationship between the input force and the ground surface vibration, called the transfer mobility, characterizes vibration propagation at a given location.

HMMH will use a purpose-built ground impacter for generating the force. A load cell and amplifier will be used to measure the force of a repeatedly-dropped weight, and ground vibration measurements will be made with high-sensitivity accelerometers mounted in the vertical direction on top of steel stakes driven into the ground. The acceleration signals will be amplified using low-noise amplifiers and recorded on a Teac Model RD-130TE 8-channel digital audio tape (DAT) recorder for subsequent analysis in the HMMH laboratory. Reference signals will be used to calibrate the load cell and accelerometers, based on their rated sensitivities.

Vibration propagation measurements to determine transfer mobility of the ground will be made at four sites with different guideway configurations: concrete beam, steel beam, hybrid beam and a switch. Following the propagation measurements at each site, vibrations from passbys of the TR08 at the exact same positions will be recorded for determination of Force Density.

---

[5] http://www.fra.dot.gov/s/env/guidance.htm

Task 3. Data Analysis

Task 3A. Noise Analysis. Noise data will be analyzed using proprietary computer processing in akustik-data laboratory in Berlin. The microphone array data are analyzed using proprietary software developed by akustik-data. A microphone array is a highly directional sound measuring instrument consisting of several closely-spaced microphones. While a single microphone measures only the total sound generated by all the sound sources on a moving train, a microphone array can be used to locate individual sound sources by properly combining the microphone output signals. When the array is located close to the sound source, where the wavefront can be considered spherical, the total sound pressure from the "focus" of the array is determined during post-processing of the recorded data by summing the individual microphone signals after correcting for the distance from each microphone to the focal point. The focal point can be electronically "steered" simply by the way in which the signals are processed using the time differences among the signals at each microphone. That is, the focus of the array can be thought of as a beam that can be moved to track a moving source with no physical adjustment of the apparatus.

The sound pressure level for each source location is determined for the position of the microphone array. Speed dependence of each source is determined by plotting sound pressure level increases with speed and performing a regression analysis on the data. Projections to source reference distances required for calculation routines are performed using akustik-data's noise model "AD-PRO 2.0." The sound energy from all sound sources measured by the array are combined using this model and compared with the actual measured time histories. The sound exposure level (SEL) of each source is determined using the same model. Time histories of each source are produced and the SEL is calculated. Further, recorded data will be analyzed into narrow band frequency spectra for source diagnostics and one-third octave-band spectra for practical application.

The results will be presented in terms of the reference quantities required for the FRA high-speed ground transportation noise and vibration guidance manual detailed noise analysis: SEL, length, and speed coefficient for each relevant source.

Task 3B. Vibration Analysis. Vibration data will be analyzed by HMMH in their laboratory in Massachusetts. Transfer mobility functions are developed from field measurements in 4 steps:
1. Analyze the field data to generate narrowband point source transfer mobilities.
2. Calculate 1/3 octave band transfer mobilities at each measurement point from the narrowband results. Because typical spectrum analyzers are not capable of obtaining 1/3 octave band transfer

functions, this processing is performed after transferring the data to a computer.
3. Calculate the transfer mobility as a function of distance for each 1/3 octave band.
4. Compute the line source transfer mobility as a function of distance in each 1/3 octave band.
1. Force density is derived from the measured vibrations of the TR08 and the transfer mobility at each of the sites. The result will be presented as Force Density spectra in one-third octave-bands

Task 4. Reporting
The results of the noise and vibration testing will be presented in a single report jointly authored by akustik-data and HMMH. It will be similar to the recent report to the FRA on the noise and vibration tests of the Acela at the Transportation Technology Center. Noise data will be interpreted by akustik-data. Results will be accompanied by figures illustrating the location of noise sources and their speed relationships. The attached figure taken from the Acela measurements (Figure 2) is a sample of how the relative sources are depicted. Pictures like this at every speed illustrate the location of noise sources. Source heights and SELs will be documented for use in prediction models. For example, source information will be presented in the appropriate format for the FRA High-speed Ground Transportation Noise and Vibration manual: SEL, effective length, speed coefficient. Differences in noise radiation among guideway types, including concrete, steel, hybrid, and switches will be documented.

The results of the vibration measurements will be presented in a form that can be used with the source levels and propagation curves in the FRA High-speed Ground Transportation Noise and Vibration manual. Force density spectra will be determined as a function of speed and type of guideway.

A summary section will compare the noise and vibration levels from the TR08 with those from other high-speed ground transportation sources.

The report will be prepared jointly by HMMH and akustik-data, with the final combined noise and vibration report edited and printed by HMMH.

KEY STAFFING

The HMMH/akustik-data team will be made up of the same members who measured and analyzed the Acela noise and vibration characteristics at the TTC. Project manager will be Dr. Carl E. Hanson, Senior Vice President and a Founder of HMMH. He will be responsible for coordinating the measurement program among team members and MAGLEV, Inc., as well as providing information as requested by the Volpe Center and the FRA. Dr. Hanson is familiar with TVE and the staff of IABG, having conducted measurements there during FRA's National Maglev Initiative Program. The akustik-data effort will be directed by Mr. Bernd Barsikow, principal of the firm. Mr. Barsikow is widely known for pioneering the development of the use of microphone arrays for diagnosing noise sources on high-speed trains. He continues to be a leading acoustical specialist in this field in Europe and is well known by the IABG acoustical specialists at TVE.

The key personnel and their respective addresses are as follows:

    Dr. Carl E. Hanson
    Harris Miller Miller & Hanson Inc.
    15 New England Executive Park
    Burlington, MA 01803
    Tel: 781-229-0707
    Fax: 781-229-7939
    Email: chanson@hmmh.com

    Dipl.-Ing. B.Barsikow
    akustik-data Engineering Office
    Kirchblick 9
    D-14129 Berlin
    Germany
    Tel: +49-30-8090 2606
    Fax: +49-30-8090 2607
    Email: info@akustik-data.de

Figure 2. Microphone Array and TR07

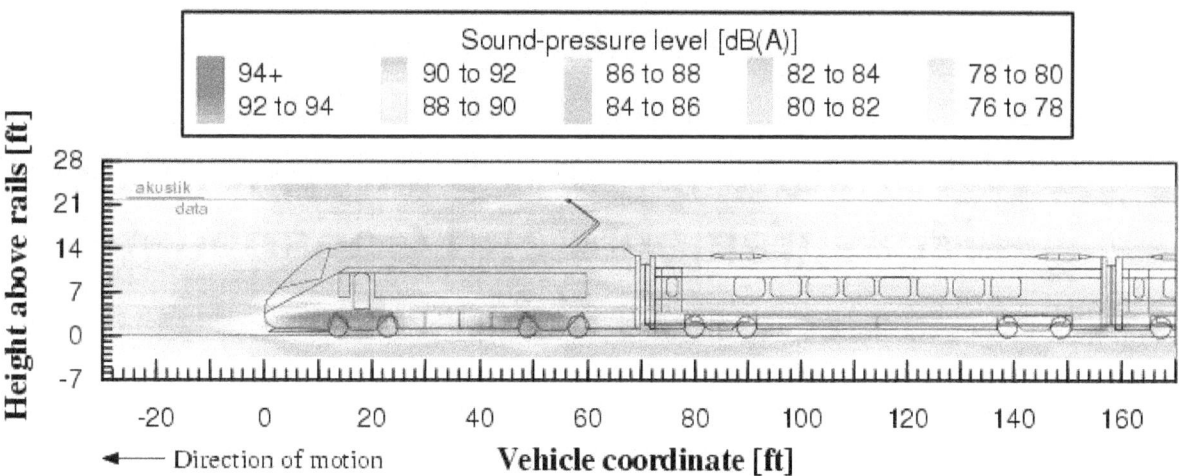

Figure 3. Noise Sources on Acela at 150 mph

Locations for Noise/Vibration/EMF Measurements and Consolidated Test Matrix

Revised May 24, 2001 To Include Noise Measurements Under the Guideway and Vibration Measurements at the Station Platform

(Appendix C in the original test plans)

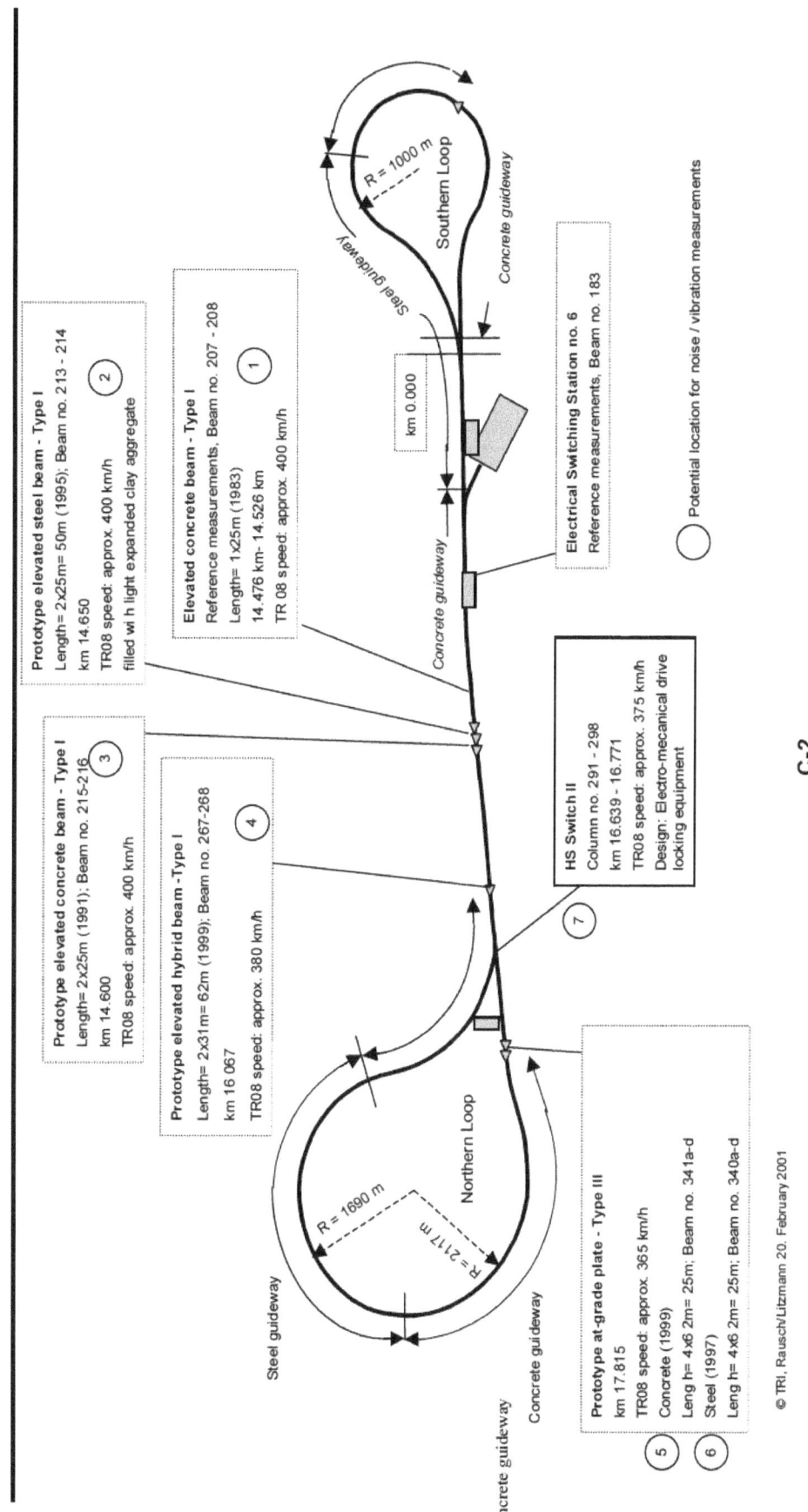

Consolidated Test Matrix Schedule of Environmental Noise, Vibration, and EMF Tests at TVE

Date: Monday, 2001-08-13

Noise Measurement Activities	Trip	Time frame	Measuring location	Beam no.	Array configuration	Single microphones	Pass-by vehicle speed [km/h]	Measuring series	Pass-by ID
Inspect equipment, assembly of electronic equipment in measuring van, installation of WX32 microphone array, mount array on IABG telescopic tower.		9:00-18:00							

Date: Monday, 2001-08-13

Vibration Measurement Activities	Trip	Time frame	Measuring location	Beam no.	Array configuration	Single microphones	Pass-by vehicle speed [km/h]	Measuring series	Pass-by ID
Arrive in Amsterdam; clear equipment through customs.		6:00 – 10:00							
Drive to Emsland, orientation at TVE.		10:00-18:00							

Date: Monday, 2001-08-13

EMF Measurement Activities	Trip	Time frame	Measuring location	Beam no.	Sensor array configuration		Vehicle operating speed profile	Measuring series	
Inspect equipment, standardise MultiWave instruments to the time reference of the TVE data acquisition system, select on-board measurement location seat numbers, install sensors on the mannequin, place instruments on charge, and stake sensor locations near the guideway to facilitate rapid relocation of the equipment between tests.		9:00-18:00							

Noise Characteristics of the Transrapid TR08 Maglev System

Date: Tuesday, 2001-08-14

Noise Measurement Activities	Trip	Time frame	Measuring location	Beam no.	Array configuration	Single microphones	Pass-by vehicle speed [km/h]	Measuring series	Pass-by ID
Install microphones on WX32 array, calibrate, elevate array to "mid" position, set up single microphones.		7:00-9:30	concrete guideway (reference type) ①	208	setup/ reconstruction	setup/ reconstruction			
Make noise measurements with WX32 array in "mid" position, make noise measurements with single microphones.	N2	09:30-10:00	concrete guideway (reference type) ①	208	WX32	6.5 m (high)	standardized speed profile (150, 200, 300, 400 km/h)	A	A-150-1 A-400-1 A-200-1 A-300-1
	N3	10:00-10:30							A-150-2 A-400-2 A-200-2 A-300-2
	N4	10:30-11:00							A-150-3 A-400-3 A-200-3 A-300-3
Reconfigure array to WX16, elevate array to "mid" position, make noise measurements with single microphones.	N5	11:00-11:30	concrete guideway (reference type) ①	208	setup/ reconstruction	6.5 m (high), 25 m, 50 ft, 100 ft, under guideway	100 km/h	A	A-100-1 A-100-2
						6.5 m (low), 25 m, 50 ft, 100 ft, under guideway	100 km/h	B	B-100-1 B-100-2
Make noise measurements with WX16 array in "mid" position, make noise measurements with single microphones.	N7	13:00-13:30	concrete guideway (reference type) ①	208	WX16	6.5 m (low), 25 m, 50 ft, 100 ft, under guideway	standardized speed profile (150, 200, 300, 400 km/h)	B	B-150-1 B-400-1 B-200-1 B-300-1
	N8	13:45-14:15							B-150-2 B-400-2 B-200-2 B-300-2
	N9 (instead of N6)	14:30-15:00							B-150-3 B-400-3 B-200-3 B-300-3

Noise Characteristics of the Transrapid TR08 Maglev System

		15:00-18:00			setup/ reconstruction	setup/ reconstruction			
Retrieve single microphones and reconstruct array to WV08/16/32.									

Date: Tuesday, 2001-08-14

Vibration Measurement Activities

	Trip	Time frame	Measuring location	Beam no.	Activity		Pass-by vehicle speed [km/h]	Measuring series	Pass-by ID
Unpack and check out equipment.		7:00-9:30			setup/ reconstruction				
Set up accelerometer, recording instruments. Calibrate.	N2	09:30-10:00	steel guideway ②	213	setup/ reconstruction		standardized speed profile (150, 200, 300, 400 km/h)	A	A-150-1 A-400-1 A-200-1 A-300-1
	N3	10:00-10:30							A-150-2 A-400-2 A-200-2 A-300-2
	N4	10:30-11:00							A-150-3 A-400-3 A-200-3 A-300-3
Record ground vibrations from TR08 at site.	N5	11:00-11:30	steel guideway ②	213	Measure		100 km/h	A	A-100-1 A-100-2
							100 km/h	B	B-100-1 B-100-2
	N7	13:00-13:30	steel guideway ②	213	Measure		standardized speed profile (150, 200, 300, 400 km/h)	B	B-150-1 B-400-1 B-200-1 B-300-1
	N8	13:45-14:15							B-150-2 B-400-2 B-200-2 B-300-2
	N9	14:30-15:00							B-150-3 B-400-3 B-200-3 B-300-3
Set up ground impacter and conduct transfer mobility test.		15:00-18:00			setup/ reconstruction/ measure				

Noise Characteristics of the Transrapid TR08 Maglev System

Date: Tuesday, 2001-08-14

EMF Measurement Activities	Trip	Time frame	Measuring location	Beam no.	Sensor array configuration	Vehicle operating speed profile	Measuring series
Transport equipment to the lead vehicle, place sensors in the first measurement location, and verify instrument performance.		8:00-9:30	Onboard lead vehicle		Mannequin positions and fixed reference		
Onboard lead vehicle. EMF and RF measurements in three adjacent seats toward the front of the vehicle and away from major on-board equipment to identify center-to-side variation in field levels.	N2	09:30-10:00	Onboard lead vehicle Aisle seat.		"	Standardized speed profile	Onboard
	N3	10:00-10:30	Onboard lead vehicle Middle seat.		"	Standardized speed profile	Onboard
	N4	10:30-11:00	Onboard lead vehicle Window seat.		"	Standardized speed profile	Onboard
Station measurements. EMF and RF measurements in the station as the vehicle departs, passes, and re-enters.	N5	11:00-11:30	Location nearest the vehicle where a passenger might wait.		"	Nonstandard speed profile required for low speed noise measurements	Station
Onboard lead vehicle. EMF and RF measurements away from major on-board equipment to identify side-to-side variation in field levels.	N7 [1]	13:00-13:30	Onboard lead vehicle Middle seat contra-lateral to seat measured on Trip N3 above.		"	Standardized speed profile	Onboard
Onboard lead vehicle. EMF and RF measurements away from major on-board equipment to identify front-to-back variation in field levels.	N8 [1]	13:45-14:15	Onboard lead vehicle Middle seat far behind seat measured on Trip N3 above but having similar proximity to onboard equipment.		"	Standardized speed profile	Onboard
Onboard lead vehicle. EMF and RF measurements away from major on-board equipment to confirm side-to-side and front-to-back variation in field levels.	N9 [1] (instead of N6)	14:30-15:00	Onboard lead vehicle Middle seat contra-lateral to seat measured on Trip N8 above.		"	Standardized speed profile	Onboard
Remove instruments from the vehicle and place them on charge. Inspect data for integrity and back up.		15:00-16:30			Instruments removed from vehicle		

Note: [1] All trips conforming to the standardized speed profile regardless of their trip number will be used for onboard measurements which will be conducted in the order given in this test plan. If additional trips are made during any given day, they will be used to accelerate the test schedule.

Noise Characteristics of the Transrapid TR08 Maglev System

Date: Wednesday, 2001-08-15

Noise Measurement Activities	Trip	Time frame	Measuring location	Beam no.	Array configuration	Single microphones	Pass-by vehicle speed [km/h]	Measuring series	Pass-by ID
Install microphones on WV08/16/32 wayside vertical array, calibrate, elevate array to "high" position.		7:00-9:30	concrete guideway (reference type) ①	208	setup/ reconstruction				
Set up single microphones at at-grade guideway.		7:30-9:30	at-grade guideway (concrete and steel) ⑤⑥	341a-d 340a-d		setup/ reconstruction			
Make noise measurements with WV08/16/32 array in "high" position.	N2	09:30-10:00	concrete guideway (reference type) ①	208	WV08/16/32 (high)	no	standardized speed profile (150, 200, 300, 400 km/h)	C	C-150-1 C-400-1 C-200-1 C-300-1
	N3	10:00-10:30							C-150-2 C-400-2 C-200-2 C-300-2
	N4	10:30-11:00							C-150-3 C-400-3 C-200-3 C-300-3
Make noise measurements with single microphones at at-grade guideway.	N2	09:30-10:00	at-grade guideway (concrete and steel) ⑤⑥	341a-d 340a-d	no	6.5 m (high and low)	standardized speed profile (300, 400 km/h)	Y	Y-400-1 Y-300-1
	N3	10:00-10:30							Y-400-2 Y-300-2
	N4	10:30-11:00							Y-400-3 Y-300-3
Lower array to "low" position.		11:00	concrete guideway (reference type) ①	208	setup/ reconstruction				
Make noise measurements with WV08/16/32 array in "low" position.	N5	11:00-11:30	concrete guideway (reference type) ①	208	WV08/16/32 (low)	no	standardized speed profile (150, 200, 300, 400 km/h)	D	D-150-1 D-400-1 D-200-1 D-300-1
	N6	12:15-12:45							D-150-2 D-400-2 D-200-2 D-300-2

Noise Characteristics of the Transrapid TR08 Maglev System

Noise Measurement Activities	Trip	Time frame	Measuring location	Beam no.	Array configuration	Single microphones	Pass-by vehicle speed [km/h]	Measuring series	Pass-by ID
	N7	13:00-13:30							D-150-3 D-400-3 D-200-3 D-300-3
Make noise measurements with single microphones at at-grade guideway.	N5	11:00-11:30	at-grade guideway (concrete and steel) ⑤⑥	341 a-d 340 a-d	no	6.5 m (high and low)	standardized speed profile (300, 400 km/h)	Y	Y-400-4 Y-300-4
	N6	12:15-12:45							Y-400-5 Y-300-5
	N7	13:00-13:30							Y-400-6 Y-300-6
Make noise measurements with WV08/16/32 array in "low" position.	N8	13:45-14:15	at-grade guideway (concrete and steel) ⑤⑥	341 a-d 340 a-d	WV08/16/32 (low)	no	100 km/h	D	D-100-1 D-100-2 D-100-3 D-100-4
Make noise measurements with single microphones at at-grade guideway.	N8	13:45-14:15	at-grade guideway (concrete and steel) ⑤⑥	341 a-d 340 a-d	no	6.5 m (high and low)	100 km/h	Y	Y-100-1 Y-100-2
Retrieve single microphones and reconstruct array to wayside horizontal configuration.		14:30-18:00			setup/ reconstruction	setup/ reconstruction			

Noise Characteristics of the Transrapid TR08 Maglev System

Date: Wednesday, 2001-08-15

Vibration Measurement Activities	Trip	Time frame	Measuring location	Beam no.	Pass-by vehicle speed [km/h]	Measuring series	Pass-by ID
Set up accelerometers, recording instrumentation, and ground impacter at site near hybrid guideway and conduct transfer mobility test.		7:00-9:30	hybrid guideway ④	267			
Record ground vibrations from TR08 at site.	N2	09:30-10:00	hybrid guideway ④	267	standardized speed profile (150, 200, 300, 400 km/h)	C	C-150-1 C-400-1 C-200-1 C-300-1
	N3	10:00-10:30					C-150-2 C-400-2 C-200-2 C-300-2
	N4	10:30-11:00					C-150-3 C-400-3 C-200-3 C-300-3
	N5	11:00-11:30				D	D-150-1 D-400-1 D-200-1 D-300-1
Move accelerometer mounts to high speed switch. Set up accelerometers, recording instrumentation at site near switch		11:30 – 13:00	HS switch II ⑦	291-298			
Record ground vibrations from TR08 at site.	N7	13:00 – 13:30			standardized speed profile (150, 200, 300, 400 km/h)	V	V-150-1 V-400-1 V-200-1 V-300-1
	N8	13:45 – 14:15					V-150-2 V-400-2 V-200-2 V-300-2
	N9	14:30 – 15:00					V-150-3 V-400-3 V-200-3 V-300-3
Set up ground impacter at site near switch and conduct transfer mobility test.		15:00 – 18:00					

Noise Characteristics of the Transrapid TR08 Maglev System

Date: Wednesday, 2001-08-15

EMF Measurement Activities	Trip	Time frame	Measuring location	Beam no.	Sensor array configuration	Vehicle operating speed profile	Measuring series
Transport equipment to the lead vehicle, place sensors in the seventh onboard measurement location, and verify instrument performance.		8:00-9:30	Onboard lead vehicle		Mannequin positions and fixed reference		
Onboard lead vehicle. EMF and RF measurements near onboard equipment.	N2	09:30-10:00	Onboard lead vehicle. Middle seat above a battery compartment.		,,	Standardized speed profile	Onboard
	N3	10:00-10:30	Onboard lead vehicle. Middle seat above battery charger and control equipment.		,,	Standardized speed profile	Onboard
	N4	10:30-11:00	Onboard lead vehicle. Middle seat above an air conditioning compressor.		,,	Standardized speed profile	Onboard
Onboard middle vehicle. EMF and RF measurements at a locations corresponding to the to test locations in the lead car to identify variation in field levels among cars. The measurement locations will correspond to three locations where measurements were made in the lead vehicle.	N5	11:00-11:30	Onboard middle vehicle. Middle seat away from onboard equipment.		,,	Standardized speed profile	Onboard
	N6	12:15-12:45	Onboard middle vehicle. Middle seat above a battery compartment.[2]		,,	Standardized speed profile	Onboard
	N7	13:00-13:30	Onboard middle vehicle. Middle seat above a battery charger and control equipment.[2]		,,	Standardized speed profile	Onboard
Station measurements. EMF and RF measurements in the station as the vehicle departs, passes, and re-enters.	N8	13:45-14:15	Location distant from the vehicle where a passenger might wait.			Nonstandard speed profile required for low speed noise measurements	Station
Remove instruments from the station and place them on charge. Inspect data for integrity and back up.		14:30-16:00			Instruments removed from station		

Note: [2] One of these locations may be changed to above the air conditioning compressor if measurements in the lead vehicle show that piece of equipment to be a significant field source.

Noise Characteristics of the Transrapid TR08 Maglev System

Date: Thursday, 2001-08-16

Noise Measurement Activities	Trip	Time frame	Measuring location	Beam no.	Array configuration	Single microphones	Pass-by vehicle speed [km/h]	Measuring series	Pass-by ID
Install microphones on WH08/16/32 wayside horizontal array, calibrate, elevate array to "low" position.		7:00-9:30	concrete guideway (reference type) ①	208	setup/ reconstruction				
Install single microphones at northern switch.		9:00-13:00	HS switch II ⑦	291-298		setup/ reconstruction			
Make noise measurements with WH08/16/32 array in "low" position.	N2	09:30-10:00	concrete guideway (reference type) ①	208	WH08/16/32	no	standardized speed profile (150, 200, 300, 400 km/h)	E	E-150-1 E-400-1 E-200-1 E-300-1
	N3	10:00-10:30							E-150-2 E-400-2 E-200-2 E-300-2
	N4	10:30-11:00							E-150-3 E-400-3 E-200-3 E-300-3
Make measurements with single microphones at switch.	N7	13:00-13:30	HS switch II ⑦	291-298	No	6.5 m (high and low) 25 m, 50 ft, 100 ft, under guideway	standardized speed profile (150, 200, 300, 400 km/h)	Z	Z-150-1 Z-400-1 Z-200-1 Z-300-1
	N8	13:45-14:15					standardized speed profile (150, 200, 300, 400 km/h)		Z-150-2 Z-400-2 Z-200-2 Z-300-2
	N9 (instead of N5/N6)	14:30-15:00					100 km/h		Z-100-1 Z-100-3 Z-100-3 Z-100-4
Retrieve single microphones and detach array from tower.		15:00-16:00			setup/ reconstruction	setup/ reconstruction			
Move tower and equipment to measuring site at steel guideway.		16:00-18:00			setup/ reconstruction				

Date: Thursday, 2001-08-16

Vibration Measurement Activities	Trip	Time frame	Measuring location	Beam no.	Pass-by vehicle speed [km/h]	Measuring series	Pass-by ID
Set up accelerometers, recording instrumentation, and ground impacter at site near concrete guideway and conduct transfer mobility test.		7:00-9:30	concrete guideway (reference type) ①	208			
Record ground vibrations from TR08 at site	N2	09:30-10:00	concrete guideway (reference type) ①		standardized speed profile (150, 200, 300, 400 km/h)	E	E-150-1 E-400-1 E-200-1 E-300-1
	N3	10:00-10:30					E-150-2 E-400-2 E-200-2 E-300-2
	N4	10:30-11:00					E-150-3 E-400-3 E-200-3 E-300-3
Move accelerometer mounts to at-grade guideway. Set up accelerometers, recording instrumentation at site.		11:00 – 13:00	at-grade guideway (concrete and steel) ⑤⑥	341a-d 340a-d			
Record ground vibrations from TR08 at site	N7	13:00-13:30	at-grade guideway (concrete and steel) ⑤⑥	341a-d 340a-d	standardized speed profile (150, 200, 300, 400 km/h)	Z	Z-150-1 Z-400-1 Z-200-1 Z-300-1
	N8	13:45-14:15			standardized speed profile (150, 200, 300, 400 km/h)		Z-150-2 Z-400-2 Z-200-2 Z-300-2
	N9 (instead of N5/N6)	14:30-15:00			100 km/h		Z-100-1 Z-100-2 Z-100-3 Z-100-4
Set up ground impacter at site near at-grade guideway and conduct transfer mobility test.		15:00-17:00					
Move accelerometers to station platform site.		17:00-18:00					

Noise Characteristics of the Transrapid TR08 Maglev System

Date:: Thursday, 2001-08-16 [3]

EMF Measurement Activities	Trip	Time frame	Measuring location	Beam no.	Sensor array configuration	Pass-by vehicle speed [km/h]	Measuring series
Transport equipment to beam 213 of the elevated guideway, place sensors beneath the guideway and at distances of 5, 20, and 35m from centerline, and verify instrument performance.		8:00-9:30	Elevated steel guideway	213	Lateral profile 1m above ground.		
Wayside. EMF and RF measurements beneath and near elevated guideway.	N2	09:30-10:00	Elevated steel guideway	213	"	Standardized speed profile (300 and 400 km/h passes recorded)	Guideway
	N3	10:00-10:30	Elevated concrete guideway	215	"	Standardized speed profile (300 and 400 km/h passes recorded)	Guideway
	N4	10:30-11:00	Elevated hybrid guideway	267-268	"		Guideway
Wayside. EMF and RF measurements near at-grade guideway.	N7[4]	13:00-13:30	Steel at-grade plate	340 a-d	"	Standardized speed profile (200 and 400 km/h passes recorded)	Guideway
	N8[4]	13:45-14:15	Concrete at-grade plate	341 a-d	"		Guideway
Wayside. EMF and RF measurements near wayside electrical switching cabinets.	N9[4] (instead of N5)	14:30-15:00	Standard switch cabinets	183	"	Standardized speed profile (300 and 400 km/h passes recorded)	Wayside equipment
	N10[4] (instead of N6)	15:00-15:30	Switch building designed for Berlin/Hamberg		"		Wayside equipment
Remove instruments from the wayside and place them on charge. Inspect data for integrity and back up.		15:30-17:30			Instruments removed from wayside		

Notes: [3] Outdoor measurements described for this day may be moved a day earlier or a day later depending on weather conditions. Non-standard trips will be used for the switching cabinet measurements if required
[4] All trips conforming to the standardized speed profile regardless of their trip number will be used for guideway measurements.

A-38

Noise Characteristics of the Transrapid TR08 Maglev System

Date: Friday, 2001-08-17

Noise Measurement Activities	Trip	Time frame	Measuring location	Beam no.	Array configuration	Single microphones	Pass-by vehicle speed [km/h]	Measuring series	Pass-by ID
Install microphones on WV08/16/32 wayside vertical array, calibrate, elevate array to "high" position, set up single microphones.		7:00-9:30	steel guideway ②		setup/ reconstruction	setup/ reconstruction			
Make noise measurements with WV08/16/32 array in "high" position, make measurements with single microphones.	N2	09:30-10:00	steel guideway ②	213	WV08/16/32 (high)	6.5 m (high)	standardized speed profile (150, 200, 300, 400 km/h)	G	G-150-1 G-400-1 G-200-1 G-300-1
	N3	10:00-10:30							G-150-2 G-400-2 G-200-2 G-300-2
	N4	10:30-11:00							G-150-3 G-400-3 G-200-3 G-300-3
Lower microphone array and 6.5 m single microphone to "low" position.		11:00			setup/ reconstruction	setup/ reconstruction			
Make noise measurements with WV08/16/32 array in "low" position, make noise measurements with single microphones.	N5	11:00-11:30	steel guideway ②	213	WV08/16/32 (low)	6.5 m (low), 25 m, 50 ft, 100 ft, under guideway	standardized speed profile (150, 200, 300, 400 km/h)	H	H-150-1 H-400-1 H-200-1 H-300-1
	N6	12:15-12:45							H-150-2 H-400-2 H-200-2 H-300-2
	N7	13:00-13:30							H-150-3 H-400-3 H-200-3 H-300-3
	N8	13:45-14:15				6.5 m (high), 25 m, 50 ft, 100 ft, under guideway	100 km/h		H-100-1 H-100-2
						6.5 m (high), 25 m, 50 ft, 100 ft, under guideway	100 km/h		H-100-3 H-100-4
Retrieve single microphones and detach array from tower.		14:30-16:30							

Noise Characteristics of the Transrapid TR08 Maglev System

Move tower to measuring site at hybrid guideway.				15:30-16:30	
Reconstruct array to WV08/16/32.				16:30-17:30	setup/ reconstruction
Pack equipment for weekend storage.				17:30-18:00	

Date: Friday, 2001-08-17

Vibration Measurement Activities	Trip	Time frame	Measuring location	Beam no.	Pass-by vehicle speed [km/h]	Measuring series	Pass-by ID
Set up accelerometers, recording instrumentation, and ground impacter at site near station platform and conduct transfer mobility test.		7:00-9:30	Station Platform				
Record ground vibrations from TR08 at site	N2	09:30-10:00	Station Platform		Low	G	G-1
	N3	10:00 – 10:30					G-2
	N4	10:30-11:00					G-3
Pack equipment and depart TVE		11:00 – 15:00					

A-40

Noise Characteristics of the Transrapid TR08 Maglev System

Date: Friday, 2001-08-17

EMF Measurement Activities	Trip	Time frame	Measuring location	Beam no.	Sensor array configuration	Vehicle operating speed profile	Measuring series
Transport equipment to the trailing vehicle, place sensors in the seventh onboard measurement location, and verify instrument performance.		8:00-9:30	Onboard trailing vehicle		Mannequin positions and fixed reference		
Onboard trailing vehicle. EMF and RF measurements at a locations corresponding to the to test locations in the lead car to identify variation in field levels among cars. The measurement locations will correspond to three locations where measurements were made in the lead vehicle.	N2	09:30-10:00	Onboard trailing vehicle. Middle seat away from onboard equipment.		"	Standardized speed profile	Onboard
	N3	10:00-10:30	Onboard middle vehicle. Middle seat above a battery compartment.[2]		"	Standardized speed profile	Onboard
	N4	10:30-11:00	Onboard middle vehicle. Middle seat above a battery charger and control equipment.[2]		"	Standardized speed profile	Onboard
Onboard leading vehicle. EMF and RF measurements at the attendant's position in the nose of the vehicle.	N5	11:00-11:30	Onboard leading vehicle. Attendant's position in the nose of the vehicle.		"	Standardized speed profile	Onboard
Onboard leading vehicle. EMF and RF measurements near onboard equipment.	N6	12:15-12:45	Onboard leading vehicle. Standing near the air handlers at the rear of the vehicle.		"	Standardized speed profile	Onboard
Contingency This trip will be used to repeat any defective onboard measurement or conduct an additional onboard measurement suggested by a review of data collected in the first two measurement days.	N7	13:00-13:30	As required.		"	Standardized speed profile	Onboard
Contingency This trip will be used to repeat any defective station measurement or conduct an additional station measurement suggested by a review of data collected in the first two measurement days.	N8	13:45-14:15	As Required.		"	Nonstandard speed profile required for low speed noise measurements	Station
Remove instruments from the station and pack for shipment. Inspect data for integrity and back up.		14:15-18:00			Instruments removed from station		

A-41

Noise Characteristics of the Transrapid TR08 Maglev System

Date: Tuesday, 2001-08-21

Noise Measurement Activities	Trip	Time frame	Measuring location	Beam no.	Array configuration	Single microphones	Pass-by vehicle speed [km/h]	Measuring series	Pass-by ID
Install microphones on WV08/16/32 wayside vertical array, calibrate, elevate array to "high" position, set up single microphones.		7:00-9:30	hybrid guideway ④	267	setup/ reconstruction	setup/ reconstruction			
Make noise measurements with WV08/16/32 array in "high" position,	N2	09:30-10:00	hybrid guideway ④	267	WV08/16/32 (high)	6.5 m (high)	standardized speed profile (150, 200, 300, 400 km/h)	I	I-150-1 I-400-1 I-200-1 I-300-1
make noise measurements with single microphones.	N3	10:00-10:30							I-150-2 I-400-2 I-200-2 I-300-2
	N4	10:30-11:00							I-150-3 I-400-3 I-200-3 I-300-3
Lower array and 6.5 m single microphone to "low" position.		11:00			setup/ reconstruction	setup/ reconstruction			
Make noise measurements with WV08/16/32 array in "low" position,	N5	11:00-11:30	hybrid guideway ④	267	WV08/16/32 (low)	6.5 m (low), 25 m, 50 ft, 100 ft, under guideway	standardized speed profile (150, 200, 300, 400 km/h)	J	J-150-1 J-400-1 J-200-1 J-300-1
make noise measurements with single microphones.	N6	12:15-12:45							J-150-2 J-400-2 J-200-2 J-300-2
	N7	13:00-13:30							J-150-3 J-400-3 J-200-3 J-300-3
	N8	13:45-14:15				6.5 m (low), 25 m, 50 ft, 100 ft, under guideway	100 km/h		J-100-1 J-100-2
						6.5 m (high), 25 m, 50 ft, 100 ft, under guideway	100 km/h		J-100-3 J-100-4
Retrieve single microphones and detach array from tower.		14:30-15:30							

Noise Characteristics of the Transrapid TR08 Maglev System

			Time	
Move tower and equipment to measuring site at concrete guideway.			15:30-16:30	
Reconstruct array to WV08/16/32.			16:30-18:00	setup/ reconstruction

Date: Tuesday, 2001-08-21

Vibration Measurement Activities	Trip	Time frame	Measuring location	Beam no.	Array configuration	Single microphones	Pass-by vehicle speed [km/h]	Measuring series	Pass-by ID
Contingency day for tests if delays.									

A-43

Noise Characteristics of the Transrapid TR08 Maglev System

Date: Wednesday, 2001-08-22

Noise Measurement Activities	Trip	Time frame	Measuring location	Beam no.	Array configuration	Single microphones	Pass-by vehicle speed [km/h]	Measuring series	Pass-by ID
Install microphones on WV08/16/32 wayside vertical array, calibrate, elevate array to "high" position, set up single microphones.		7:00-9:30	prototype concrete guideway ③	215	setup/ reconstruction	setup/ reconstruction			
Make noise measurements with WV08/16/32 array in "high" position,	N2	09:30-10:00	prototype concrete guideway ③	215	WV08/16/32 (high)	6.5 m (high)	standardized speed profile (150, 200, 300, 400 km/h)	K	K-150-1 K-400-1 K-200-1 K-300-1
make noise measurements with single microphones.	N3	10:00-10:30							K-150-2 K-400-2 K-200-2 K-300-2
	N4	10:30-11:00							K-150-3 K-400-3 K-200-3 K-300-3
Lower array and 6.5 m single microphone to "low" position.		11:00			setup/ reconstruction	setup/ reconstruction			
Make noise measurements with WV08/16/32 array in "low" position,	N5	11:00-11:30	prototype concrete guideway ③	215	WV08/16/32 (low)	6.5 m (low), 25 m, 50 ft, 100 ft, under guideway	standardized speed profile (150, 200, 300, 400 km/h)	L	L-150-1 L-400-1 L-200-1 L-300-1
make noise measurements with single microphones.	N6	12:15-12:45							L-150-2 L-400-2 L-200-2 L-300-2
	N7	13:00-13:30							L-150-3 L-400-3 L-200-3 L-300-3
	N8	13:45-14:15				6.5 m (low), 25 m, 50 ft, 100 ft, under guideway	100 km/h		L-100-1 L-100-2

Noise Characteristics of the Transrapid TR08 Maglev System

			6.5 m (high), 25 m, 50 ft, 100 ft, under guideway	100 km/h		L-100-3 L-100-4
Retrieve single microphones and detach array from tower. Pack equipment.		14:30-18:00				

Date: Wednesday, 2001-08-22

Vibration Measurement Activities	Trip	Time frame	Measuring location	Beam no.	Array configuration	Single microphones	Pass-by vehicle speed [km/h]	Measuring series	Pass-by ID
Contingency day for tests if delays encountered.									

Date: Thursday, 2001-08-22

Noise Measurement Activities	Trip	Time frame	Measuring location	Beam no.	Array configuration	Single microphones	Pass-by vehicle speed [km/h]	Measuring series	Pass-by ID
Additional contingency day due to weather problems or other unforeseen events.									

A-45

APPENDIX B. RESULTS OF SINGLE-MICROPHONE MEASUREMENTS AT THE VARIOUS GUIDEWAY TYPES

B.1 Introduction

In the appendix, selected time histories of the A-weighted sound-pressure level (SPL) measured with the single microphones are shown for the various guideway types and microphone positions. For each guideway type and each microphone position, time histories are to be illustrated by representative examples at about 100, 150, 200, 300, and 400 km/h (62, 93, 124, 186, and 249 mph). Time histories during passbys of the TR08 traveling on the reference concrete guideway, the prototype steel and prototype concrete guideway, the hybrid guideway, the North switch, and the at-grade guideway are given for lateral distances of 30.5, 25.0, 15.2, and 6.5 m (100.0, 82.0, 50.0, and 21.3 ft) as well as for the microphone beneath the elevated guideways.

For each guideway type and each microphone position, the single-number levels calculated from the time histories are listed in a table containing the date, time, and vehicle speed of each test run of the TR08 together with the maximum SPL $L_{Amax,fast}$, the event level $L_{Aeq,E}$, the event time t_E, the sound-exposure level (SEL), and the one-hour equivalent level $L_{Aeq,1h}$ (see definitions in Section 2.5.1).

In some of the following figures, it appears as if the background noise for these measurements was rather high, i.e., approx. 67 A-weighted decibels (dBA) in Figure B-149, which, however, was not the case. The constant levels that sometimes occur when the TR08 is approaching or leaving, are produced by the analogous measuring amplifier, B&K type 2610, whose output section has a dynamic range of only 50 dB. On the other hand, this amplifier was necessary for A-weighting the sound pressures and also averaging them using the "fast" mode. Nevertheless, the noise floor was so low that it did not influence the single-number levels that were calculated from the time histories shown.

As can be seen in Table B-28 and Table B-29, the measured single-number levels at the highest vehicle speeds are marked. These values appear to be questionable, since in the corresponding time histories high peak levels occur that cannot be explained (see Figures B139 and B-144). Measurements for the other guideway types carried out at the same microphone positions (6.5 m (21.3 ft) distance from track centerline) showed significantly lower peak levels.

B.2 Reference Concrete Guideway

B.2.1 Microphone at 30.5 m (100.0 ft) distance from track centerline

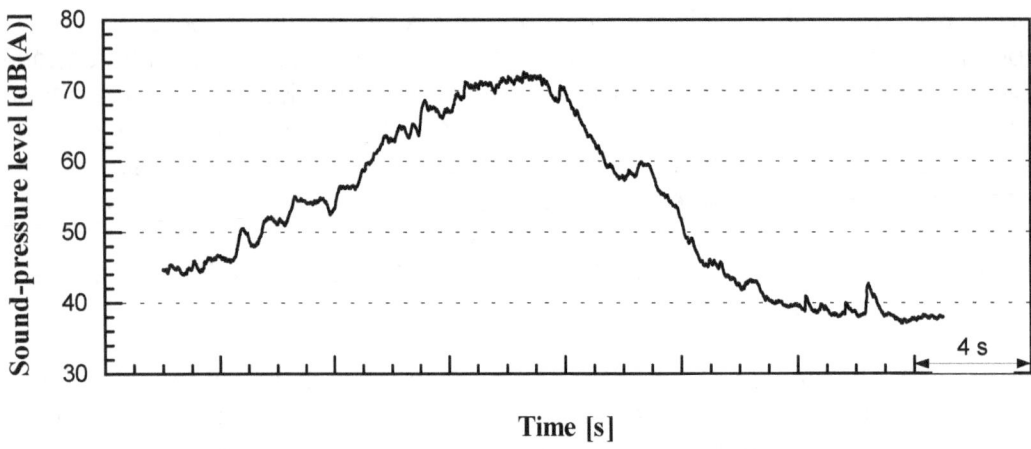

Figure B-1. Time history of the A-weighted SPL during a passby of the TR08 travelling on the reference concrete guideway at about 100 km/h (62 mph) measured at 30.5 m *(100 ft) distance from track centerline and 1.2 m (4.0 ft) above the grund.

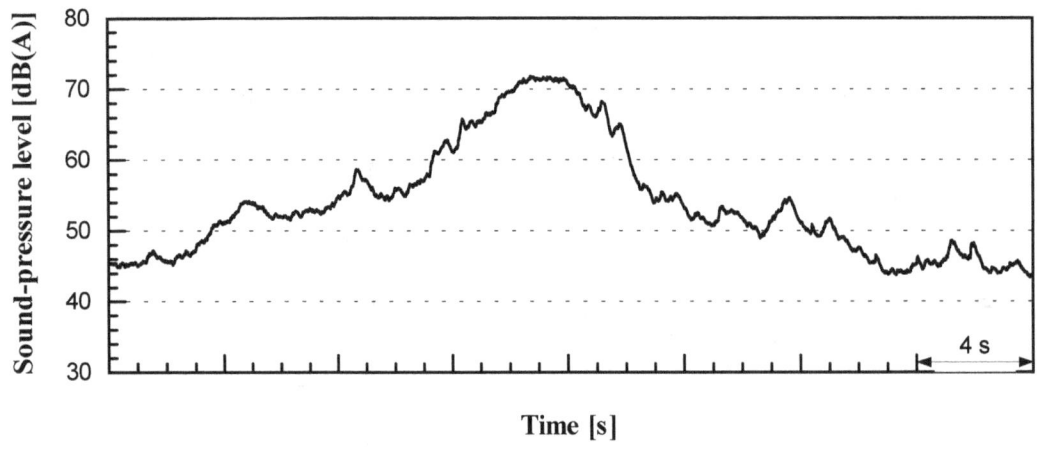

Figure B-2. Time history of the A-weighted SPL during a passby of the TR08 travelling on the reference concrete guideway at about 150 km/h (93 mph) measured at 30.5 m (100.0 ft) distance from track centerline and 1.2 m (4.0 ft) above the ground.

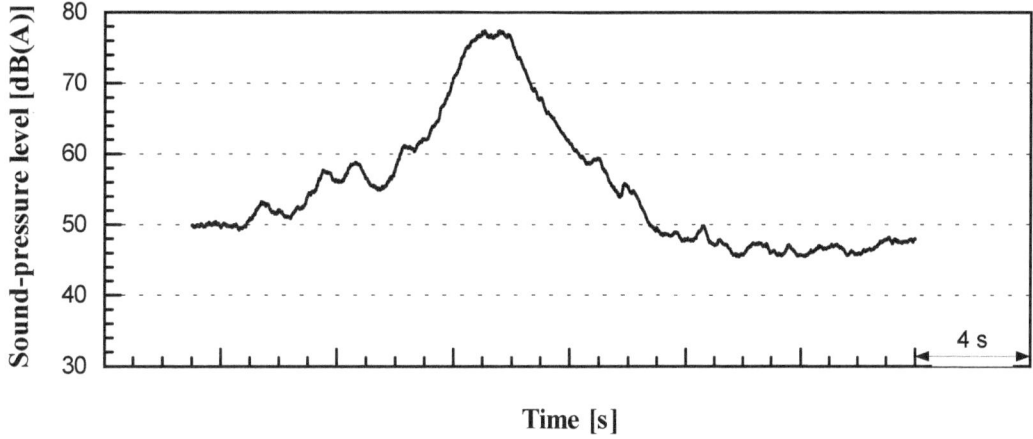

Figure B-3. Time history of the A-weighted SPL during a passby of the TR08 travelling on the reference concrete guideway at about 200 km/h (124 mph) measured at 30.5 m (100.0 ft) distance from track centerline and 1.2 m (4.0 ft) above the ground.

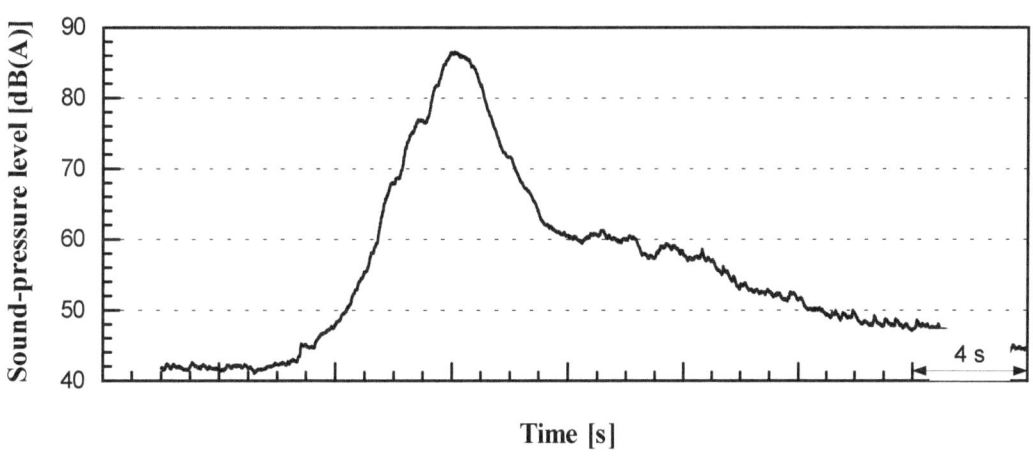

Figure B-4. Time history of the A-weighted SPL during a passby of the TR08 travelling on the reference concrete guideway at about 300 km/h (186 mph) measured at 30.5 m (100.0 ft) distance from track centerline and 1.2 m (4.0 ft) above the ground.

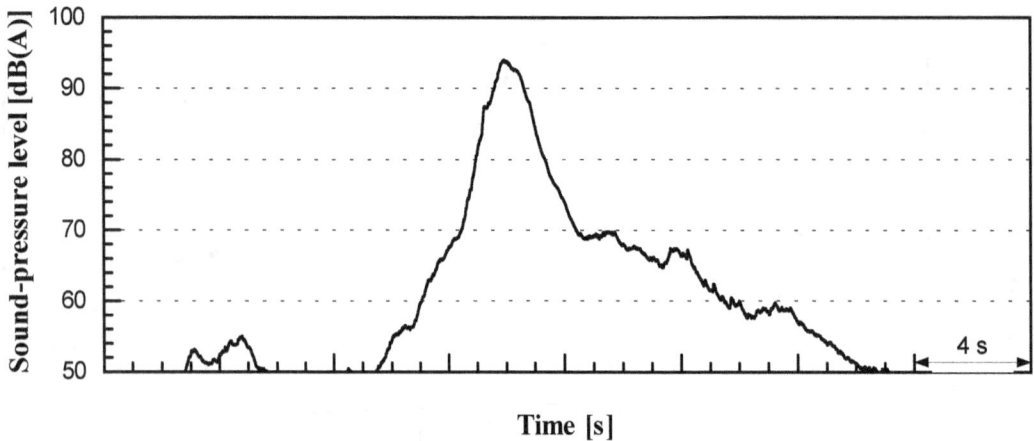

Figure B-5. Time history of the A-weighted SPL during a passby of the TR08 travelling on the reference concrete guideway at about 400 km/h (249 mph) measured at 30.5 m (100.0 ft) distance from track centerline and 1.2 m (4.0 ft) above the ground.

Table B-1. Results of the microphone positioned close to the reference concrete guideway at 30.5 m (100.0 ft) distance from track centerline and 1.2 m (4.0 ft) above the ground (measuring series A/B).

Date	Time	Vehicle speed [km/h (mph)]	$L_{Amax,\,fast}$ [dB(A)]	$L_{Aeq,E}$ [dB(A)]	t_E [s]	SEL [dB(A)]	$L_{Aeq,1h}$ [dB(A)]
2002-05-17	12:57	100.0 (62.1)	73.3	67.7	11.75	78.4	42.8
2002-05-17	13:02	99.9 (62.1)	72.6	67.3	12.02	78.1	42.5
2002-05-17	13:10	100.1 (62.2)	73.6	67.6	12.19	78.4	42.9
2002-05-17	13:37	100.1 (62.2)	72.8	67.3	11.49	77.9	42.3
2001-08-16	10:38	149.8 (93.1)	72.2	65.8	10.96	76.2	40.6
2001-08-16	11:30	149.7 (93.0)	71.5	66.8	7.84	75.8	40.2
2001-08-16	13:28	149.7 (93.0)	71.7	66.0	10.21	76.1	40.6
2001-08-16	14:23	149.7 (93.0)	71.7	65.5	13.90	76.9	41.4
2001-08-16	12:56	199.6 (124.0)	78.2	71.6	7.95	80.6	45.0
2001-08-16	13:39	199.8 (124.1)	77.8	72.0	6.94	80.4	44.9
2001-08-16	14:35	199.9 (124.2)	77.3	71.5	7.30	80.1	44.6
2001-08-16	10:55	299.6 (186.2)	86.3	80.7	4.71	87.5	51.9
2001-08-16	11:46	299.6 (186.2)	86.1	81.1	4.44	87.6	52.0
2001-08-16	13:44	299.6 (186.2)	86.5	81.3	4.39	87.7	52.2
2001-08-16	14:40	299.8 (186.3)	86.4	80.7	4.91	87.6	52.1
2001-08-16	09:44	370.0 (229.9)	91.0	85.3	3.99	91.3	55.8
2001-08-16	10:14	401.0 (249.2)	93.8	88.5	3.47	93.9	58.3
2001-08-16	10:43	401.5 (249.5)	93.5	88.3	3.50	93.7	58.2
2001-08-16	11:35	400.9 (249.1)	94.0	89.1	3.08	94.0	58.5
2001-08-16	13:33	401.2 (249.3)	93.6	88.3	3.60	93.8	58.3
2001-08-16	14:28	400.9 (249.1)	93.9	88.7	3.36	94.0	58.4

B.2.2 Microphone at 25.0 m (82.0 ft) distance from track centerline

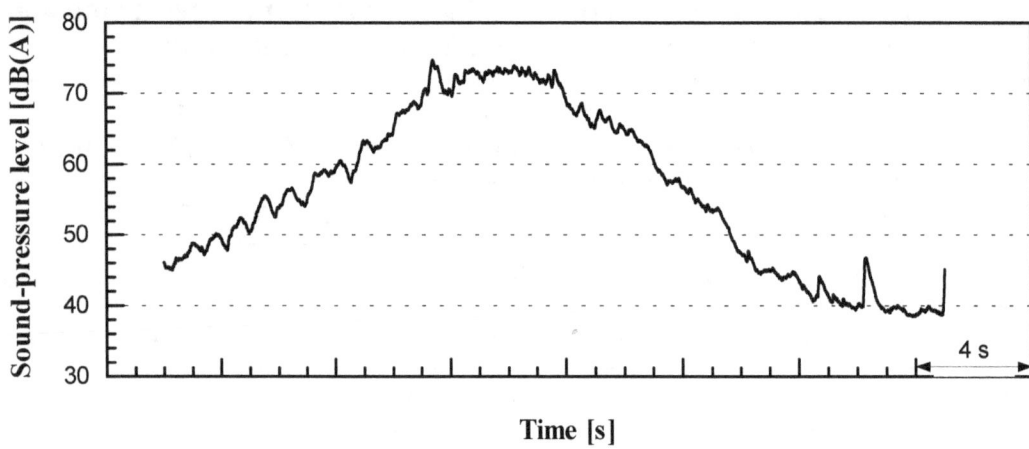

Figure B-6. Time history of the A-weighted SPL during a passby of the TR08 travelling on the reference concrete guideway at about 100 km/h (62 mph) measured at 25.0 m (82 ft) distance from track centerline and 3.5 m (11.5 ft) above the ground.

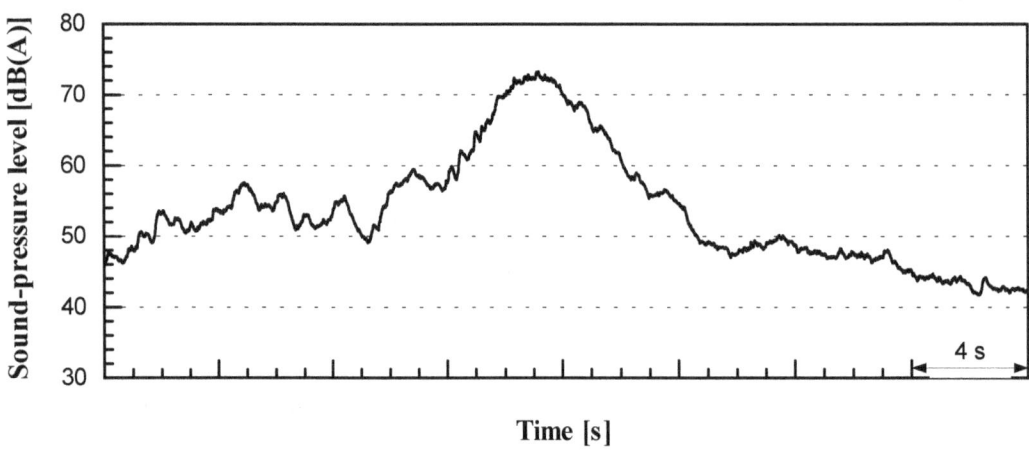

Figure B-7. Time history of the A-weighted SPL during a passby of the TR08 travelling on the reference concrete guideway at about 150 km/h (93 mph) measured at 25.0 m (82.0 ft) distance from track centerline and 3.5 m (11.5 ft) above the ground.

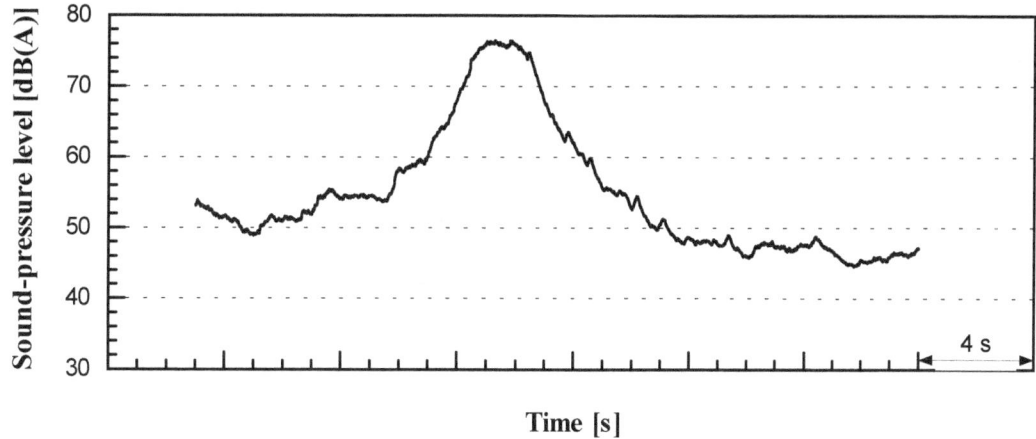

Figure B-8. Time history of the A-weighted SPL during a passby of the TR08 travelling on the reference concrete guideway at about 200 km/h (124 mph) measured at 25.0 m (82.0 ft) distance from track centerline and 3.5 m (11.5 ft) above the ground.

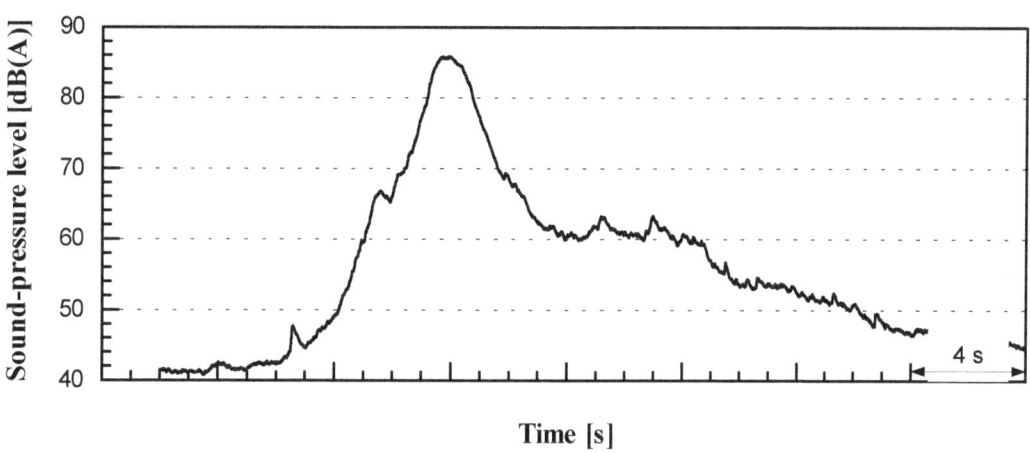

Figure B-9. Time history of the A-weighted SPL during a passby of the TR08 travelling on the reference concrete guideway at about 300 km/h (186 mph) measured at 25.0 m (82.0 ft) distance from track centerline and 3.5 m (11.5 ft) above the ground.

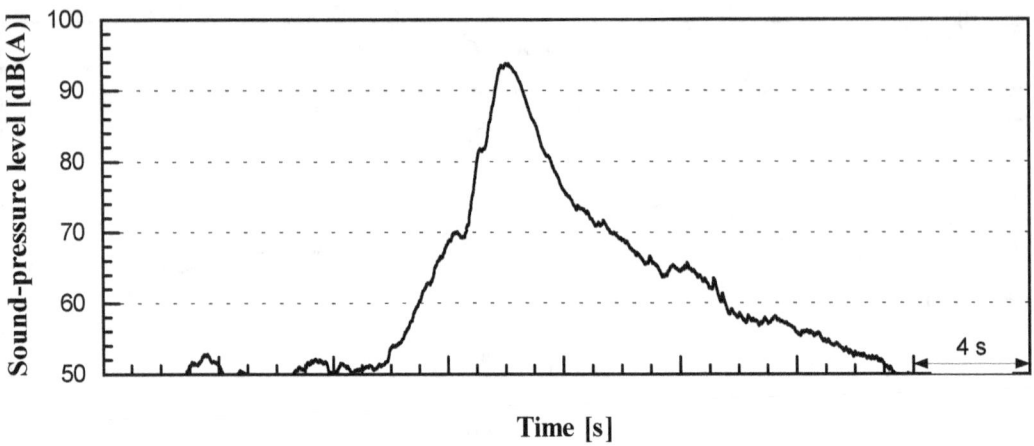

Figure B-10. Time history of the A-weighted SPL during a passby of the TR08 travelling on the reference concrete guideway at about 400 km/h (249 mph) measured at 25.0 m (82.0 ft) distance from track centerline and 3.5 m (11.5 ft) above the ground.

Table B-2. Results of the microphone positioned close to the reference concrete guideway at 25.0 m (82.0 ft) distance from track centerline and 3.5 m (11.5 ft) above the ground (measuring series A/B).

Date	Time	Vehicle speed [km/h (mph)]	$L_{Amax,\,fast}$ [dB(A)]	$L_{Aeq,E}$ [dB(A)]	t_E [s]	SEL [dB(A)]	$L_{Aeq,1h}$ [dB(A)]
2002-05-17	12:57	100.0 (62.1)	73.7	68.1	14.85	79.8	44.2
2002-05-17	13:02	99.9 (62.1)	74.7	69.1	13.42	80.4	44.8
2002-05-17	13:10	100.1 (62.2)	74.1	68.8	13.37	80.0	44.5
2002-05-17	13:37	100.1 (62.2)	73.2	67.5	14.41	79.1	43.5
2001-08-16	10:38	149.8 (93.1)	72.7	66.3	10.93	76.6	41.1
2001-08-16	11:30	149.7 (93.0)	72.7	66.8	9.45	76.5	41.0
2001-08-16	13:28	149.7 (93.0)	72.9	66.9	9.73	76.8	41.2
2001-08-16	14:23	149.7 (93.0)	73.3	66.7	10.56	77.0	41.4
2001-08-16	12:56	199.6 (124.0)	76.2	69.8	9.97	79.3	43.7
2001-08-16	13:39	199.8 (124.1)	76.1	70.2	8.02	79.3	43.7
2001-08-16	14:35	199.9 (124.2)	76.3	71.0	7.07	79.5	43.9
2001-08-16	10:55	299.6 (186.2)	85.5	80.0	4.97	87.0	51.4
2001-08-16	11:46	299.6 (186.2)	85.6	80.0	4.66	86.7	51.2
2001-08-16	13:44	299.6 (186.2)	85.7	80.1	4.55	86.7	51.2
2001-08-16	14:40	299.8 (186.3)	85.7	80.2	4.57	86.8	51.2
2001-08-16	09:44	370.0 (229.9)	91.2	85.1	4.34	91.4	55.9
2001-08-16	10:14	401.0 (249.2)	93.7	88.3	3.30	93.5	57.9
2001-08-16	10:43	401.5 (249.5)	93.4	87.4	3.89	93.3	57.8
2001-08-16	11:35	400.9 (249.1)	93.8	88.7	3.17	93.7	58.2
2001-08-16	13:33	401.2 (249.3)	94.1	88.1	3.50	93.6	58.0
2001-08-16	14:28	400.9 (249.1)	93.7	88.2	3.55	93.7	58.1

B.2.3 Microphone at 15.2 m (50.0 ft) distance from track centerline

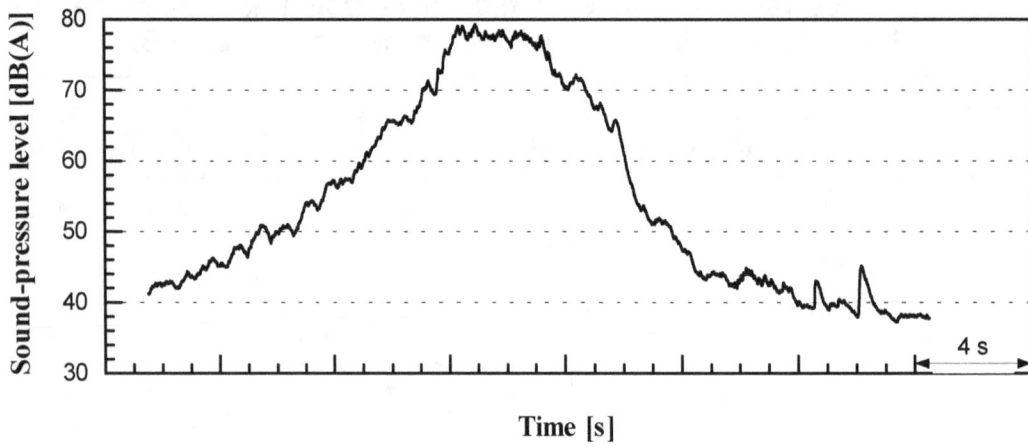

Figure B-11. Time history of the A-weighted SPL during a passby of the Tr08 travelling on the reference concrete guideway at about 100 km/h (62 mph) measured at 15.2 m (50.0 ft) distance from track centerline and 1.5 m (5.0 ft) above the ground.

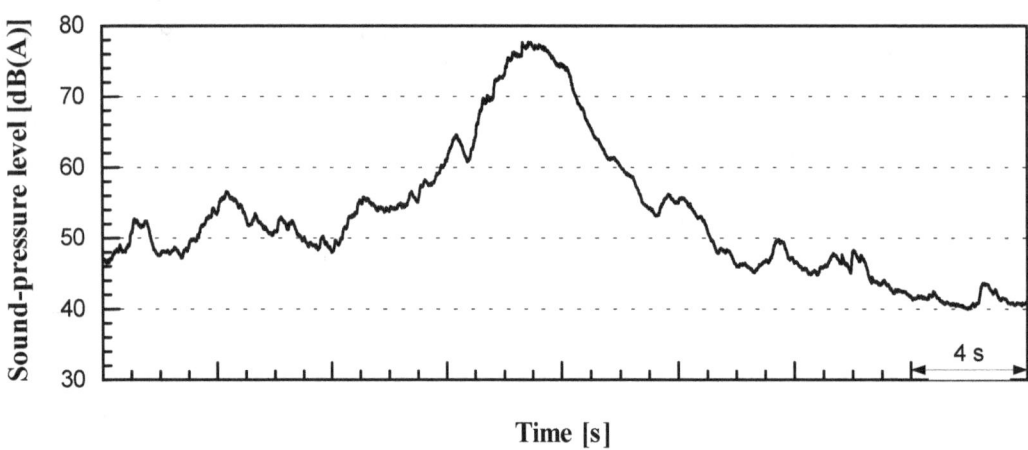

Figure B-12. Time history of the A-weighted SPL during a passby of the TR08 travelling on the reference concrete guideway at about 150 km/h (93 mph) measured at 15.2 m (50.0 ft) distance from track centerline and 1.5 m (5.0 ft) above the ground.

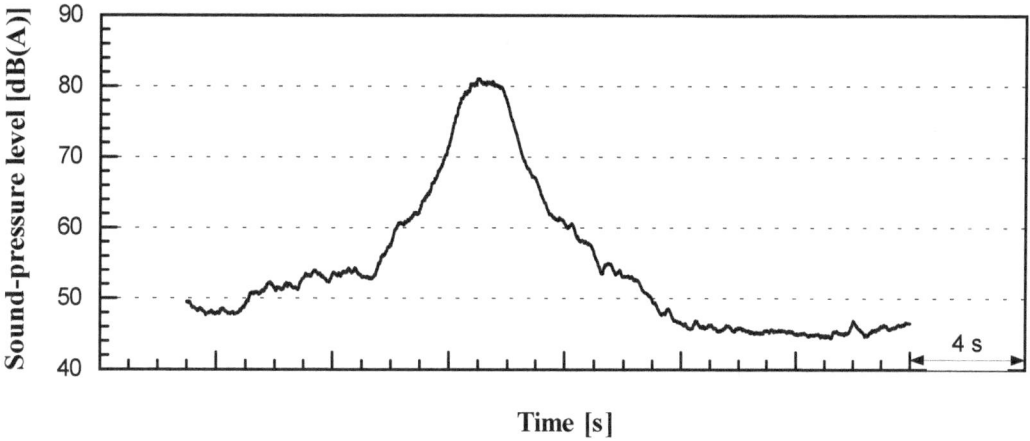

Figure B-13. Time history of the A-weighted SPL during a passby of the TR08 travelling on the reference concrete guideway at about 200 km/h (124 mph) measured at 15.2 m (50.0 ft) distance from track centerline and 1.5 m (5.0 ft) above the ground.

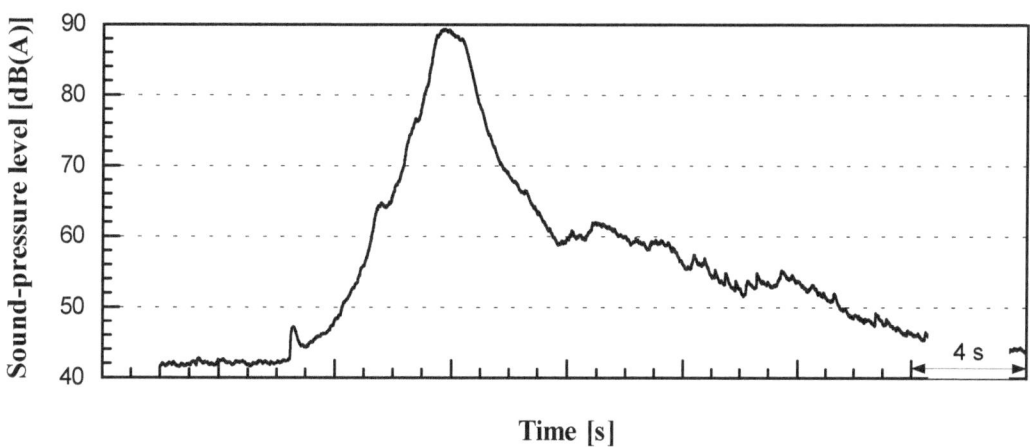

Figure B-14. Time history of the A-weighted SPL during a passby of the TR08 travelling on the reference concrete guideway at about 300 km/h (186 mph) measured at 15.2 m (50.0 ft) distance from track centerline and 1.5 m (5.0 ft) above the ground.

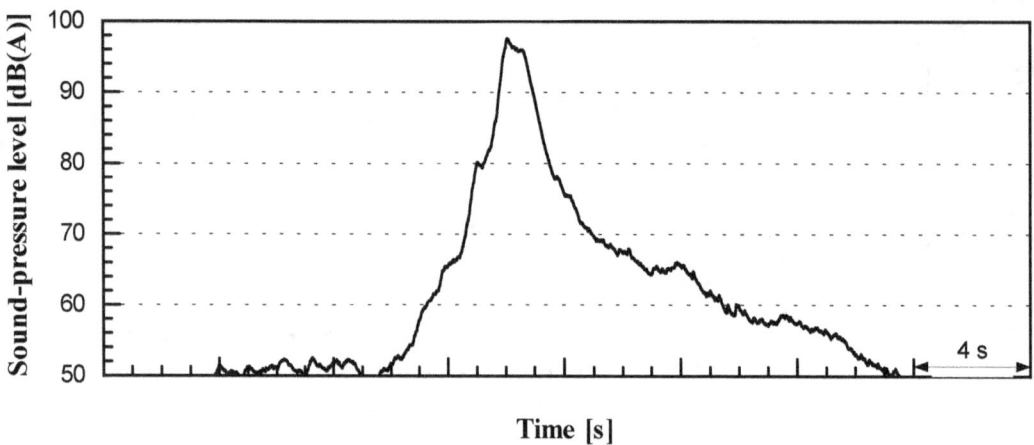

Figure B-15. Time history of the A-weighted SPL during a passby of the TR08 travelling on the reference concrete guideway at about 400 km/h (249 mph) measured at 15.2 m (50.0 ft) distance from track centerline and 1.5 m (5.0 ft) above the ground.

Table B-3. Results of the microphone positioned close to the reference concrete guideway at 15.2 m (50.0 ft) distance from track centerline and 1.5 m (5.0 ft) above the ground (measuring series A/B).

Date	Time	Vehicle speed [km/h (mph)]	$L_{Amax, fast}$ [dB(A)]	$L_{Aeq,E}$ [dB(A)]	t_E [s]	SEL [dB(A)]	$L_{Aeq,1h}$ [dB(A)]
2002-05-17	12:57	100.0 (62.1)	79.1	74.6	8.38	83.8	48.3
2002-05-17	13:02	99.9 (62.1)	79.2	74.2	9.15	83.8	48.3
2002-05-17	13:10	100.1 (62.2)	79.6	74.4	8.70	83.8	48.3
2002-05-17	13:37	100.1 (62.2)	78.6	74.0	8.95	83.5	47.9
2001-08-16	10:38	149.8 (93.1)	78.0	72.3	6.71	80.6	45.0
2001-08-16	11:30	149.7 (93.0)	77.4	71.6	7.41	80.3	44.7
2001-08-16	13:28	149.7 (93.0)	77.6	71.7	7.39	80.4	44.8
2001-08-16	14:23	149.7 (93.0)	77.6	72.0	7.03	80.5	44.9
2001-08-16	12:56	199.6 (124.0)	80.4	75.3	5.31	82.6	47.0
2001-08-16	13:39	199.8 (124.1)	80.7	75.2	5.76	82.8	47.2
2001-08-16	14:35	199.9 (124.2)	81.0	75.7	5.13	82.8	47.2
2001-08-16	10:55	299.6 (186.2)	89.4	84.4	3.57	89.9	54.4
2001-08-16	11:46	299.6 (186.2)	89.1	84.3	3.53	89.8	54.3
2001-08-16	13:44	299.6 (186.2)	89.2	84.1	3.74	89.8	54.2
2001-08-16	14:40	299.8 (186.3)	89.3	84.3	3.54	89.8	54.3
2001-08-16	09:44	370.0 (229.9)	94.8	89.5	2.91	94.1	58.6
2001-08-16	10:14	401.0 (249.2)	97.1	92.3	2.52	96.3	60.8
2001-08-16	10:43	401.5 (249.5)	96.6	92.1	2.58	96.2	60.6
2001-08-16	11:35	400.9 (249.1)	97.1	92.3	2.65	96.6	61.0
2001-08-16	13:33	401.2 (249.3)	97.2	91.7	2.95	96.4	60.9
2001-08-16	14:28	400.9 (249.1)	97.6	91.7	2.94	96.4	60.8

B.2.4 Microphone at 6.5 m (21.3 ft) distance from track centerline (high position)

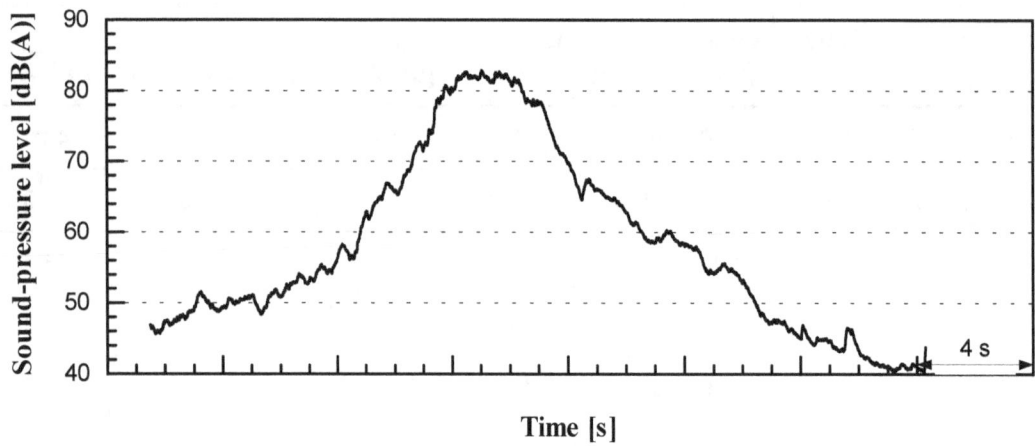

Figure B-16. Time history of the A-weighted SPL during a passby of the TR08 travelling on the reference concrete guideway at about 100 km/h (62 mph) measured at 6.5 m (21.3 ft) distance from track centerline and the height of the upper surface of the guideway.

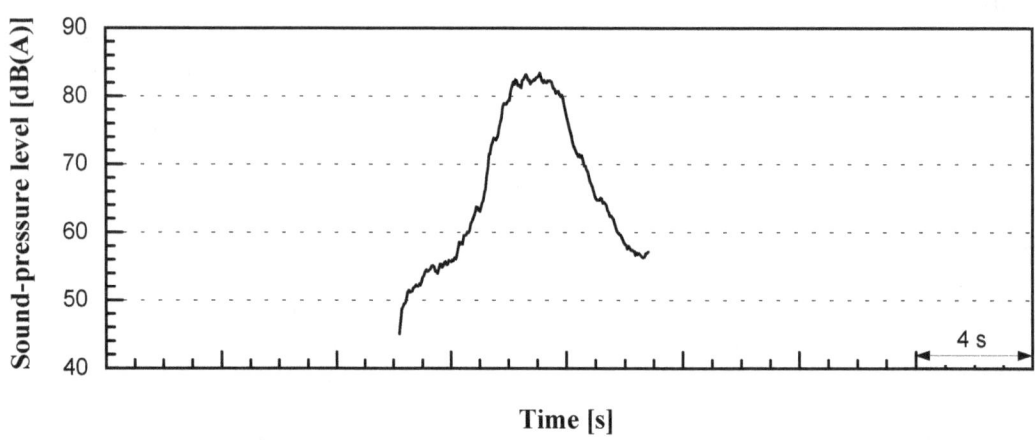

Figure B-17. Time history of the A-weighted SPL during a passby of the TR08 travelling on the reference concrete guideway at about 150 km/h (93 mph) measured at 6.5 m (21.3 ft) distance from track centerline and the height of the upper surface of the guideway.

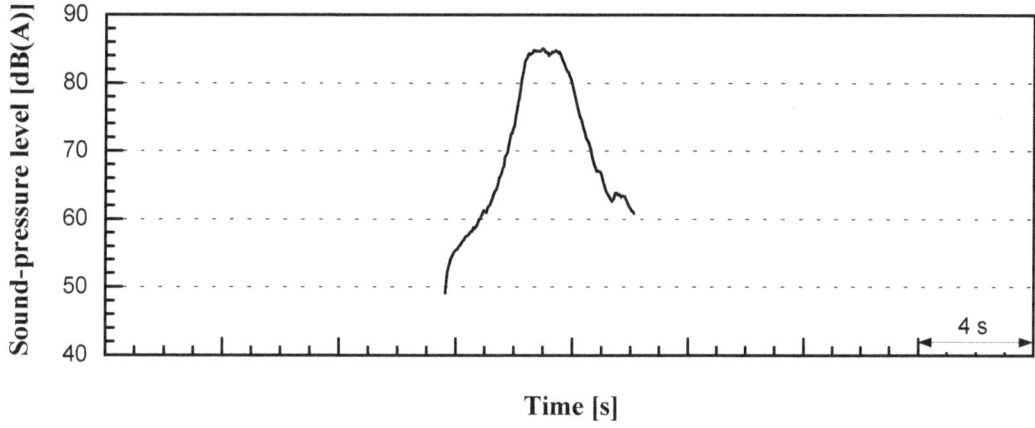

Figure B-18. Time history of the A-weighted SPL during a passby of the TR08 travelling on the reference concrete guideway at about 200 km/h (124 mph) measured at 6.5 m (21.3 ft) distance from track centerline and the height of the upper surface of the guideway.

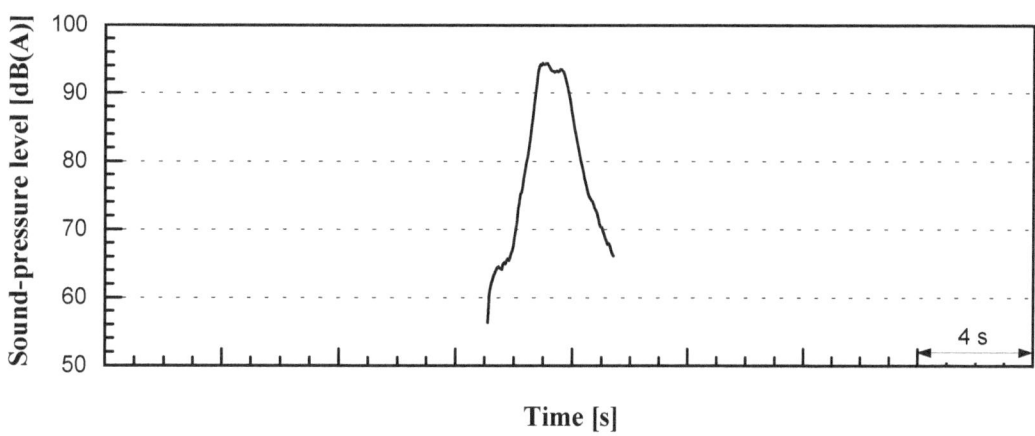

Figure B-19. Time history of the A-weighted SPL during a passby of the TR08 travelling on the reference concrete guideway at about 300 km/h (186 mph) measured at 6.5 m (21.3 ft) distance from track centerline and the height of the upper surface of the guideway.

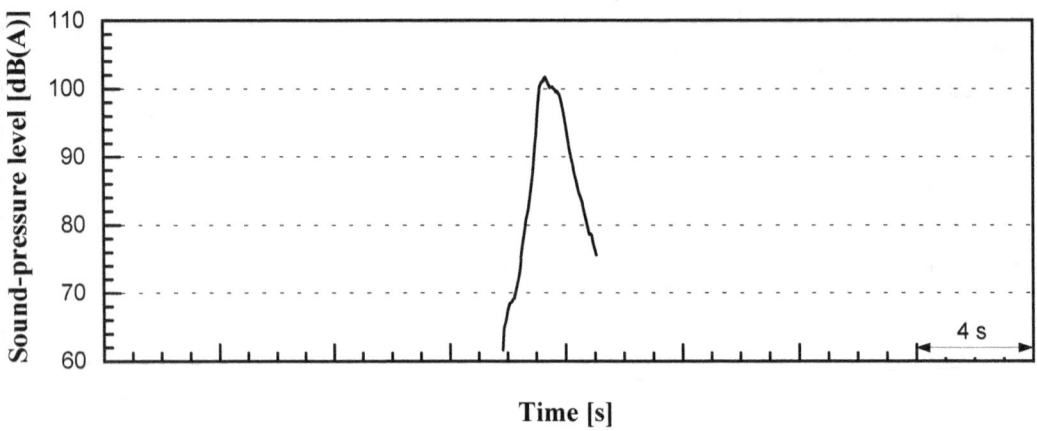

Figure B-20. Time history of the A-weighted SPL during a passby of the TR08 travelling on the reference concrete guideway at about 400 km/h (249 mph) measured at 6.5 m (21.3 ft) distance from track centerline and the height of the upper surface of the guideway.

Table B-4. Results of the microphone positioned close to the reference concrete guideway at 6.5 m (21.3 ft) distance from track centerline and the height of the upper surface of the guideway (measuring series A).

Date	Time	Vehicle speed [km/h (mph)]	$L_{Amax,\,fast}$ [dB(A)]	$L_{Aeq,E}$ [dB(A)]	t_E [s]	SEL [dB(A)]	$L_{Aeq,1h}$ [dB(A)]
2002-05-17	12:57	100.0 (62.1)	83.7	80.8	4.06	86.9	51.3
2002-05-17	13:02	99.9 (62.1)	82.9	80.5	4.37	86.9	51.3
2002-05-17	13:10	100.1 (62.2)	83.6	80.7	4.02	86.8	51.2
2002-05-17	13:37	100.1 (62.2)	84.3	80.7	3.86	86.6	51.0
2001-08-15	09:22	149.8 (93.1)	83.6	81.3	2.92	86.0	50.4
2001-08-15	10:15	149.6 (93.0)	83.4	80.8	2.80	85.2	49.7
2001-08-15	11:13	149.6 (93.0)	82.8	80.2	2.76	84.6	49.1
2001-08-15	09:40	199.7 (124.1)	85.5	83.6	2.28	87.2	51.6
2001-08-15	10:26	199.6 (124.0)	85.1	83.1	2.24	86.6	51.0
2001-08-15	11:25	199.4 (123.9)	85.1	82.9	2.24	86.4	50.9
2001-08-15	09:27	299.7 (186.2)	94.3	92.2	1.64	94.3	58.8
2001-08-15	10:31	299.6 (186.2)	94.8	92.3	1.56	94.3	58.7
2001-08-15	11:30	299.6 (186.2)	94.3	92.3	1.56	94.2	58.6
2001-08-15	10:40	400.8 (249.0)	101.9	99.4	1.28	100.5	64.9
2001-08-15	13:01	400.0 (248.5)	101.7	99.1	1.20	99.9	64.3
2001-08-15	14:28	400.0 (248.5)	101.2	98.7	1.28	99.7	64.2

B.2.5 Microphone at 6.5 m (21.3 ft) distance from track centerline (low position)

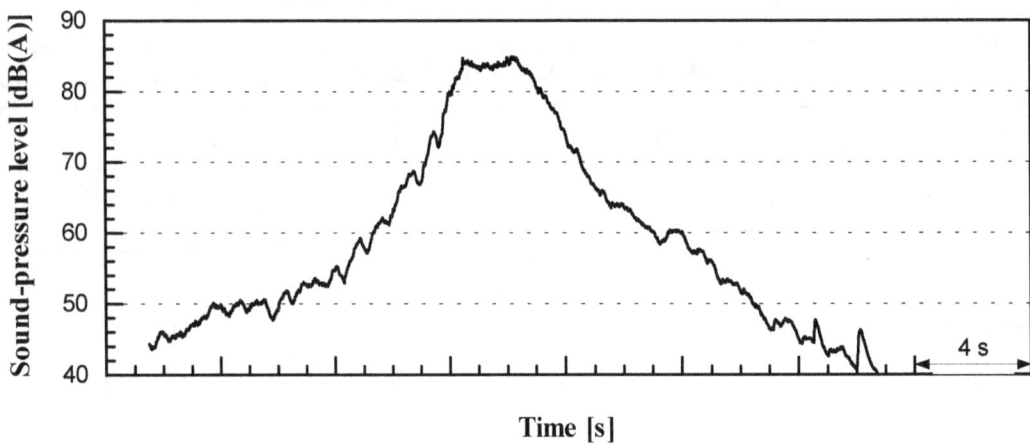

Figure B-21. Time history of the A-weighted SPL during a passby of the TR08 travelling on the reference concrete guideway at about 100 km/h (62 mph) measured at 6.5 m (21.3 ft) distance from track centerline and 1.5 m (5.0 ft) below the upper surface of the guideway.

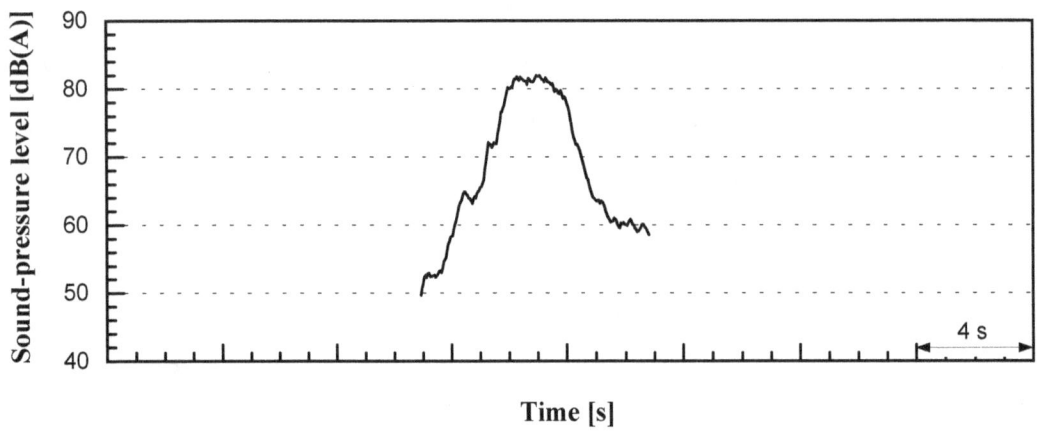

Figure B-22. Time history of the A-weighted SPL during a passby of the TR08 travelling on the reference concrete guideway at about 150 km/h (93 mph) measured at 6.5 m (21.3 ft) distance from track centerline and 1.5 m (5.0 ft) below the upper surface of the guideway.

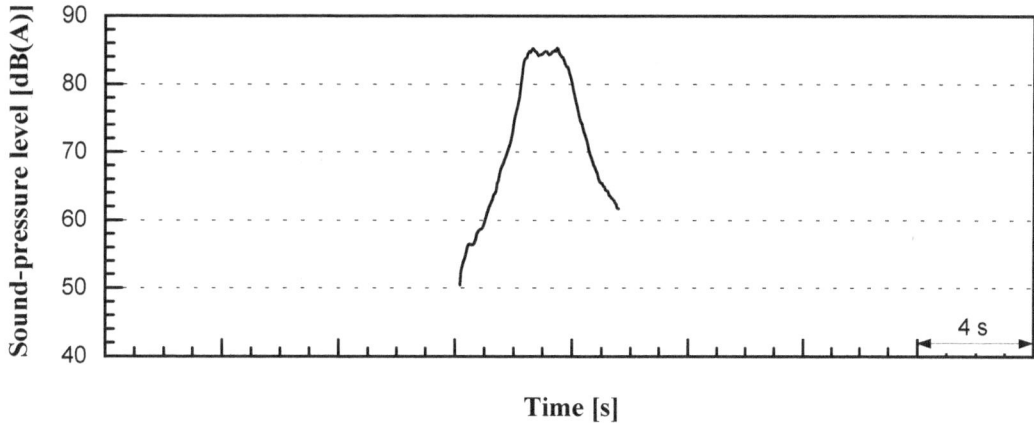

Figure B-23. Time history of the A-weighted SPL during a passby of the TR08 travelling on the reference concrete guideway at about 200 km/h (124 mph) measured at 6.5 m (21.3 ft) distance from track centerline and 1.5 m (5.0 ft) below the upper surface of the guideway.

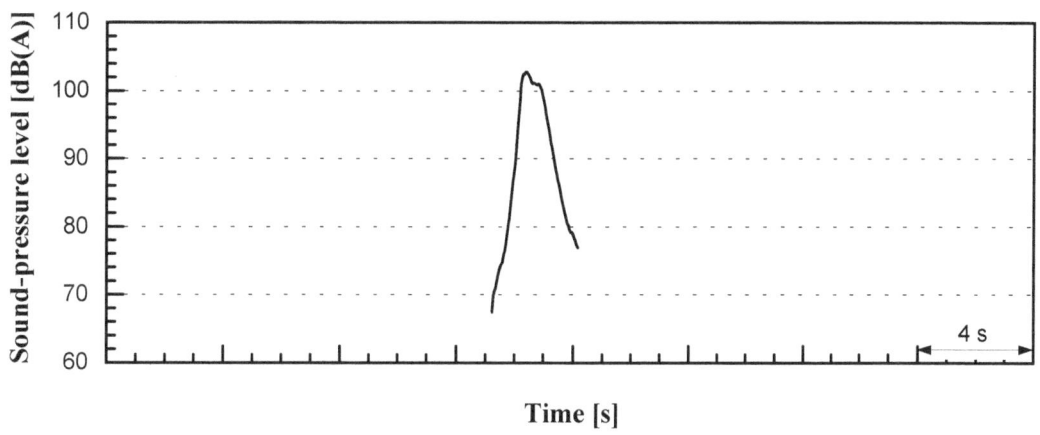

Figure B-24. Time history of the A-weighted SPL during a passby of the TR08 travelling on the reference concrete guideway at about 400 km/h (249 mph) measured at 6.5 m (21.3 ft) distance from track centerline and 1.5 m (5.0 ft) below the upper surface of the guideway.

Table B-5. Results of the microphone positioned close to the reference concrete guideway at 6.5 m (21.3 ft) distance from track centerline and 1.5 m (5.0 ft) below the upper surface of the guideway (measuring series B).

Date	Time	Vehicle speed [km/h (mph)]	$L_{Amax,\,fast}$ [dB(A)]	$L_{Aeq,E}$ [dB(A)]	t_E [s]	SEL [dB(A)]	$L_{Aeq,1h}$ [dB(A)]
2002-05-17	12:57	100.0 (62.1)	85.4	82.8	3.71	88.5	52.9
2002-05-17	13:02	99.9 (62.1)	84.9	82.5	4.10	88.6	53.1
2002-05-17	13:10	100.1 (62.2)	85.3	82.7	3.68	88.3	52.8
2002-05-17	13:37	100.1 (62.2)	85.4	82.7	3.62	88.3	52.7
2001-08-14	12:49	149.7 (93.0)	83.3	80.9	2.84	85.4	49.8
2001-08-14	13:23	151.0 (93.8)	86.3	83.7	2.68	87.9	52.4
2001-08-14	14:25	149.8 (93.1)	82.0	80.0	2.80	84.4	48.9
2001-08-14	13:35	199.9 (124.2)	85.2	83.3	2.24	86.8	51.2
2001-08-14	13:46	199.5 (124.0)	85.3	83.2	2.28	86.8	51.2
2001-08-14	14:37	199.6 (124.0)	85.4	83.2	2.28	86.8	51.2
2001-08-14	14:42	199.5 (124.0)	86.5	84.0	2.24	87.5	52.0
2001-08-14	12:54	400.8 (249.0)	102.4	99.9	1.28	101.0	65.5
2001-08-14	13:28	400.7 (249.0)	102.8	100.2	1.24	101.1	65.6
2001-08-14	13:50	401.1 (249.2)	102.3	99.9	1.28	101.0	65.4
2001-08-14	14:30	400.9 (249.1)	101.8	99.6	1.28	100.7	65.1

B.2.6 Microphone beneath guideway centerline

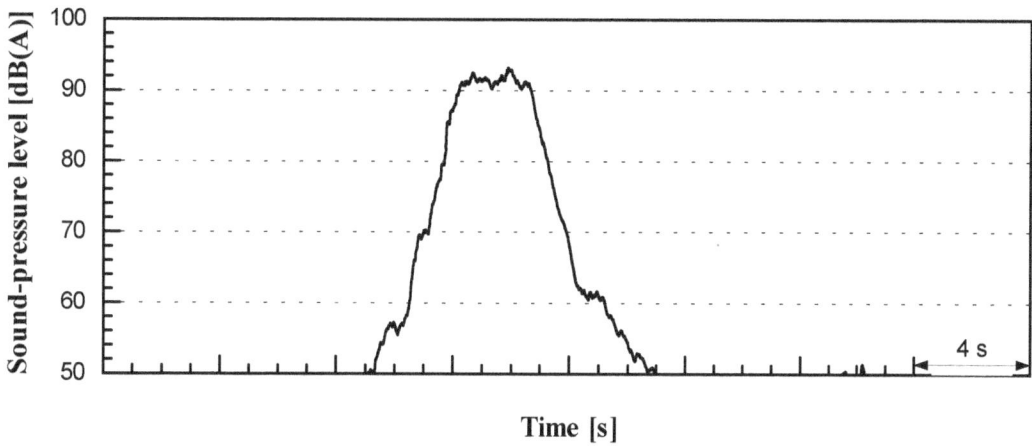

Figure B-25. Time history of the A-weighted SPL during a passby of the TR08 travelling on the reference concrete guideway at about 100 km/h (62 mph) measured beneath the guidway centerline at a height of 1.5 m (5.0 ft) above the ground.

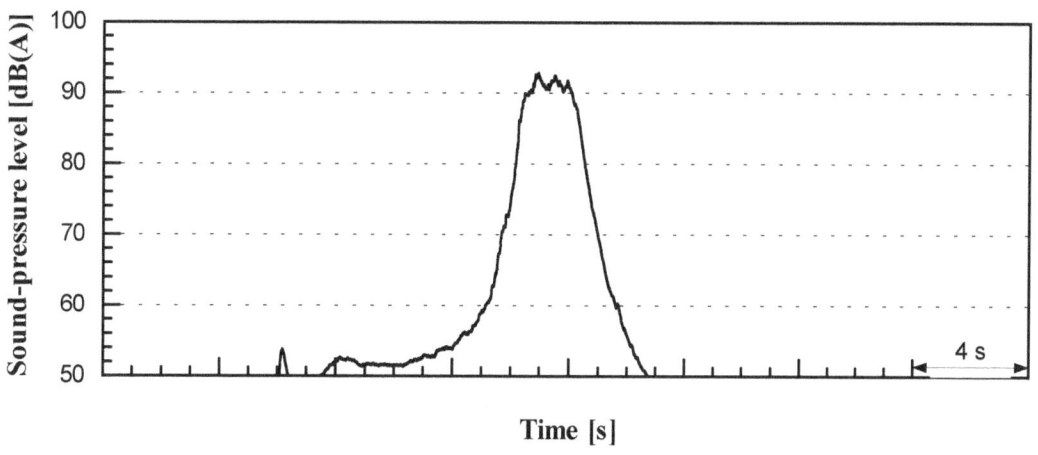

Figure B-26. Time history of the A-weighted SPL during a passby of the TR08 travelling on the reference concrete guideway at about 150 km/h (93 mph) measured beneath guideway centerline at a height of 1.5 m (5.0 ft) above the ground.

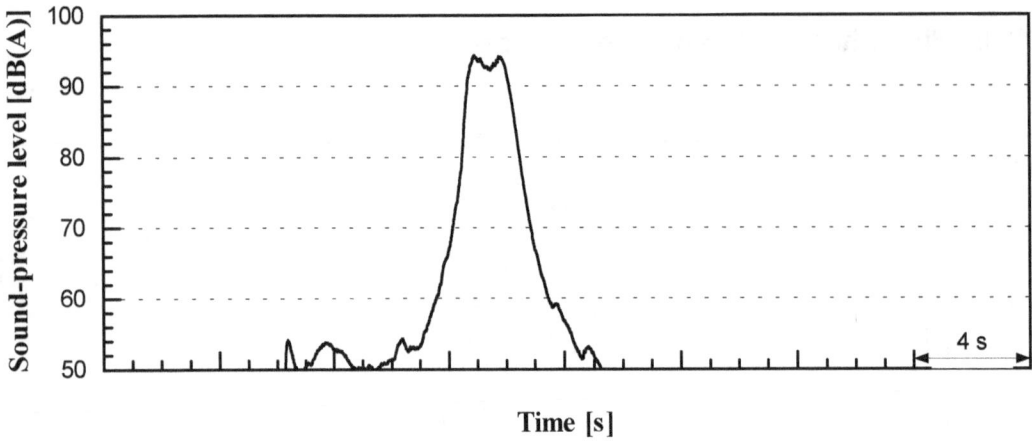

Figure B-27. Time history of the A-weighted SPL during a passby of the TR08 travelling on the reference concrete guideway at about 200 km/h (124 mph) measured beneath guideway centerline at a height of 1.5 m (5.0 ft) above the ground.

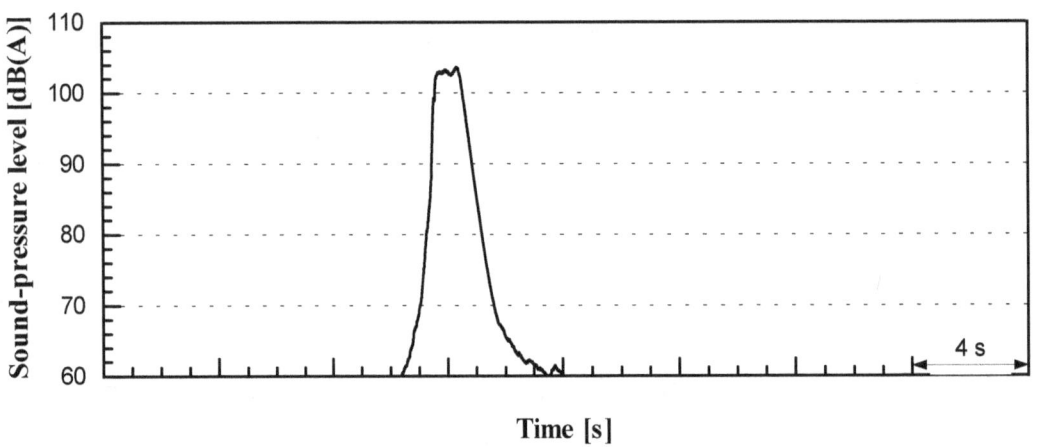

Figure B-28. Time history of the A-weighted SPL during a passby of the TR08 travelling on the reference concrete guideway at about 300 km/h (186 mph) measured beneath guideway centerline at a height of 1.5 m (5.0 ft) above the ground.

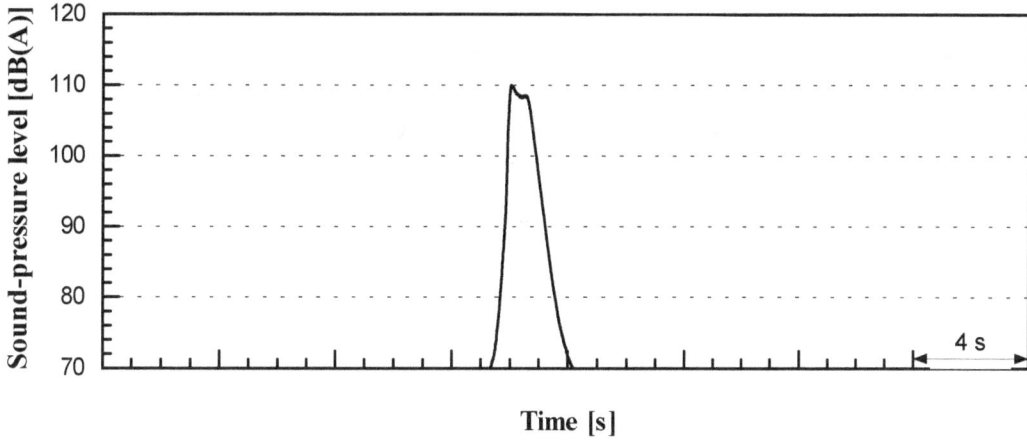

Figure B-29. Time history of the A-weighted SPL during a passby of the TR08 travelling on the reference concrete guideway at about 400 km/h (249 mph) measured beneath guideway centerline at a height of 1.5 m (5.0 ft) above the ground.

Table B-6. Results of the microphone positioned beneath the centerline of the reference concrete guideway at a height of 1.5 m (5.0 ft) above the ground (measuring series A/B).

Date	Time	Vehicle speed [km/h (mph)]	$L_{Amax, fast}$ [dB(A)]	$L_{Aeq,E}$ [dB(A)]	t_E [s]	SEL [dB(A)]	$L_{Aeq,1h}$ [dB(A)]
2002-05-17	12:57	100.0 (62.1)	92.7	89.2	4.41	95.6	60.1
2002-05-17	13:02	99.9 (62.1)	93.2	89.5	4.36	95.9	60.4
2002-05-17	13:10	100.1 (62.2)	92.9	89.1	4.45	95.6	60.0
2002-05-17	13:37	100.1 (62.2)	92.6	89.1	4.53	95.6	60.1
2001-08-16	10:38	149.8 (93.1)	93.3	89.6	2.87	94.1	58.6
2001-08-16	11:30	149.7 (93.0)	92.8	89.3	2.84	93.8	58.3
2001-08-16	13:28	149.7 (93.0)	92.6	89.1	2.90	93.7	58.2
2001-08-16	14:23	149.7 (93.0)	92.7	89.2	2.91	93.8	58.3
2001-08-16	12:56	199.6 (124.0)	94.3	91.4	2.38	95.1	59.6
2001-08-16	13:39	199.8 (124.1)	94.1	91.2	2.37	94.9	59.4
2001-08-16	14:35	199.9 (124.2)	94.2	91.1	2.38	94.9	59.3
2001-08-16	10:55	299.6 (186.2)	103.0	100.1	1.72	102.4	66.9
2001-08-16	11:46	299.6 (186.2)	103.0	100.1	1.72	102.5	66.9
2001-08-16	13:44	299.6 (186.2)	103.4	100.1	1.71	102.4	66.9
2001-08-16	14:40	299.8 (186.3)	103.5	100.5	1.71	102.9	67.3
2001-08-16	09:44	370.0 (229.9)	107.8	104.1	1.50	105.9	70.3
2001-08-16	10:14	401.0 (249.2)	109.8	106.2	1.41	107.7	72.1
2001-08-16	10:43	401.5 (249.5)	109.4	106.0	1.42	107.6	72.0
2001-08-16	11:35	400.9 (249.1)	109.2	105.8	1.42	107.3	71.7
2001-08-16	13:33	401.2 (249.3)	109.8	106.1	1.42	107.7	72.1
2001-08-16	14:28	400.9 (249.1)	109.8	106.3	1.40	107.8	72.2

B.3 Prototype Steel Guideway

B.3.1 Microphone at 30.5 m (100.0 ft) distance from track centerline

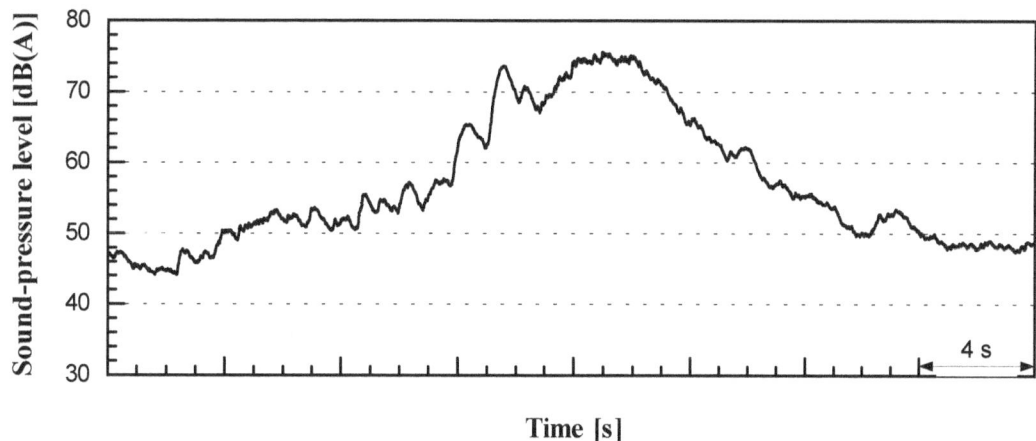

Figure B-30. Time history of the A-weighted SPL during a passby of the TR08 travelling on the prototype steel guideway at about 100 km/h (62 mph) measured at 30.5 m (100.0 ft) distanace from track centerline and 1.2 m (4.0 ft) above the ground.

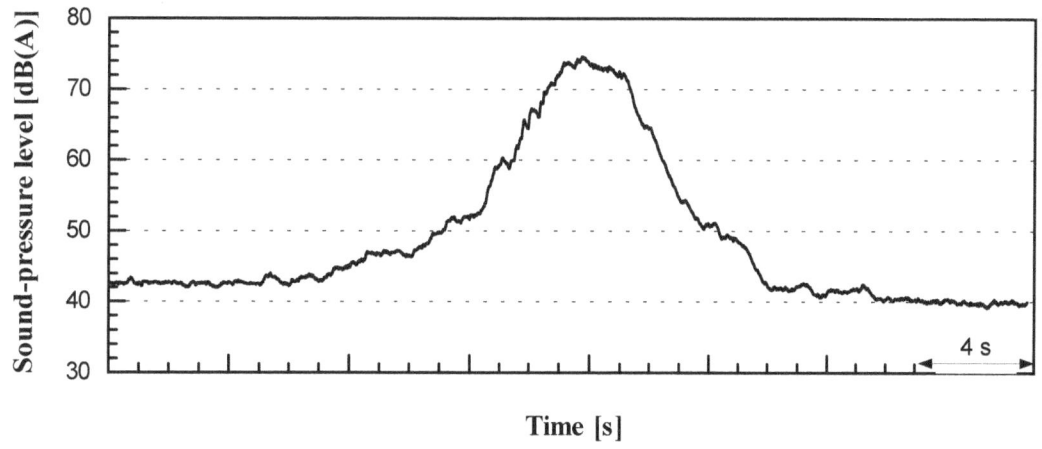

Figure B-31. Time history of the A-weighted SPL during a passby of the TR08 travelling on the prototype steel guideway at about 150 km/h (93 mph) measured at 30.5 m (100.0 ft) distance from track centerline and 1.2 m (4.0 ft) above the ground.

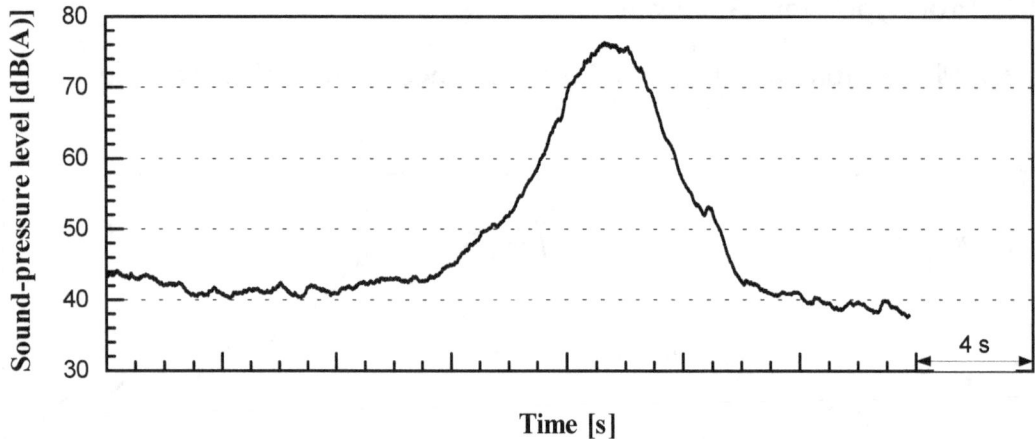

Figure B-32. Time history of the A-weighted SPL during a passby of the TR08 travelling on the prototype steel guideway at about 200 km/h (124 mph) measured at 30.5 m (100.0 ft) distance from track centerline and 1.2 m (4.0 ft) above the ground.

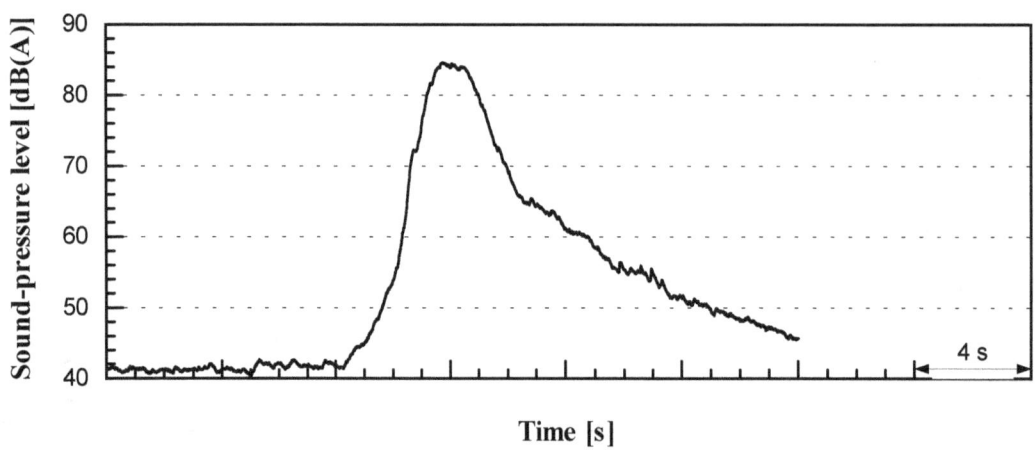

Figure B-33. Time history of the A-weighted SPL during a passby of the TR08 travelling on the prototype steel guideway at about 300 km/h (186 mph) measured at 30.5 m (100.0 ft) distance from track centerline and 1.2 m (4.0 ft) above the ground.

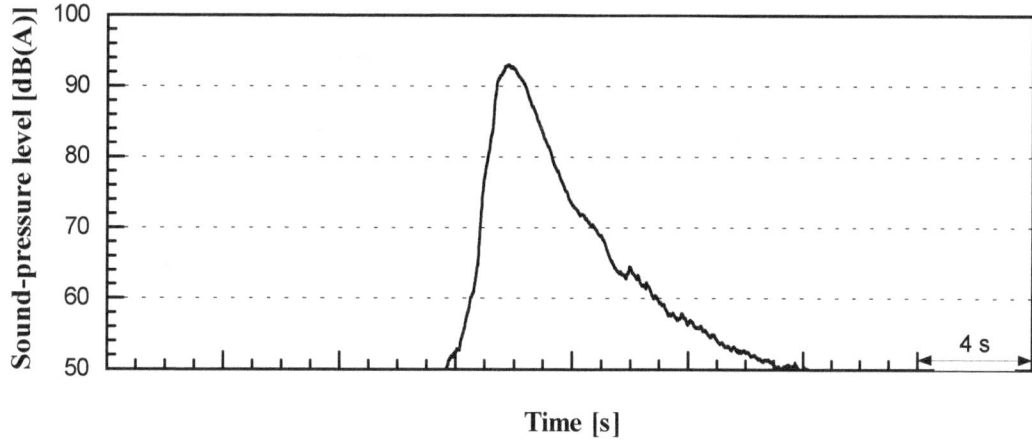

Figure B-34. Time history of the A-weighted SPL during a passby of the TR08 travelling on the prototype steel guideway at about 400 km/h (249 mph) measured at 30.5 m (100.0 ft) distance from track centerline and 1.2 m (4.0 ft) above the ground.

Table B-7. Results of the microphone positioned close to the prototype steel guideway at 30.5 m (100.0 ft) distance from track centerline and 1.2 m (4.0 ft) above the ground (measuring series H).

Date	Time	Vehicle speed [km/h (mph)]	$L_{Amax, fast}$ [dB(A)]	$L_{Aeq,E}$ [dB(A)]	t_E [s]	SEL [dB(A)]	$L_{Aeq,1h}$ [dB(A)]
2002-05-15	10:40	100.2 (62.3)	76.2	69.9	13.35	81.1	45.6
2002-05-15	10:51	100.3 (62.3)	75.6	70.1	12.42	81.0	45.5
2002-05-15	10:56	99.8 (62.0)	76.2	69.9	12.45	80.8	45.3
2002-05-15	11:03	100.1 (62.2)	76.0	70.5	11.90	81.3	45.7
2001-08-23	10:35	149.5 (92.9)	78.2	73.2	7.48	81.9	46.4
2001-08-23	11:21	149.8 (93.1)	75.5	70.7	6.36	78.7	43.1
2001-08-23	12:17	149.5 (92.9)	74.6	69.8	6.48	78.0	42.4
2001-08-23	13:06	149.7 (93.0)	74.7	70.2	6.55	78.3	42.8
2001-08-23	10:53	199.4 (123.9)	78.2	72.9	5.47	80.3	44.7
2001-08-23	11:33	199.5 (124.0)	76.3	71.7	5.44	79.0	43.5
2001-08-23	13:17	199.5 (124.0)	76.0	70.8	5.84	78.5	42.9
2001-08-23	14:09	199.7 (124.1)	76.7	71.6	5.41	78.9	43.4
2001-08-23	10:41	299.3 (186.0)	84.2	79.3	4.84	86.2	50.6
2001-08-23	12:23	299.7 (186.3)	84.6	80.1	4.28	86.4	50.8
2001-08-23	13:23	299.4 (186.1)	85.2	80.3	4.17	86.5	51.0
2001-08-23	11:26	399.0 (248.0)	93.0	88.2	3.28	93.3	57.8
2001-08-23	11:37	398.1 (247.4)	93.0	88.2	3.16	93.1	57.5
2001-08-23	13:10	398.3 (247.5)	92.5	87.9	2.98	92.6	57.1
2001-08-23	14:14	398.4 (247.6)	92.1	86.9	3.52	92.4	56.8

B.3.2 Microphone at 25.0 m (82.0 ft) distance from track centerline

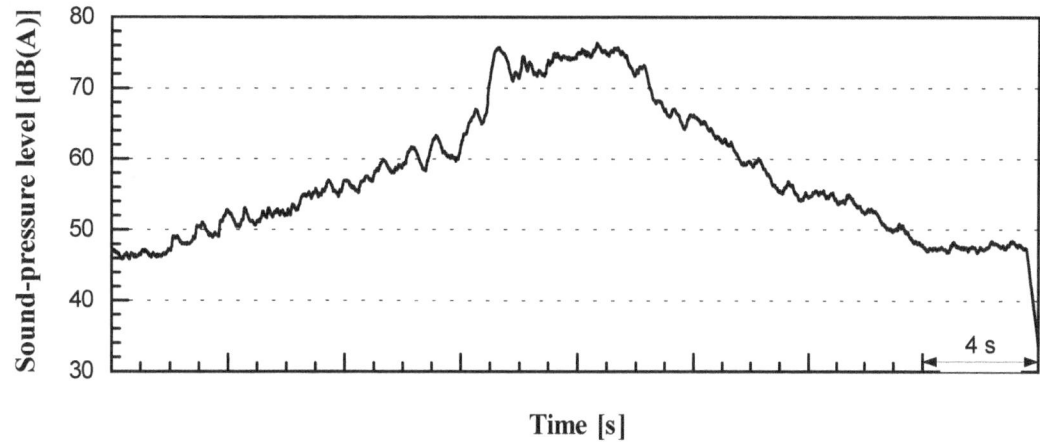

Figure B-35. Time history of the A-weighted SPL during a passby of the TR08 travelling on the prototype steel guideway at aobut 100 km/h (62 mph) measured at 25.0 m (82.0 ft) distance from track centerline and 3.5 m (11.5 ft) above the ground.

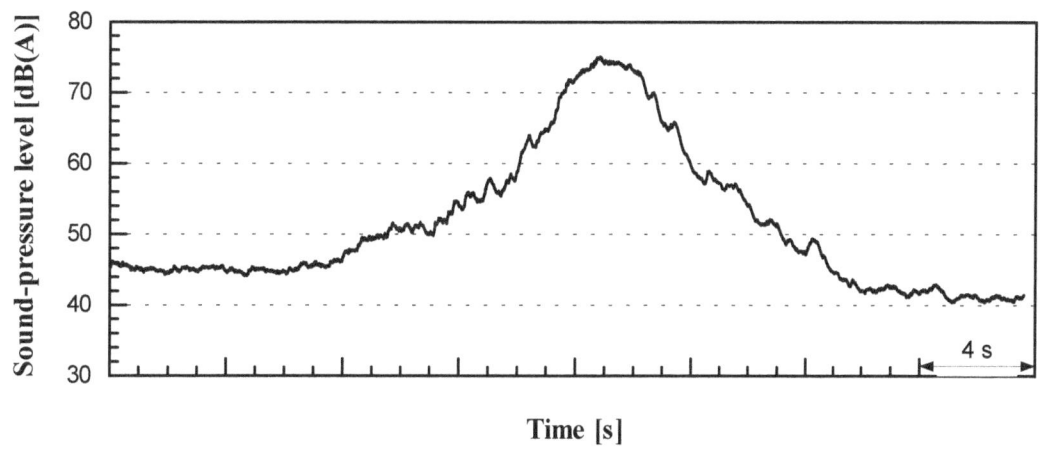

Figure B-36. Time history of the A-weighted SPL during a passby of the TR08 travelling on the prototype steel guideway at about 150 km/h (93 mph) measured at 25.0 m (82.0 ft) distance from track centerline and 3.5 m (11.5 ft) above the ground.

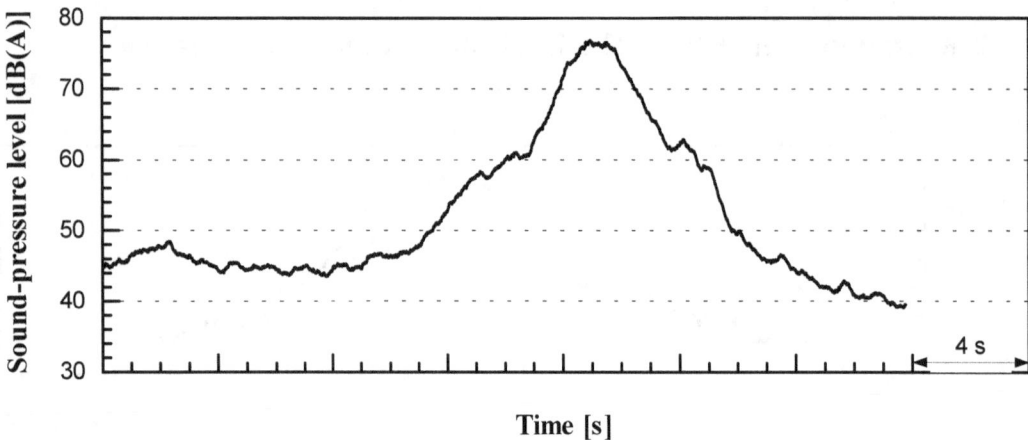

Figure B-37. Time history of the A-weighted SPL during a passby of the TR08 travelling on the prototype steel guideway at about 200 km/h (124 mph) measured at 25.0 m (82.0 ft) distance from track centerline and 3.5 m (11.5 ft) above the ground.

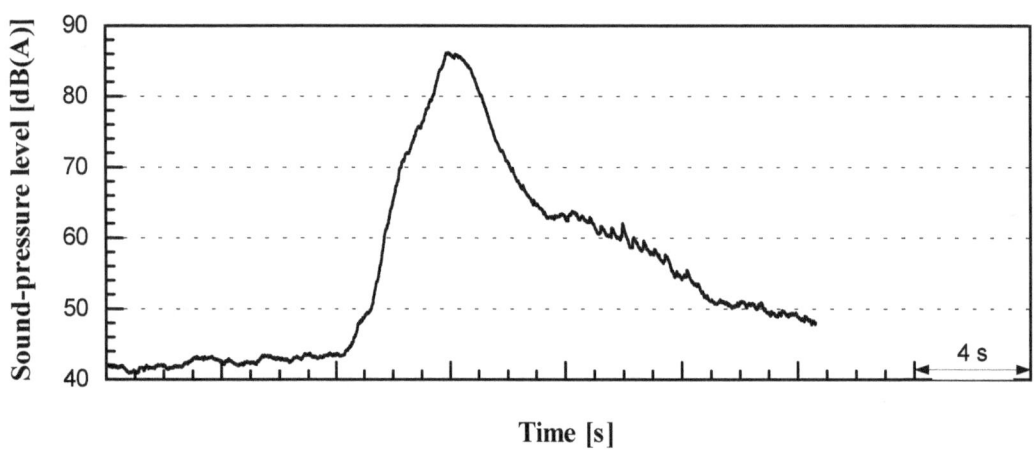

Figure B-38. Time history of the A-weighted SPL during a passby of the TR08 travelling on the prototype steel guideway at about 300 km/h (186 mph) measured at 25.0 m (82.0 ft) distance from track centerline and 3.5 m (11.5 ft) above the ground.

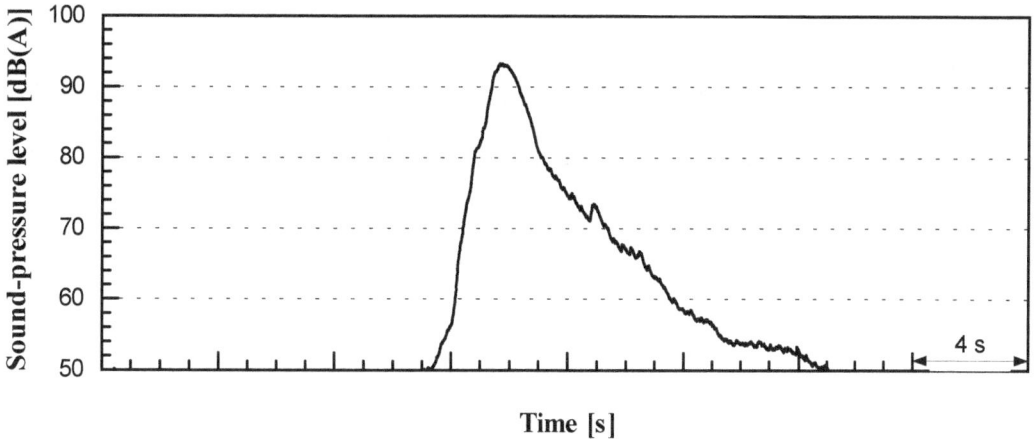

Figure B-39. Time history of the A-weighted SPL during a passby of the TR08 travelling on the prototype steel guideway at about 400 km/h (249 mph) measured at 25.0 m (82.0 ft) distance from track centerline and 3.5 m (11.5 ft) above the ground.

Table B-8. Results of the microphone positioned close to the prototype steel guideway at 25.0 m (82.0 ft) distance from track centerline and 3.5 m (11.5 ft) above the ground (measuring series H).

Date	Time	Vehicle speed [km/h (mph)]	$L_{Amax, fast}$ [dB(A)]	$L_{Aeq,E}$ [dB(A)]	t_E [s]	SEL [dB(A)]	$L_{Aeq,1h}$ [dB(A)]
2002-05-15	10:40	100.2 (62.3)	76.6	70.1	16.96	82.4	46.8
2002-05-15	10:51	100.3 (62.3)	76.3	70.4	14.22	82.0	46.4
2002-05-15	10:56	99.8 (62.0)	75.6	70.0	15.21	81.8	46.3
2002-05-15	11:03	100.1 (62.2)	76.2	70.8	12.85	81.9	46.3
2001-08-23	10:35	149.5 (92.9)	78.4	72.6	9.30	82.3	46.8
2001-08-23	11:21	149.8 (93.1)	75.7	70.4	8.27	79.6	44.0
2001-08-23	12:17	149.5 (92.9)	75.1	69.3	8.99	78.8	43.2
2001-08-23	13:06	149.7 (93.0)	75.2	69.2	9.89	79.2	43.6
2001-08-23	10:53	199.4 (123.9)	77.2	71.0	7.88	80.0	44.4
2001-08-23	11:33	199.5 (124.0)	76.8	70.2	8.47	79.5	44.0
2001-08-23	13:17	199.5 (124.0)	76.1	69.9	8.25	79.0	43.5
2001-08-23	14:09	199.7 (124.1)	76.2	70.6	6.69	78.9	43.3
2001-08-23	10:41	299.3 (186.0)	85.6	79.8	5.24	87.0	51.4
2001-08-23	12:23	299.7 (186.3)	86.1	80.6	4.63	87.3	51.7
2001-08-23	13:23	299.4 (186.1)	85.7	80.6	4.59	87.3	51.7
2001-08-23	11:26	399.0 (248.0)	93.1	87.7	3.49	93.1	57.6
2001-08-23	11:37	398.1 (247.4)	93.2	87.5	3.81	93.3	57.7
2001-08-23	13:10	398.3 (247.5)	93.3	87.5	3.75	93.2	57.7
2001-08-23	14:14	398.4 (247.6)	92.8	87.0	3.66	92.7	57.1

B.3.3 Microphone at 15.2 m (50.0 ft) distance from track centerline

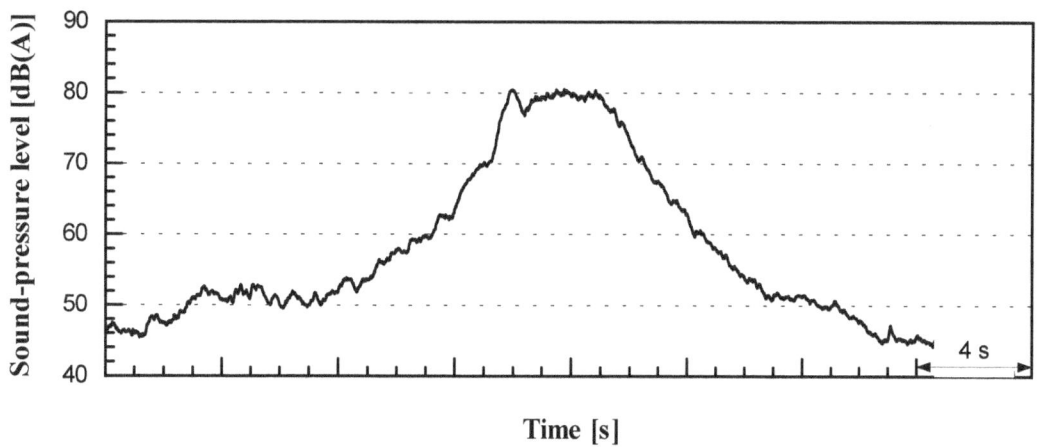

Figure B-40. Time history of the A-weighted SPL during a pssby of the TR08 travelling on the prototype steel guideway at about 100 km/h (62 mph) measured at 15.2 m (50.0 ft) distance from track centerline and 1.5 m (5.0 ft) above the ground.

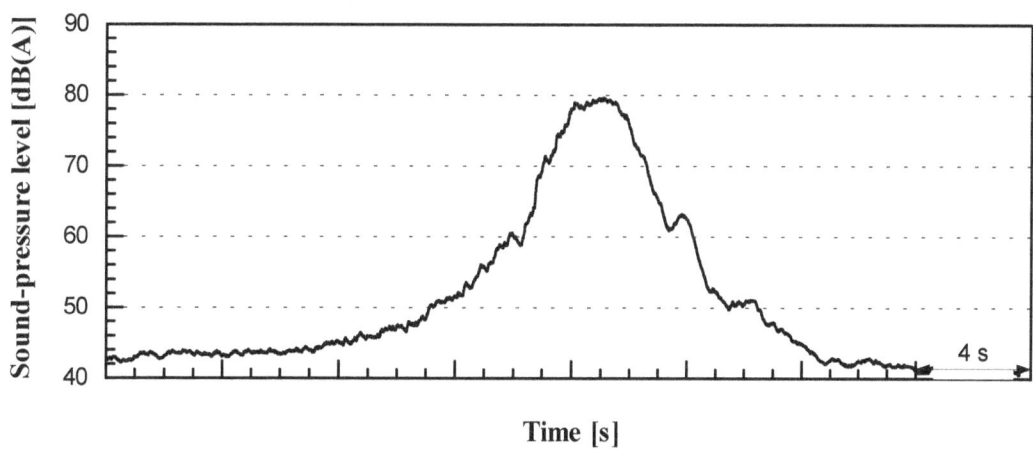

Figure B-41. Time history of the A-weighted SPL during a passby of the TR08 travelling on the prototype steel guideway at about 150 km/h (93 mph) measured at 15.2 m (50.0 ft) distance from track centerline and 1.5 m (5.0 ft) above the ground.

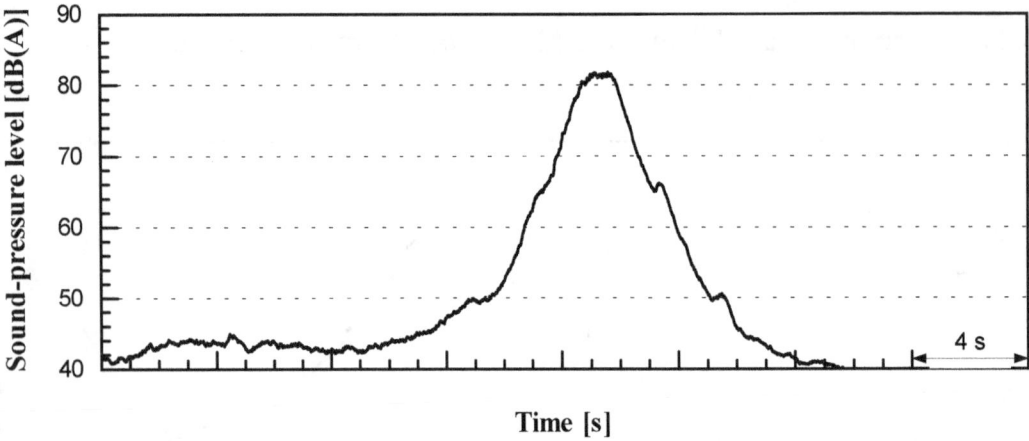

Figure B-42. Time history of the A-weighted SPL during a passby of the TR08 travelling on the prototype steel guideway at about 200 km/h (124 mph) measured at 15.2 m (50.0 ft) distance from track centerline and 1.5 m (5.0 ft) above the ground.

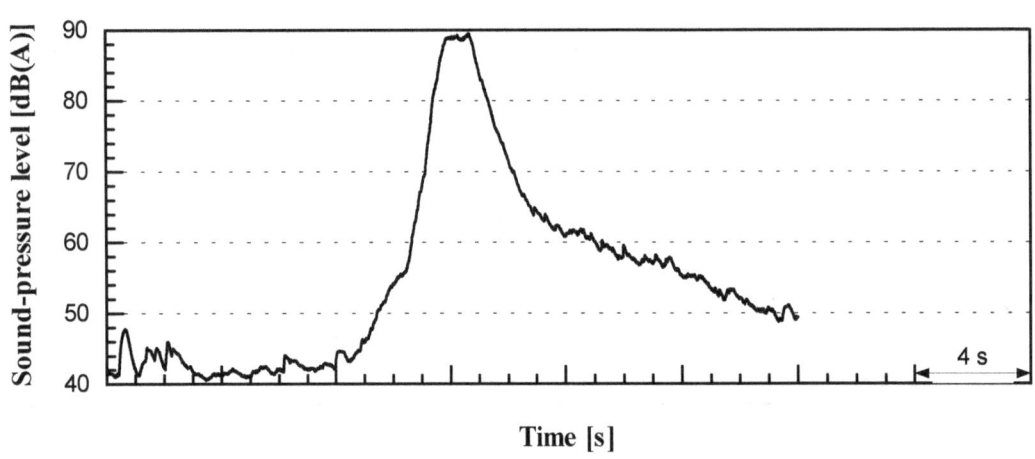

Figure B-43. Time history of the A-weighted SPL during a passby of the TR08 travelling on the prototype steel guideway at about 300 km/h (186 mph) measured at 15.2 m (50.0 ft) distance from track centerline and 1.5 m (5.0 ft) above the ground.

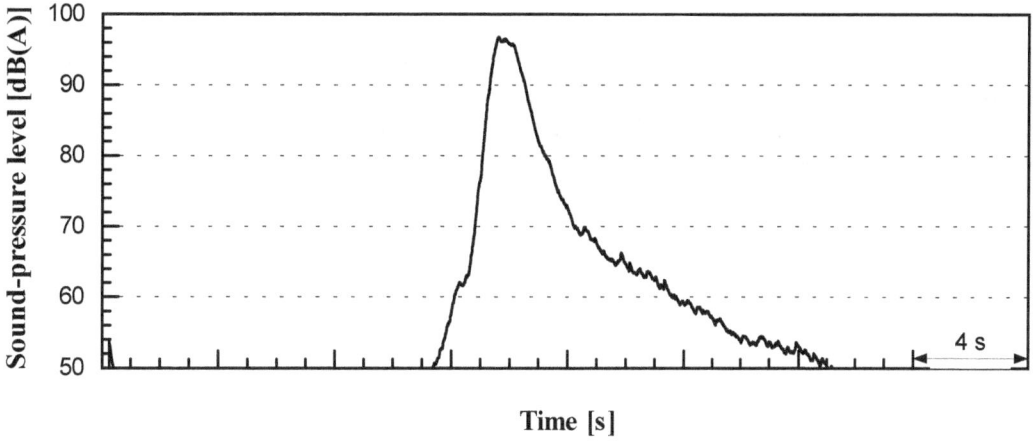

Figure B-44. Time history of the A-weighted SPL during a passby of the TR08 travelling on the prototype steel guideway at about 400 km/h (249 mph) measured at 15.2 m (50.0 ft) distance from track centerline and 1.5 m (5.0 ft) above the ground.

Table B-9. Results of the microphone positioned close to the prototype steel guideway at 15.2 m (50.0 ft) distance from track centerline and 1.5 m (5.0 ft) above the ground (measuring series H).

Date	Time	Vehicle speed [km/h (mph)]	$L_{Amax, fast}$ [dB(A)]	$L_{Aeq,E}$ [dB(A)]	t_E [s]	SEL [dB(A)]	$L_{Aeq,1h}$ [dB(A)]
2002-05-15	10:40	100.2 (62.3)	81.4	76.9	8.47	86.2	50.7
2002-05-15	10:51	100.3 (62.3)	80.5	76.2	8.91	85.7	50.1
2002-05-15	10:56	99.8 (62.0)	81.2	76.4	7.97	85.5	49.9
2002-05-15	11:03	100.1 (62.2)	80.9	76.1	9.09	85.7	50.1
2001-08-23	10:35	149.5 (92.9)	83.2	78.3	6.24	86.2	50.7
2001-08-23	11:21	149.8 (93.1)	80.7	76.4	4.91	83.3	47.7
2001-08-23	12:17	149.5 (92.9)	79.7	75.1	5.95	82.9	47.3
2001-08-23	13:06	149.7 (93.0)	79.8	76.0	5.31	83.3	47.7
2001-08-23	10:53	199.4 (123.9)	82.6	77.5	4.55	84.0	48.5
2001-08-23	11:33	199.5 (124.0)	81.9	76.8	4.87	83.7	48.1
2001-08-23	13:17	199.5 (124.0)	81.8	76.8	4.67	83.5	47.9
2001-08-23	14:09	199.7 (124.1)	81.4	76.1	5.21	83.2	47.7
2001-08-23	10:41	299.3 (186.0)	89.4	84.8	3.40	90.1	54.5
2001-08-23	12:23	299.7 (186.3)	89.7	85.4	3.14	90.4	54.8
2001-08-23	13:23	299.4 (186.1)	89.6	85.2	3.40	90.5	55.0
2001-08-23	11:26	399.0 (248.0)	96.6	91.5	2.88	96.1	60.5
2001-08-23	11:37	398.1 (247.4)	96.9	92.3	2.54	96.3	60.8
2001-08-23	13:10	398.3 (247.5)	96.6	91.8	2.52	95.9	60.3
2001-08-23	14:14	398.4 (247.6)	96.3	91.4	2.86	96.0	60.4

B.3.4 Microphone at 6.5 m (21.3 ft) distance from track centerline (high position)

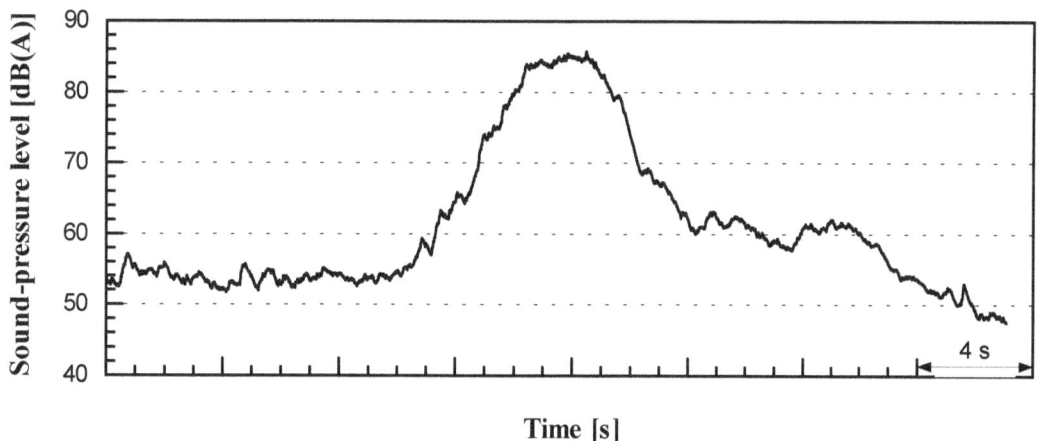

Figure B-45. Time history of the A-weighted SPL during a passby of the TR08 travelling on the prototype steel guideway at about 100 km/h (62 mph) measured at 6.5 m (21.3 ft) distance from track centerline and the height of the upper surface of the guideway.

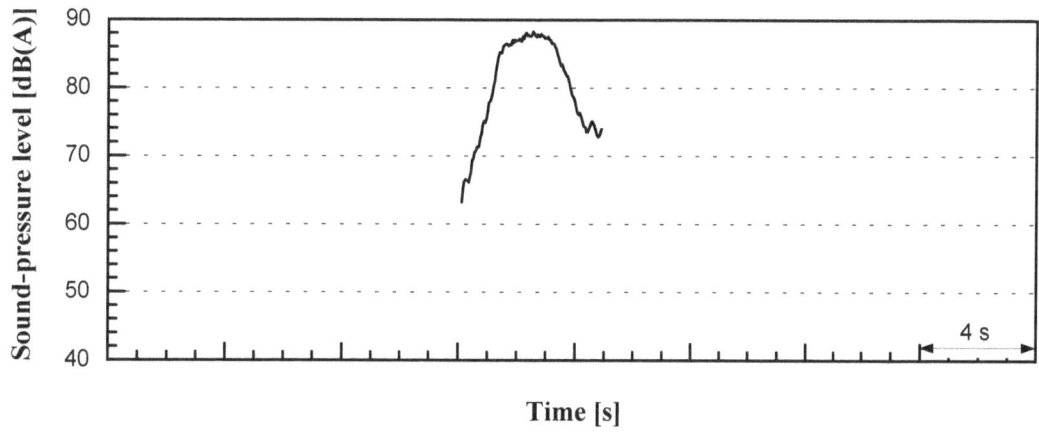

Figure B-46. Time history of the A-weighted SPL during a passby of the TR08 travelling on the prototype steel guideway at about 150 km/h (93 mph) measured at 6.5 m (21.3 ft) distance from track centerline and the height of the upper surface of the guideway.

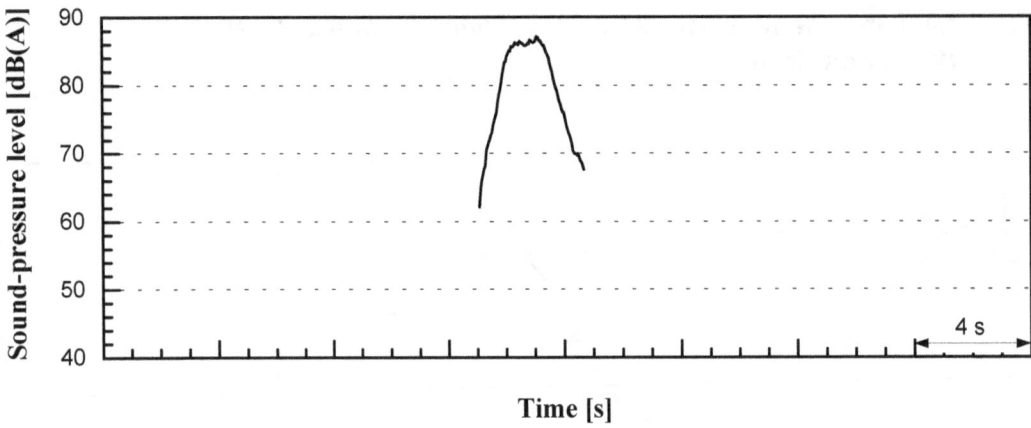

Figure B-47. Time history of the A-weighted SPL during a passby of the TR08 travelling on the prototype steel guideway at about 200 km/h (124 mph) measured at 6.5 m (21.3 ft) distance from track centerline and the height of the upper surface of the guideway.

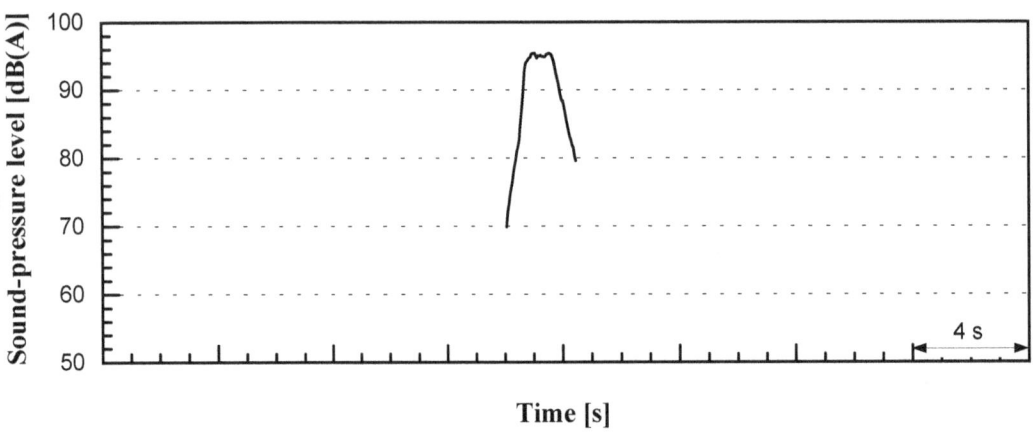

Figure B-48. Time history of the A-weighted SPL during a passby of the TR08 travelling on the prototype steel guideway at about 300 km/h (186 mph) measured at 6.5 m (21.3 ft) distance from track centerline and the height of the upper surface of the guideway.

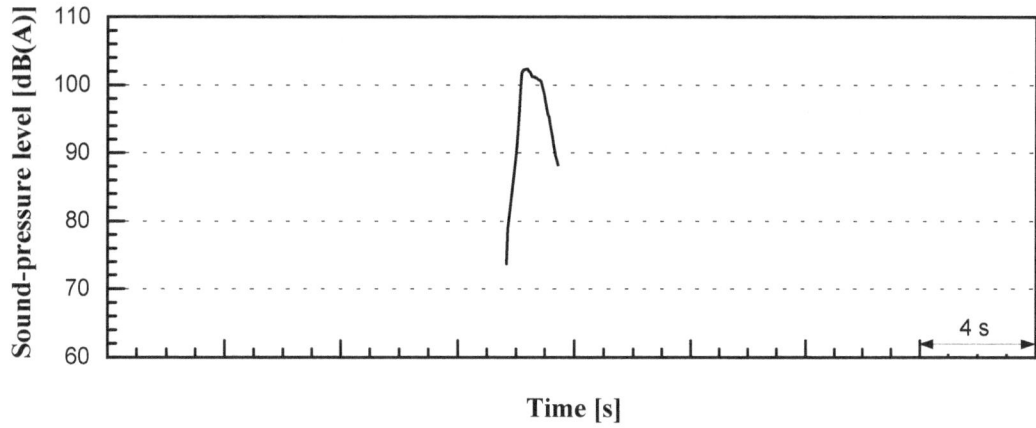

Figure B-49. Time history of the A-weighted SPL during a passby of the TR08 travelling on the prototype steel guideway at about 400 km/h (249 mph) measured at 6.5 m (21.3 ft) distance from track centerline and the height of the upper surface of the guideway.

Table B-10. Results of the microphone positioned close to the prototype steel guideway at 6.5 m (21.3 ft) distance from track centerline and the height of the upper surface of the guideway (measuring series G/H).

Date	Time	Vehicle speed [km/h (mph)]	$L_{Amax, fast}$ [dB(A)]	$L_{Aeq,E}$ [dB(A)]	t_E [s]	SEL [dB(A)]	$L_{Aeq,1h}$ [dB(A)]
2002-05-15	10:40	100.2 (62.3)	86.1	83.6	4.34	90.0	54.5
2002-05-15	10:51	100.3 (62.3)	85.8	83.0	4.31	89.4	53.8
2002-05-15	10:56	99.8 (62.0)	86.7	83.6	3.97	89.5	54.0
2002-05-15	11:03	100.1 (62.2)	86.0	83.4	4.31	89.8	54.2
2001-08-21	9:19	149.8 (93.1)	86.5	84.2	2.88	88.8	53.3
2001-08-21	9:55	149.7 (93.0)	88.3	85.9	2.96	90.6	55.1
2001-08-21	10:34	149.5 (92.9)	88.3	85.6	3.08	90.5	54.9
2001-08-21	8:56	199.5 (124.0)	88.3	86.2	2.24	89.7	54.2
2001-08-21	11:14	199.5 (124.0)	86.6	84.5	2.36	88.2	52.7
2001-08-21	11:26	199.4 (123.9)	87.2	84.7	2.28	88.3	52.7
2001-08-21	8:43	299.2 (186.0)	95.0	93.0	1.64	95.2	59.6
2001-08-21	9:24	299.6 (186.2)	95.5	93.6	1.64	95.7	60.2
2001-08-21	11:19	299.7 (186.3)	95.4	93.5	1.68	95.7	60.2
2001-08-21	10:39	399.7 (248.4)	102.8	100.3	1.28	101.4	65.8
2001-08-21	12:27	398.9 (247.9)	101.9	99.7	1.28	100.8	65.2
2001-08-21	13:12	398.6 (247.7)	102.3	99.9	1.32	101.1	65.6

B.3.5 Microphone at 6.5 m (21.3 ft) distance from track centerline (low position)

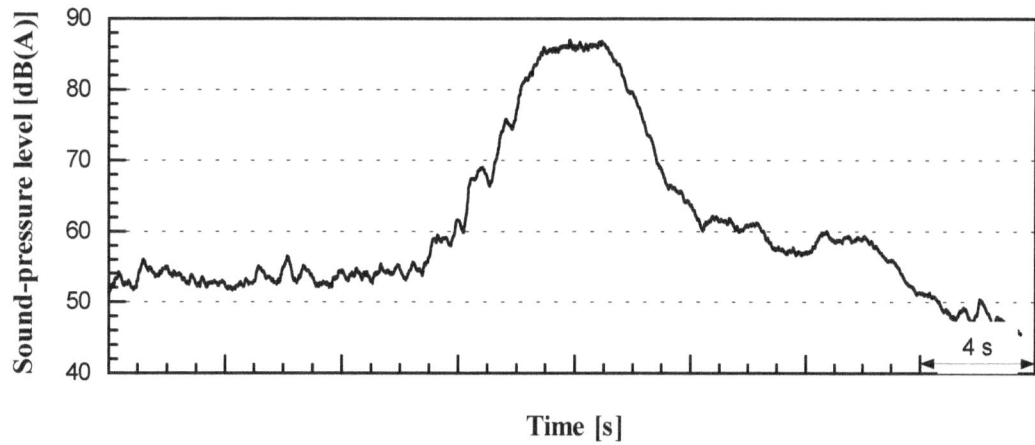

Figure B-50. Time history of the A-weighted SPL during a passby of the TR08 travelling on the prototype steel guideway at about 100 km/h (62 mph) measured at 6.5 m (21.3 ft) distance from track centerline and 1.5 m (5.0 ft) below the upper surface of the guideway.

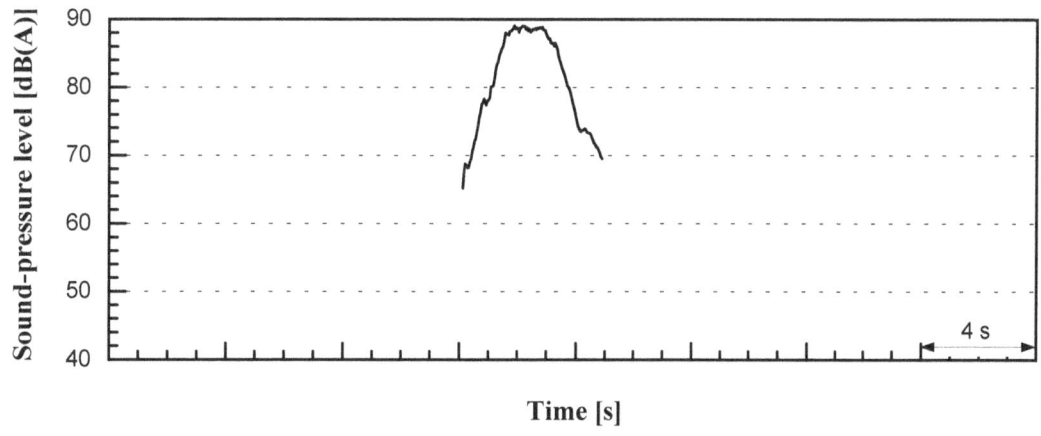

Figure B-51. Time history of the A-weighted SPL during a passby of the TR08 travelling on the prototype steel guideway at about 150 km/h (93 mph) measured at 6.5 m (21.3 ft) distance from track centerline and 1.5 m (5.0 ft) below the upper surface of the guideway.

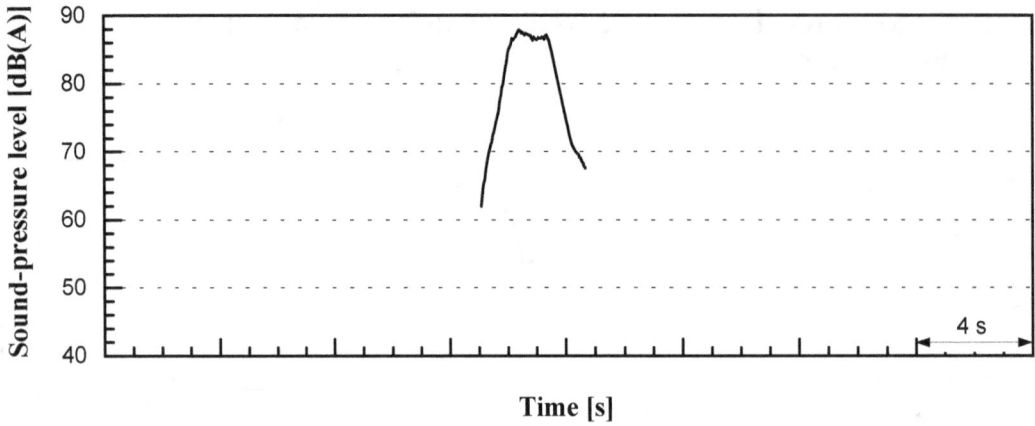

Figure B-52. Time history of the A-weighted SPL during a passby of the TR08 travelling on the prototype steel guideway at about 200 km/h (124 mph) measured at 6.5 m (21.3 ft) distance from track centerline and 1.5 m (5.0 ft) below the upper surface of the guideway.

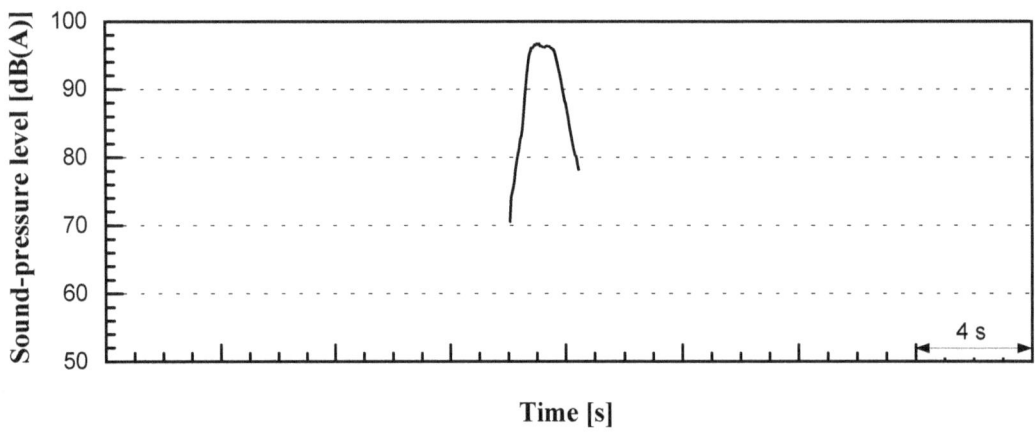

Figure B-53. Time history of the A-weighted SPL during a passby of the TR08 travelling on the prototype steel guideway at about 300 km/h (186 mph) measured at 6.5 m (21.3 ft) distance from track centerline and 1.5 m (5.0 ft) below the upper surface of the guideway.

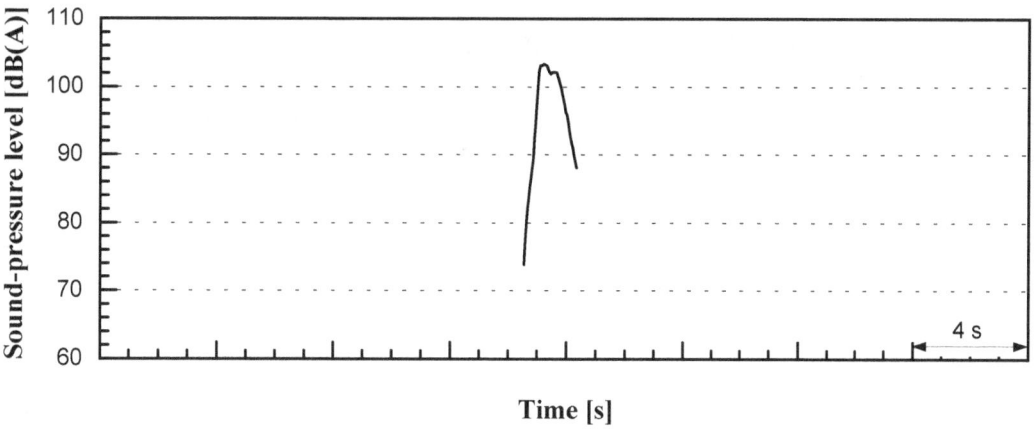

Figure B-54. Time history of the A-weighted SPL during a passby of the TR08 travelling on the prototype steel guideway at about 400 km/h (249 mph) measured at 6.5 m (21.3 ft) distance from track centerline and 1.5 m (5.0 ft) below the upper surface of the guideway.

Table B-11. Results of the microphone positioned close to the prototype steel guideway at 6.5 m (21.3 ft) distance from track centerline and 1.5 m (5.0 ft) below the upper surface of the guideway (measuring series H).

Date	Time	Vehicle speed [km/h (mph)]	$L_{Amax, fast}$ [dB(A)]	$L_{Aeq,E}$ [dB(A)]	t_E [s]	SEL [dB(A)]	$L_{Aeq,1h}$ [dB(A)]
2002-05-15	10:40	100.2 (62.3)	87.8	85.3	4.25	91.6	56.0
2002-05-15	10:51	100.3 (62.3)	87.0	84.6	4.27	90.9	55.4
2002-05-15	10:56	99.8 (62.0)	87.7	85.5	3.84	91.3	55.8
2002-05-15	11:03	100.1 (62.2)	87.3	85.0	4.19	91.2	55.6
2001-08-22	8:59	149.5 (92.9)	89.0	86.9	2.84	91.4	55.9
2001-08-22	11:08	149.3 (92.8)	89.4	86.8	2.64	91.0	55.5
2001-08-22	11:42	149.7 (93.0)	86.4	84.2	2.80	88.6	53.1
2001-08-22	10:33	199.5 (124.0)	87.7	85.8	2.28	89.4	53.8
2001-08-22	12:34	199.6 (124.1)	87.7	85.4	2.28	89.0	53.4
2001-08-22	12:46	199.5 (124.0)	87.8	85.7	2.20	89.1	53.5
2001-08-22	9:04	299.6 (186.2)	96.9	94.8	1.56	96.7	61.1
2001-08-22	10:16	299.5 (186.1)	96.7	94.7	1.60	96.8	61.2
2001-08-22	10:38	299.6 (186.2)	96.5	94.4	1.56	96.4	60.8
2001-08-22	11:47	399.3 (248.2)	103.3	100.9	1.28	102.0	66.5
2001-08-22	12:39	398.6 (247.7)	103.5	101.0	1.28	102.0	66.5
2001-08-22	12:50	397.8 (247.2)	103.3	100.9	1.28	102.0	66.4

B.3.6 Microphone beneath guideway centerline

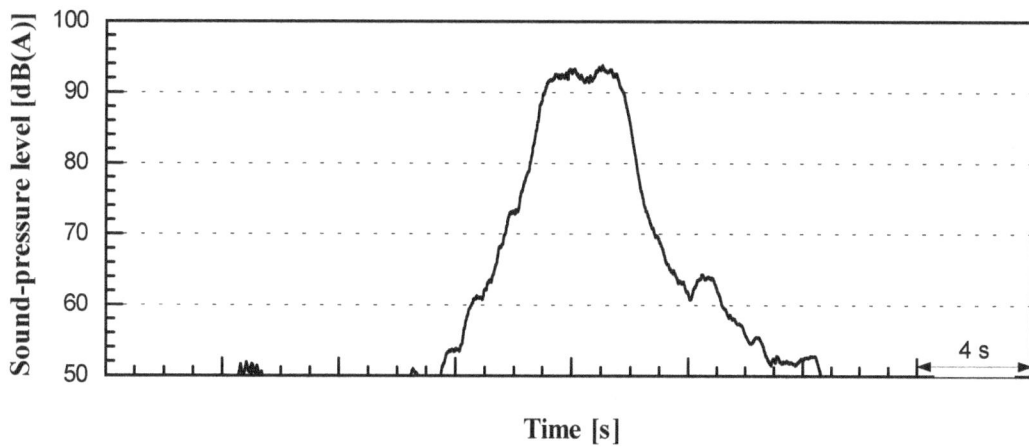

Figure B-55. Time history of the A-weighted SPL during a passby of the TR08 travelling on the prototype steel guideway at about 100 km/h (62 mph) measured beneath the guideway centerline at a height of 1.5 m (5.0 ft) above the ground.

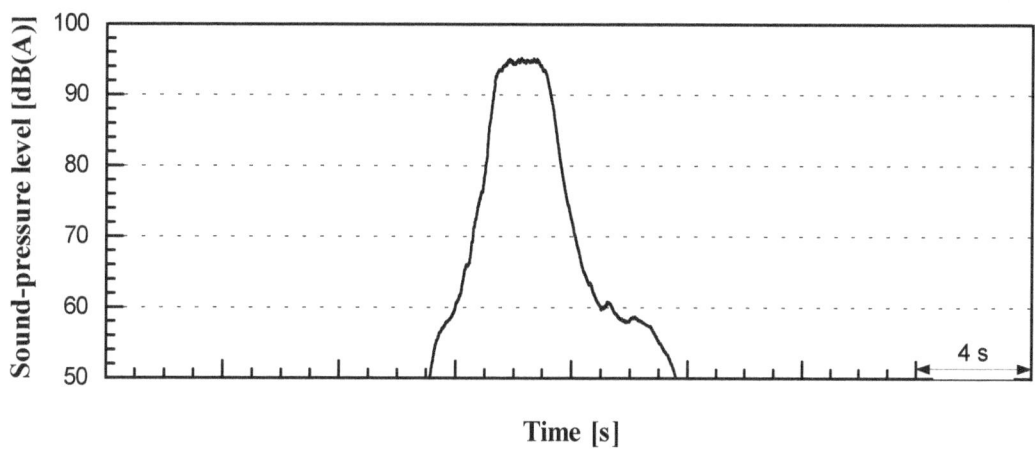

Figure B-56. Time history of the A-weighted SPL during a passby of the TR08 travelling on the prototype steel guideway at about 150 km/h (93 mph) measured beneath guideway centerline at a height of 1.5 m (5.0 ft) above the ground.

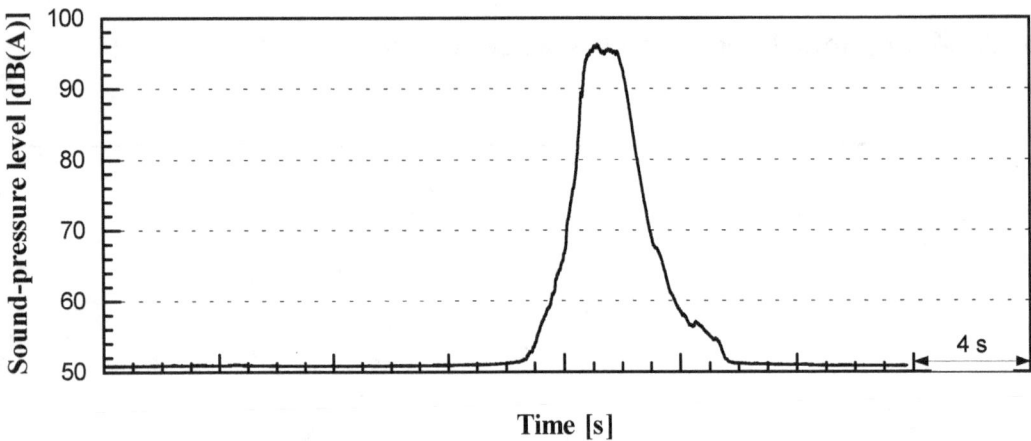

Figure B-57. Time history of the A-weighted SPL during a passby of the TR08 travelling on the prototype steel guideway at about 200 km/h (124 mph) measured beneath guideway centerline at a height of 1.5 m (5.0 ft) above the ground.

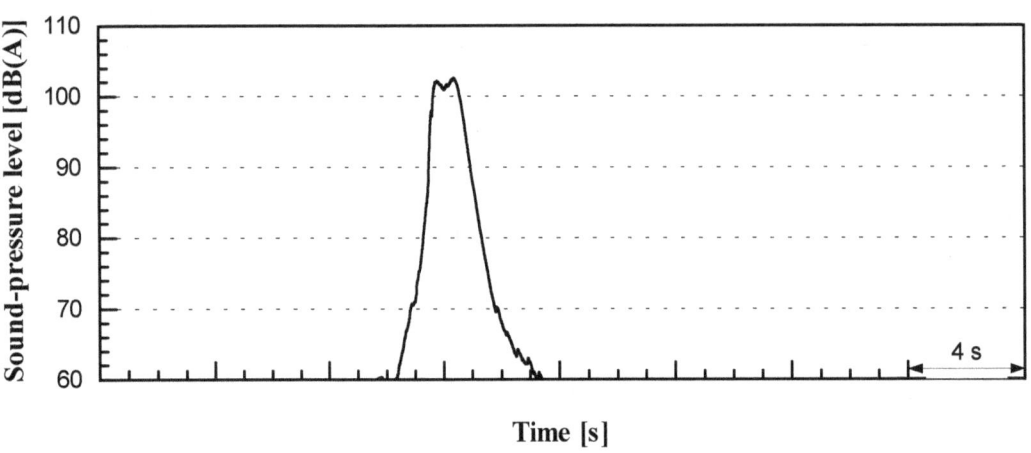

Figure B-58. Time history of the A-weighted SPL during a passby of the TR08 travelling on the prototype steel guideway at about 300 km/h (186 mph) measured beneath guideway centerline at a height of 1.5 m (5.0 ft) above the ground.

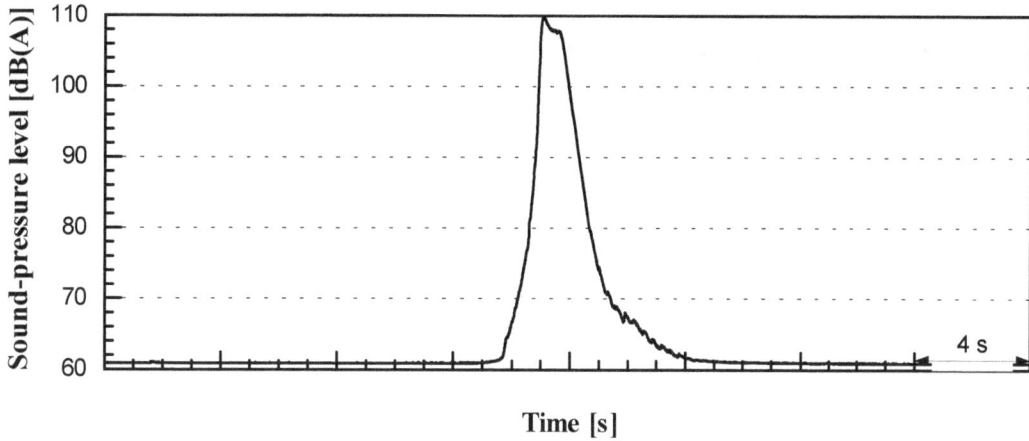

Figure B-59. Time history of the A-weighted SPL during a passby of the TR08 travelling on the prototype steel guideway at about 400 km/h (249 mph) measured beneath guideway centerline at a height of 1.5 m (5.0 ft) above the ground.

Table B-12. Results of the microphone positioned beneath the centerline of the prototype steel guideway at a height of 1.5 m (5.0 ft) above the ground (measuring series H).

Date	Time	Vehicle speed [km/h (mph)]	$L_{Amax, fast}$ [dB(A)]	$L_{Aeq,E}$ [dB(A)]	t_E [s]	SEL [dB(A)]	$L_{Aeq,1h}$ [dB(A)]
2002-05-15	10:40	100.2 (62.3)	94.5	90.9	4.36	97.3	61.7
2002-05-15	10:51	100.3 (62.3)	93.8	90.6	4.33	97.0	61.4
2002-05-15	10:56	99.8 (62.0)	93.8	90.2	4.55	96.7	61.2
2002-05-15	11:03	100.1 (62.2)	93.9	90.9	4.37	97.3	61.7
2001-08-23	10:35	149.5 (92.9)	98.6	95.4	2.94	100.0	64.5
2001-08-23	11:21	149.8 (93.1)	95.9	92.7	3.06	97.5	61.9
2001-08-23	12:17	149.5 (92.9)	95.1	92.5	3.04	97.3	61.7
2001-08-23	13:06	149.7 (93.0)	95.2	92.6	3.01	97.4	61.8
2001-08-23	10:53	199.4 (123.9)	95.9	92.8	2.42	96.6	61.1
2001-08-23	11:33	199.5 (124.0)	96.2	93.1	2.35	96.8	61.3
2001-08-23	13:17	199.5 (124.0)	95.9	92.9	2.40	96.7	61.1
2001-08-23	14:09	199.7 (124.1)	96.5	93.3	2.33	97.0	61.5
2001-08-23	10:41	299.3 (186.0)	103.0	99.9	1.85	102.5	67.0
2001-08-23	12:23	299.7 (186.3)	102.6	99.3	1.88	102.1	66.5
2001-08-23	13:23	299.4 (186.1)	102.5	99.2	1.88	102.0	66.4
2001-08-23	11:26	399.0 (248.0)	109.9	105.9	1.51	107.7	72.2
2001-08-23	11:37	398.1 (247.4)	109.7	105.9	1.51	107.7	72.1
2001-08-23	13:10	398.3 (247.5)	109.6	105.8	1.50	107.5	72.0
2001-08-23	14:14	398.4 (247.6)	109.8	105.8	1.48	107.5	72.0

B.4 Prototype Concrete Guideway

B.4.1 Microphone at 30.5 m (100.0 ft) distance from track centerline

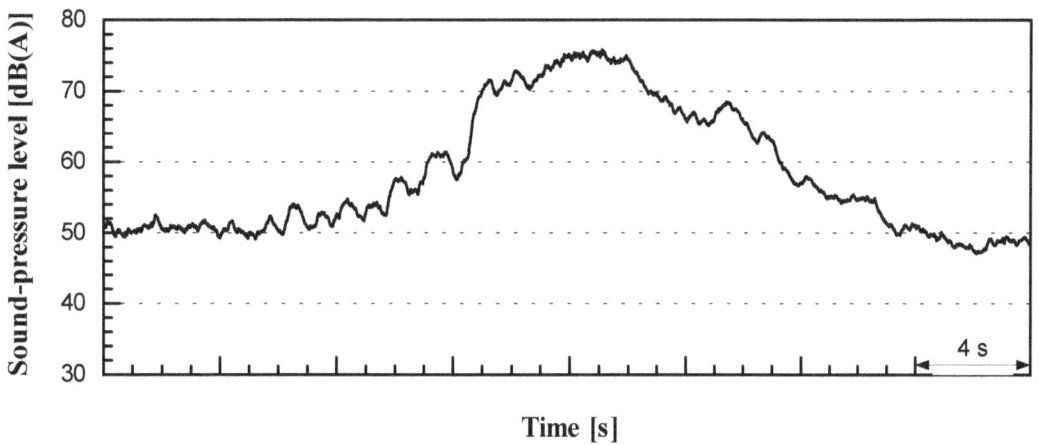

Figure B-60. Time history of the A-weighted SPL during a passby of the TR08 travelling on the prototype concrete guideway at about 100 km/h (62 mph) measured at 30.5 m (100.0 ft) distance from track centeriline and 1.2 m (4.0 ft) above the ground.

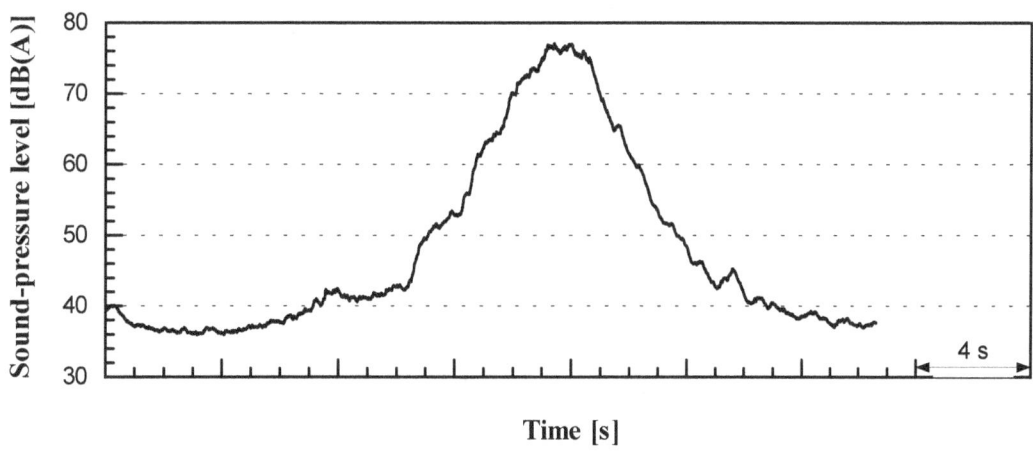

Figure B-61. Time history of the A-weighted SPL during a passby of the TR08 travelling on the prototype concrete guideway at about 150 km/h (93 mph) measured at 30.5 m (100.0 ft) distance from track centerline and 1.2 m (4.0 ft) above the ground.

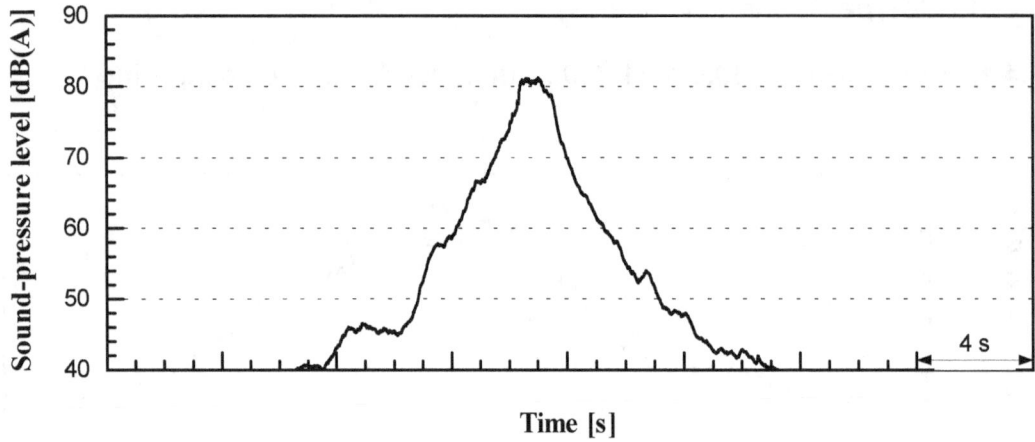

Figure B-62. Time history of the A-weighted SPL during a passby of the TR08 travelling on the prototype concrete guideway at about 200 km/h (124 mph) measured at 30.5 m (100.0 ft) distance from track centerline and 1.2 m (4.0 ft) above the ground.

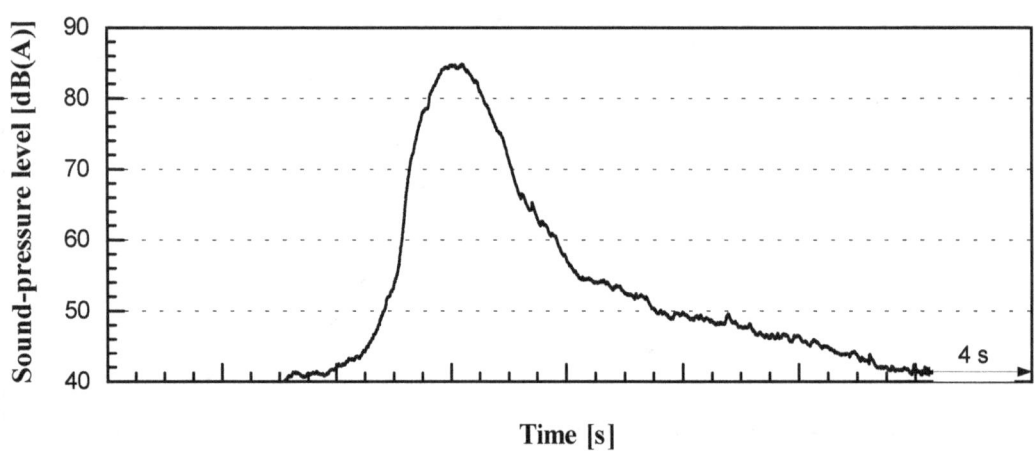

Figure B-63. Time history of the A-weighted SPL during a passby of the TR08 travelling on the prototype concrete guideway at about 300 km/h (186 mph) measured at 30.5 m (100.0 ft) distance from track centerline and 1.2 m (4.0 ft) above the ground.

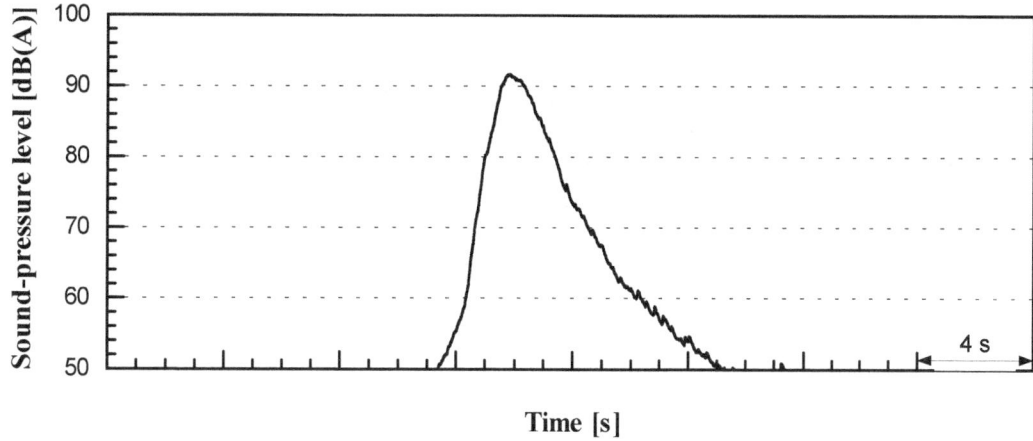

Figure B-64. Time history of the A-weighted SPL during a passby of the TR08 travelling on the prototype concrete guideway at about 400 km/h (249 mph) measured at 30.5 m (100.0 ft) distance from track centerline and 1.2 m (4.0 ft) above the ground.

Table B-13. Results of the microphone positioned close to the prototype concrete guideway at 30.5 m (100.0 ft) distance from track centerline and 1.2 m (4.0 ft) above the ground (measuring series L)

Date	Time	Vehicle speed [km/h (mph)]	$L_{Amax, fast}$ [dB(A)]	$L_{Aeq,E}$ [dB(A)]	t_E [s]	SEL [dB(A)]	$L_{Aeq,1h}$ [dB(A)]
2002-05-15	13:31	100.0 (62.1)	75.5	70.4	12.23	81.3	45.7
2002-05-15	13:36	99.9 (62.1)	77.9	72.4	13.03	83.5	48.0
2002-05-15	13:42	100.1 (62.2)	75.9	70.4	13.81	81.8	46.2
2002-05-15	13:47	100.1 (62.2)	77.0	71.6	12.22	82.5	46.9
2001-08-21	09:55	149.8 (93.1)	79.5	74.5	6.31	82.5	46.9
2001-08-21	10:35	149.6 (93.0)	78.9	73.9	6.23	81.8	46.2
2001-08-21	13:53	150.0 (93.2)	77.2	72.2	6.03	80.0	44.5
2001-08-21	14:10	149.7 (93.0) (N→S)	77.2	71.2	6.39	79.2	43.7
2001-08-21	11:15	199.6 (124.1)	81.4	75.7	4.35	82.1	46.5
2001-08-21	11:27	199.4 (123.9)	80.8	74.4	4.61	81.0	45.4
2001-08-21	12:23	199.5 (124.0)	82.5	76.2	4.52	82.8	47.2
2001-08-21	13:08	199.6 (124.1)	81.3	75.6	4.69	82.3	46.7
2001-08-21	14:03	199.6 (124.1)	81.4	75.2	4.68	81.9	46.4
2001-08-21	09:25	299.7 (186.3)	84.8	80.0	4.70	86.7	51.1
2001-08-21	10:00	299.3 (186.0)	n/a	n/a	n/a	n/a	n/a
2001-08-21	11:20	299.7 (186.3)	84.9	80.6	4.26	86.9	51.3
2001-08-21	10:41	399.7 (248.4)	92.4	87.8	3.36	93.0	57.5
2001-08-21	12:28	400.1 (248.7)	92.6	87.3	3.58	92.8	57.3
2001-08-21	13:13	399.9 (248.5)	91.7	86.7	3.60	92.2	56.7
2001-08-21	13:58	400.0 (248.6)	93.4	88.8	2.94	93.5	57.9

B.4.2 Microphone at 25.0 m (82.0 ft) distance from track centerline

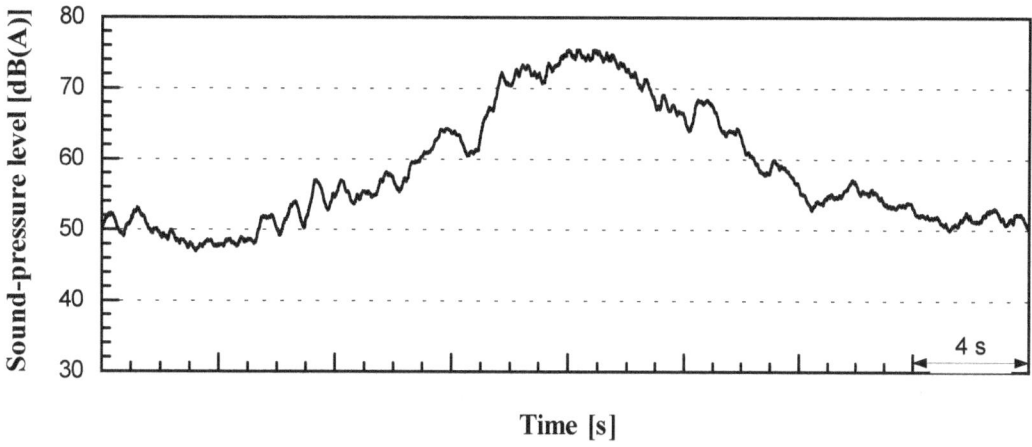

Figure B-65. Time history of the A-weighted SPL during a passby of the TR08 travelling on the prototype concrete guideway at about 100 km/h (62 mph) measured at 25.0 m (82.0 ft) distance from track centerline and 3.5 m (11.5 ft) above the ground.

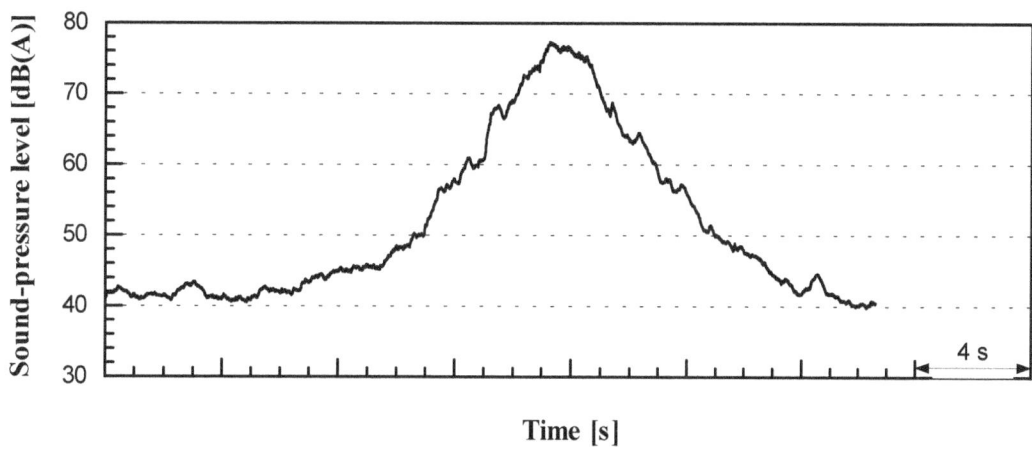

Figure B-66. Time history of the A-weighted SPL during a passby of the TR08 travelling on the prototype concrete guideway at about 150 km/h (93 mph) measured at 25.0 m (82.0 ft) distance from track centerline and 3.5 m (11.5 ft) above the ground.

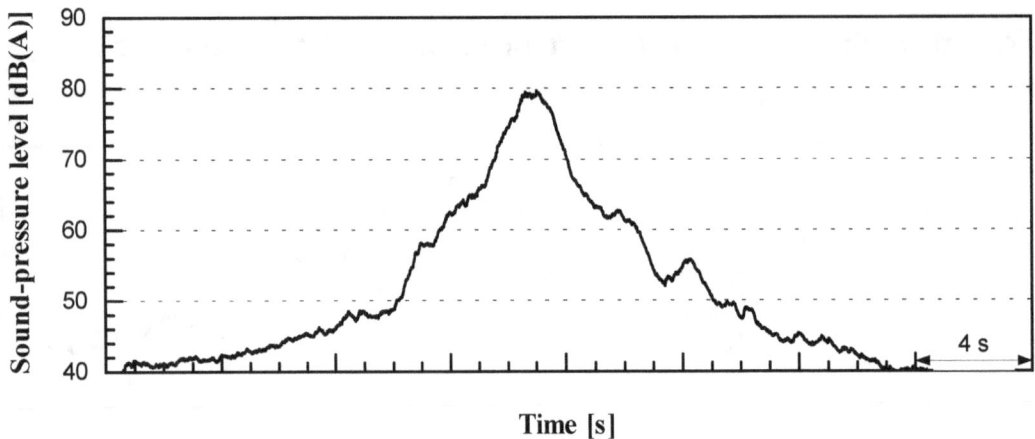

Figure B-67. Time history of the A-weighted SPL during a passby of the TR08 travelling on the prototype concrete guideway at about 200 km/h (124 mph) measured at 25.0 m (82.0 ft) distance from track centerline and 3.5 m (11.5 ft) above the ground.

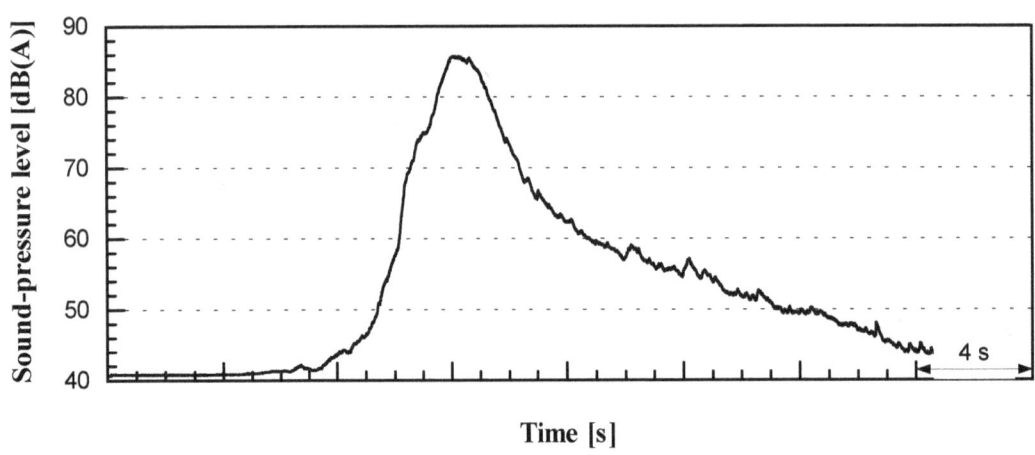

Figure B-68. Time history of the A-weighted SPL during a passby of the TR08 travelling on the prototype concrete guideway at about 300 km/h (186 mph) measured at 25.0 m (82.0 ft) distance from track centerline and 3.5 m (11.5 ft) above the ground.

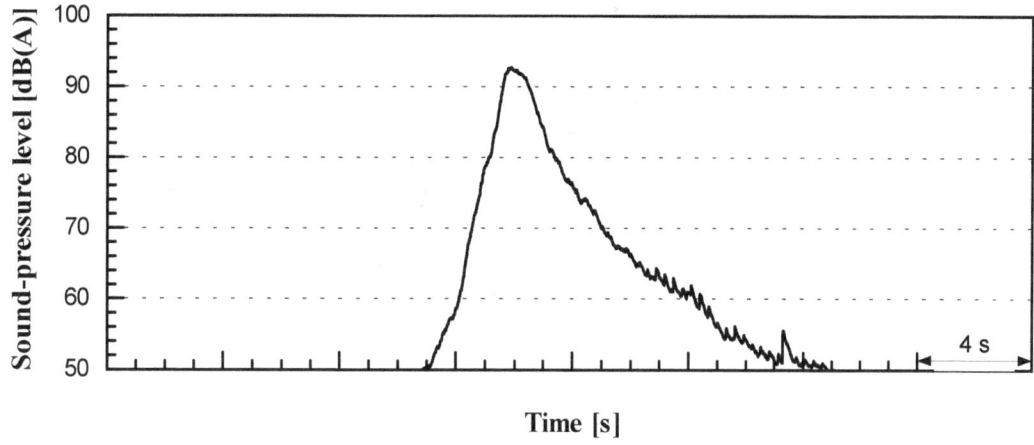

Figure B-69. Time history of the A-weighted SPL during a passby of the TR08 travelling on the prototype concrete guideway at about 400 km/h (249 mph) measured at 25.0 m (82.0 ft) distance from track centerline and 3.5 m (11.5 ft) above the ground.

Table B-14. Results of the microphone positioned close to the prototype concrete guideway at 25.0 m (82.0 ft) distance from track centerline and 3.5 m (11.5 ft) above the ground (measuring series L).

Date	Time	Vehicle speed [km/h (mph)]	$L_{Amax, fast}$ [dB(A)]	$L_{Aeq,E}$ [dB(A)]	t_E [s]	SEL [dB(A)]	$L_{Aeq,1h}$ [dB(A)]
2002-05-15	13:31	100.0 (62.1)	75.1	69.6	14.49	81.3	45.7
2002-05-15	13:36	99.9 (62.1)	78.9	72.7	11.60	83.3	47.8
2002-05-15	13:42	100.1 (62.2)	75.4	69.8	13.82	81.2	45.6
2002-05-15	13:47	100.1 (62.2)	76.4	70.6	12.81	81.7	46.1
2001-08-21	09:55	149.8 (93.1)	78.3	72.5	8.52	81.8	46.2
2001-08-21	10:35	149.6 (93.0)	77.9	72.0	9.07	81.5	46.0
2001-08-21	13:53	150.0 (93.2)	77.3	71.4	7.31	80.1	44.5
2001-08-21	14:10	149.7 (93.0) (N→S)	75.7	70.1	7.98	79.2	43.6
2001-08-21	11:15	199.6 (124.1)	80.7	74.1	5.11	81.2	45.6
2001-08-21	11:27	199.4 (123.9)	79.7	72.9	6.13	80.8	45.2
2001-08-21	12:23	199.5 (124.0)	80.5	73.9	5.65	81.4	45.9
2001-08-21	13:08	199.6 (124.1)	79.7	72.7	6.88	81.1	45.5
2001-08-21	14:03	199.6 (124.1)	80.1	72.9	6.04	80.7	45.2
2001-08-21	09:25	299.7 (186.3)	85.8	80.1	5.39	87.5	51.9
2001-08-21	10:00	299.3 (186.0)	85.8	80.0	5.49	87.4	51.8
2001-08-21	11:20	299.7 (186.3)	85.8	80.7	4.63	87.4	51.8
2001-08-21	10:41	399.7 (248.4)	93.3	87.6	3.94	93.5	58.0
2001-08-21	12:28	400.1 (248.7)	93.6	87.6	3.74	93.3	57.7
2001-08-21	13:13	399.9 (248.5)	92.7	86.9	3.97	92.9	57.3
2001-08-21	13:58	400.0 (248.6)	93.9	88.5	3.44	93.9	58.3

B.4.3 Microphone at 15.2 m (50.0 ft) distance from track centerline

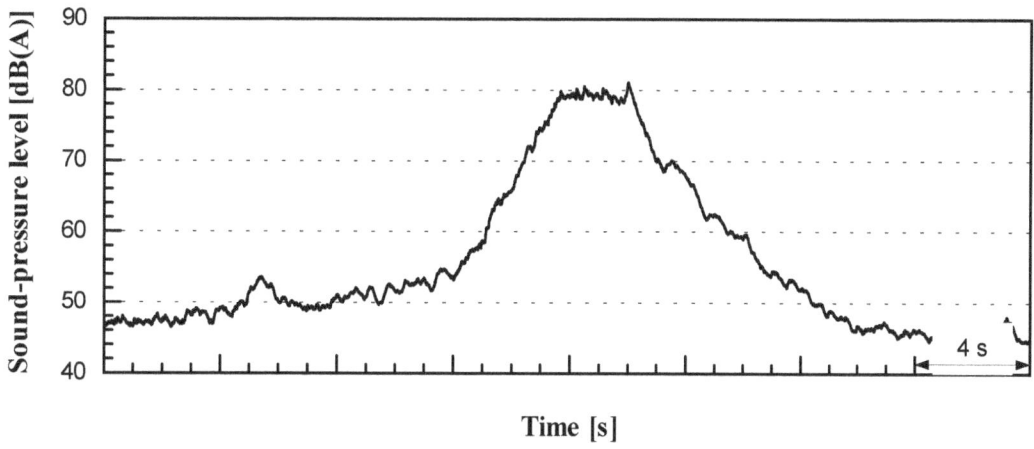

Figure B-70. Time history of the A-weighted SPL during a passby of the the TR08 travelling on the prototype concrete guideway at about 100 km/h (62 mph) measured at 15.2 m (50.0 ft) distance from track centerline and 1.5 m (5.0 ft) above the ground.

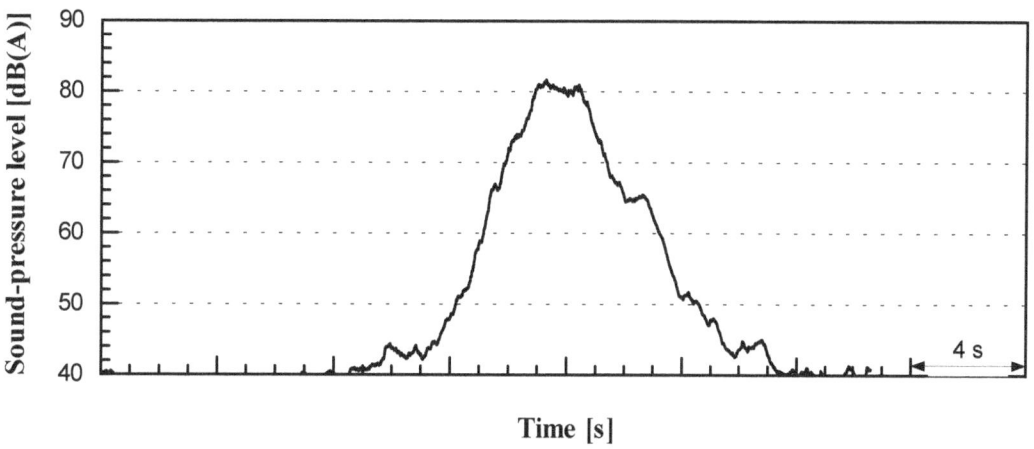

Figure B-71. Time history of the A-weighted SPL during a passby of the TR08 travelling on the prototype concrete guideway at about 150 km/h (93 mph) measured at 15.2 m (50.0 ft) distance from track centerline and 1.5 m (5.0 ft) above the ground.

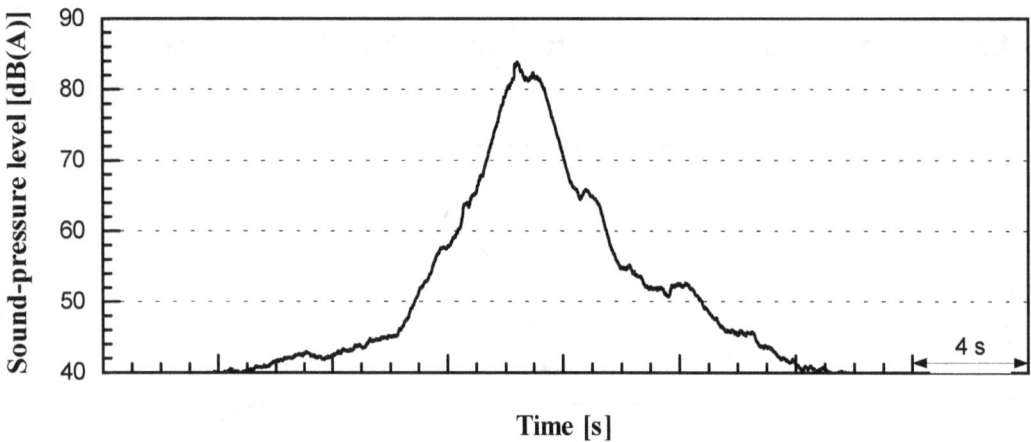

Figure B-72. Time history of the A-weighted SPL during a passby of the TR08 travelling on the prototype concrete guideway at about 200 km/h (124 mph) measured at 15.2 m (50.0 ft) distance from track centerline and 1.5 m (5.0 ft) above the ground.

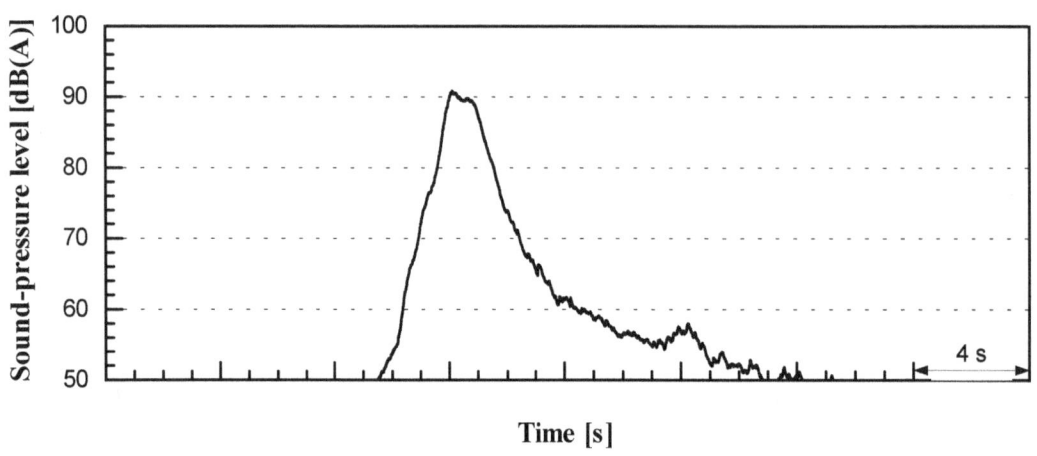

Figure B-73. Time history of the A-weighted SPL during a passby of the TR08 travelling on the prototype concrete guideway at about 300 km/h (186 mph) measured at 15.2 m (50.0 ft) distance from track centerline and 1.5 m (5.0 ft) above the ground.

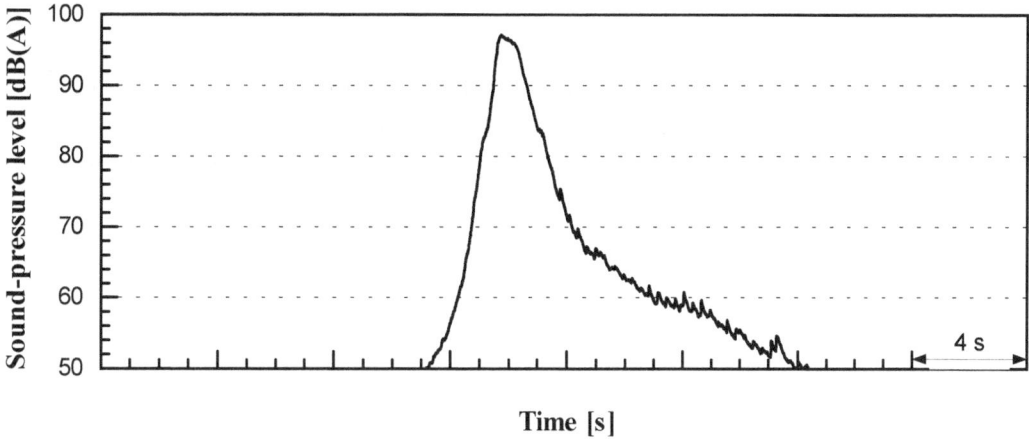

Figure B-74. Time history of the A-weighted SPL during a passby of the TR08 travelling on the prototype concrete guideway at about 400 km/h (249 mph) measured at 15.2 m (50.0 ft) distance from track centerline and 1.5 m (5.0 ft) above the ground.

Table B-15. Results of the microphone positioned close to the prototype concrete guideway at 15.2 m (50.0 ft) distance from track centerline and 1.5 m (5.0 ft) above the ground (measuring series L).

Date	Time	Vehicle speed [km/h (mph)]	$L_{Amax,\,fast}$ [dB(A)]	$L_{Aeq,E}$ [dB(A)]	t_E [s]	SEL [dB(A)]	$L_{Aeq,1h}$ [dB(A)]
2002-05-15	13:31	100.0 (62.1)	80.3	75.5	7.81	84.5	48.9
2002-05-15	13:36	99.9 (62.1)	82.0	77.5	7.78	86.4	50.9
2002-05-15	13:42	100.1 (62.2)	81.0	75.6	8.09	84.7	49.2
2002-05-15	13:47	100.1 (62.2)	81.5	76.5	8.09	85.5	50.0
2001-08-21	09:55	149.8 (93.1)	84.1	78.4	5.91	86.2	50.6
2001-08-21	10:35	149.6 (93.0)	84.5	79.5	5.51	86.7	51.3
2001-08-21	13:53	150.0 (93.2)	81.7	76.4	5.85	84.0	48.5
2001-08-21	14:10	149.7 (93.0) (N→S)	82.1	76.9	5.16	84.1	48.5
2001-08-21	11:15	199.6 (124.1)	85.1	79.9	3.70	85.6	50.0
2001-08-21	11:27	199.4 (123.9)	84.6	79.1	3.79	84.9	49.3
2001-08-21	12:23	199.5 (124.0)	83.7	77.7	4.71	84.5	48.9
2001-08-21	13:08	199.6 (124.1)	83.8	77.7	4.43	84.2	48.6
2001-08-21	14:03	199.6 (124.1)	83.3	77.6	4.34	84.0	48.4
2001-08-21	09:25	299.7 (186.3)	90.7	85.7	3.47	91.1	55.6
2001-08-21	10:00	299.3 (186.0)	90.4	85.5	3.44	90.9	55.3
2001-08-21	11:20	299.7 (186.3)	90.7	85.7	3.38	91.0	55.5
2001-08-21	10:41	399.7 (248.4)	97.5	92.8	2.58	96.9	61.3
2001-08-21	12:28	400.1 (248.7)	97.3	92.4	2.59	96.6	61.0
2001-08-21	13:13	399.9 (248.5)	97.1	92.2	2.62	96.4	60.8
2001-08-21	13:58	400.0 (248.6)	97.9	93.1	2.45	97.0	61.4

B.4.4 Microphone at 6.5 m (21.3 ft) distance from track centerline (high position)

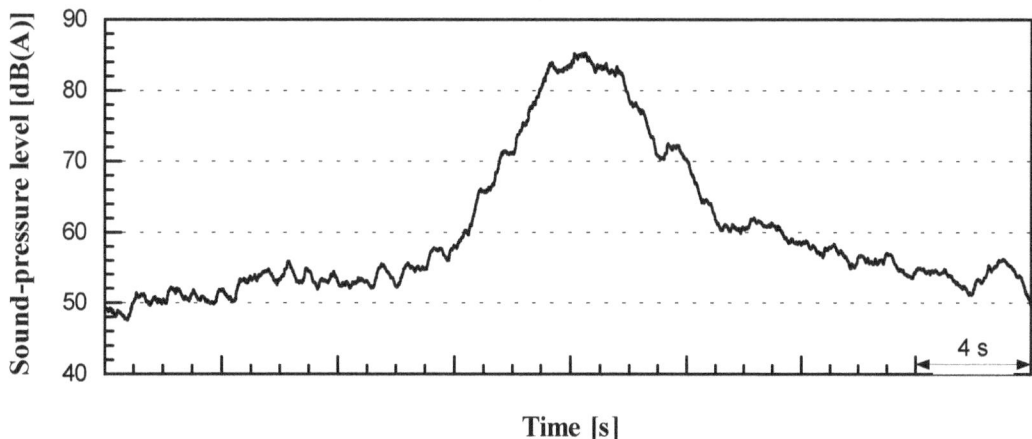

Figure B-75. Time history of the A-weighted SPL during a passby of the TR08 travelling on the prototype concrete guideway at about 100 km/h (62 mph) measured at 6.5 m (21.3 ft) distanace from track centerline and the height of the upper surface of the guideway.

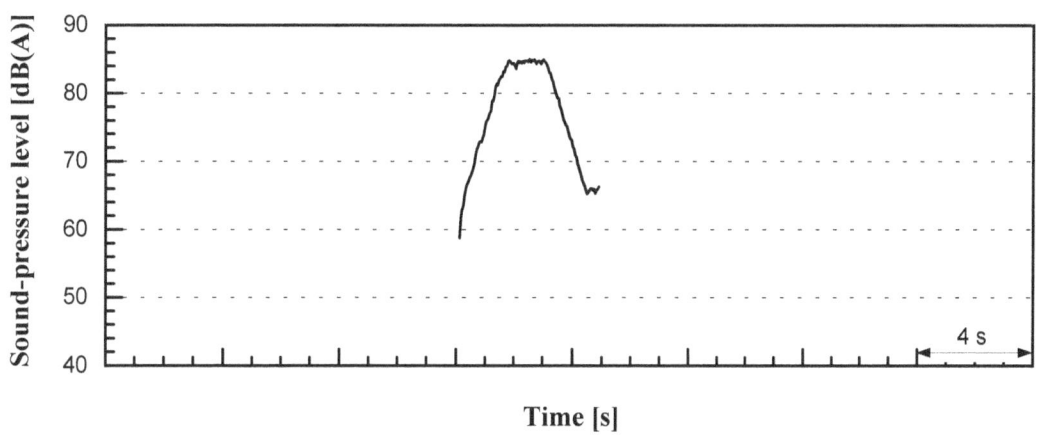

Figure B-76. Time history of the A-weighted SPL during a passby of the TR08 travelling on the prototype concrete guideway at about 150 km/h (93 mph) measured at 6.5 m (21.3 ft) distance from track centerline and the height of the upper surface of the guideway.

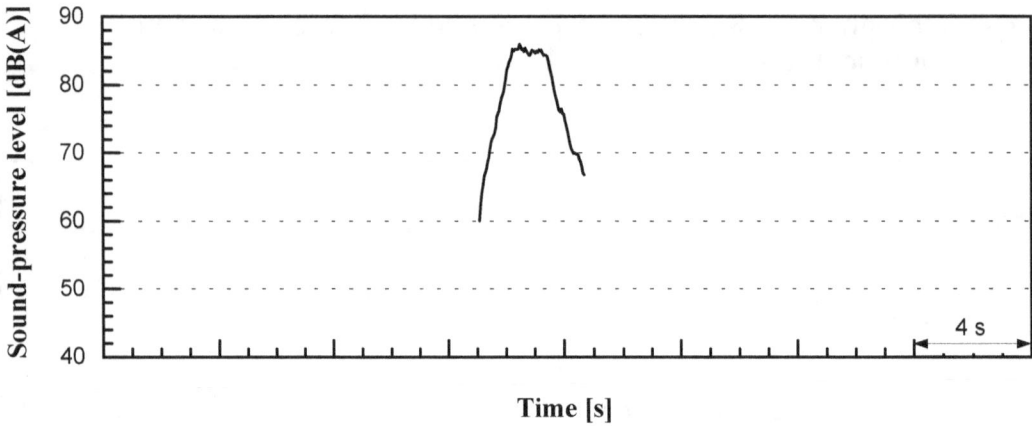

Figure B-77. Time history of the A-weighted SPL during a passby of the TR08 travelling on the prototype concrete guideway at about 200 km/h (124 mph) measured at 6.5 m (21.3 ft) distance from track centerline and the height of the upper surface of the guideway.

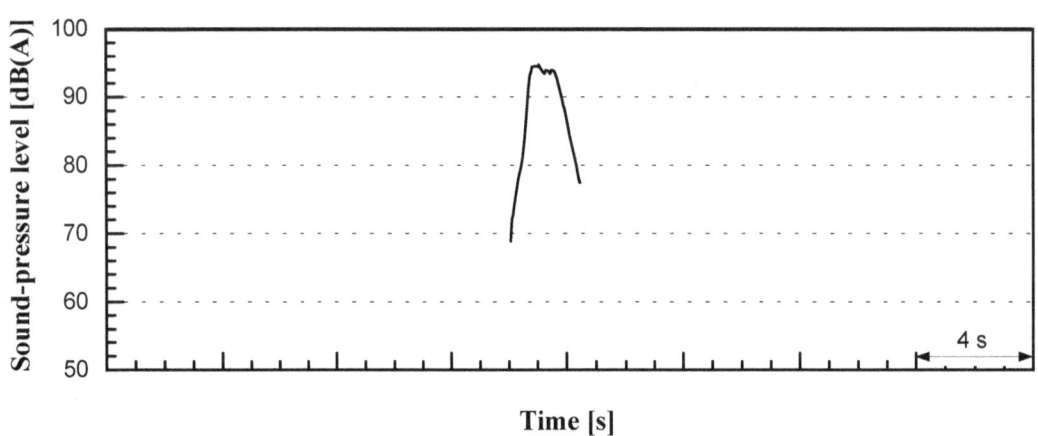

Figure B-78. Time history of the A-weighted SPL during a passby of the TR08 travelling on the prototype concrete guideway at about 300 km/h (186 mph) measured at 6.5 m (21.3 ft) distance from track centerline and the height of the upper surface of the guideway.

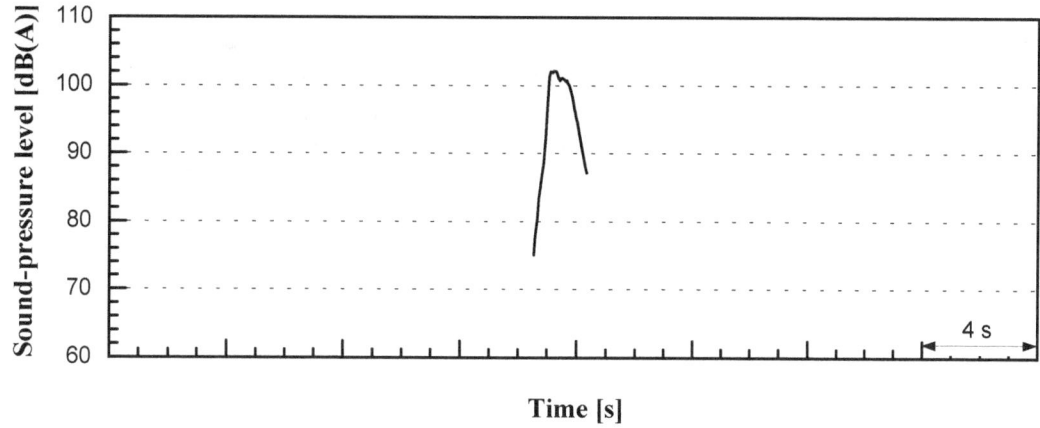

Figure B-79. Time history of the A-weighted SPL during a passby of the TR08 travelling on the prototype concrete guideway at about 400 km/h (249 mph) measured at 6.5 m (21.3 ft) distance from track centerline and the height of the upper surface of the guideway.

Table B-16. Results of the microphone positioned close to the prototype concrete guideway at 6.5 m (21.3 ft) distance from track centerline and the height of the upper surface of the guideway (measuring series K/L).

Date	Time	Vehicle speed [km/h (mph)]	$L_{Amax, fast}$ [dB(A)]	$L_{Aeq,E}$ [dB(A)]	t_E [s]	SEL [dB(A)]	$L_{Aeq,1h}$ [dB(A)]
2002-05-15	13:31	100.0 (62.1)	85.8	82.7	4.02	88.8	53.2
2002-05-15	13:36	99.9 (62.1)	87.1	83.8	4.28	90.1	54.6
2002-05-15	13:42	100.1 (62.2)	85.3	82.3	4.13	88.5	52.9
2002-05-15	13:47	100.1 (62.2)	86.4	83.3	3.94	89.2	53.7
2001-08-23	10:35	149.5 (92.9)	88.6	86.1	2.80	90.5	55.0
2001-08-23	11:20	149.9 (93.2)	85.0	82.9	2.96	87.6	52.0
2001-08-23	12:17	149.7 (93.0)	84.6	82.1	2.88	86.7	51.1
2001-08-23	10:53	199.2 (123.8)	86.8	84.8	2.20	88.2	52.7
2001-08-23	11:31	199.5 (124.0)	86.5	84.2	2.20	87.7	52.1
2001-08-23	14:09	199.7 (124.1)	85.9	83.4	2.28	87.0	51.4
2001-08-23	10:40	299.3 (186.0)	94.8	92.5	1.68	94.8	59.2
2001-08-23	12:22	299.8 (186.3)	94.7	92.5	1.64	94.6	59.1
2001-08-23	13:23	299.4 (186.1)	94.6	92.5	1.64	94.6	59.0
2001-08-23	11:25	399.0 (248.0)	102.1	99.7	1.28	100.8	65.2
2001-08-23	11:36	399.4 (248.2)	102.3	100.0	1.28	101.0	65.5
2001-08-23	13:10	399.6 (248.4)	102.2	99.5	1.28	100.6	65.1

B.4.5 Microphone at 6.5 m (21.3 ft) distance from track centerline (low position)

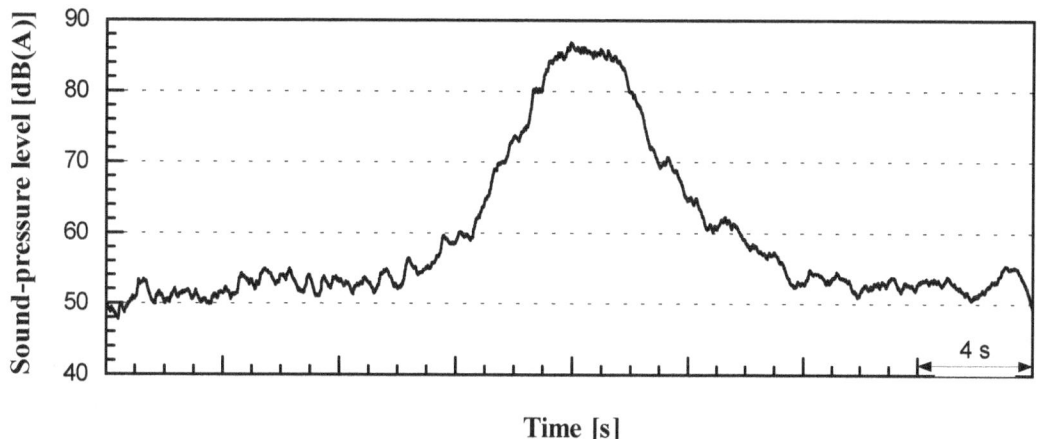

Figure B-80. Time history of the A-weighted SPL during a passby of the TR08 travelling on the prototype concrete guideway at about 100 km/h (62 mph) measured at 6.5 m (21.3 ft) distance from track centerline and 1.5 m (5.0 ft) below the upper surface of the guideway.

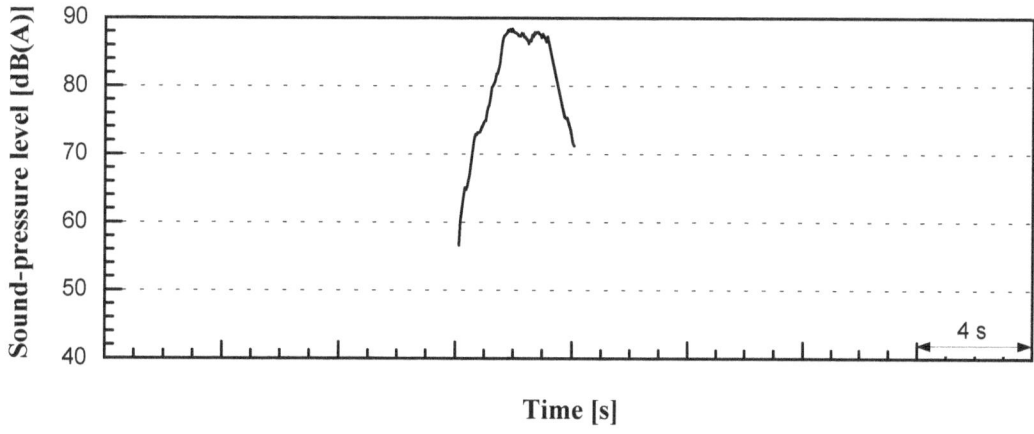

Figure B-81. Time history of the A-weighted SPL during a passby of the TR08 travelling on the prototype concrete guideway at about 150 km/h (93 mph) measured at 6.5 m (21.3 ft) distance from track centerline and 1.5 m (5.0 ft) below the upper surface of the guideway.

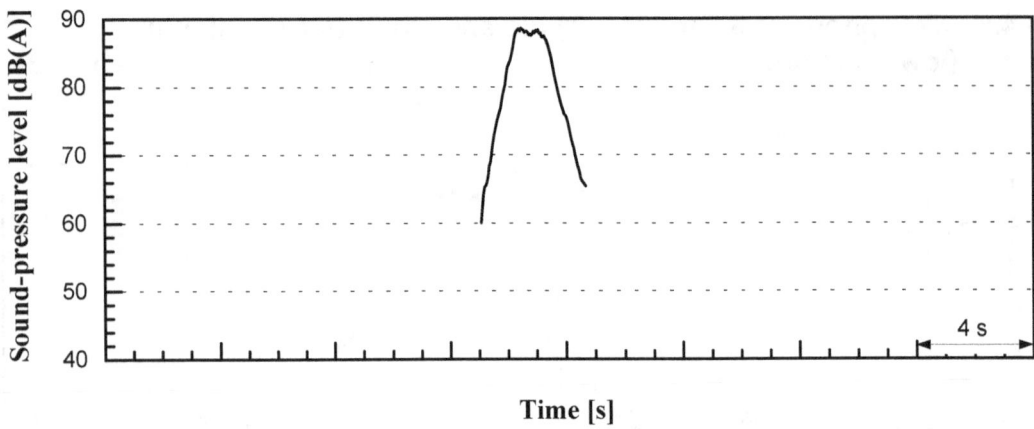

Figure B-82. Time history of the A-weighted SPL during a passby of the TR08 travelling on the prototype concrete guideway at about 200 km/h (124 mph) measured at 6.5 m (21.3 ft) distance from track centerline and 1.5 m (5.0 ft) below the upper surface of the guideway.

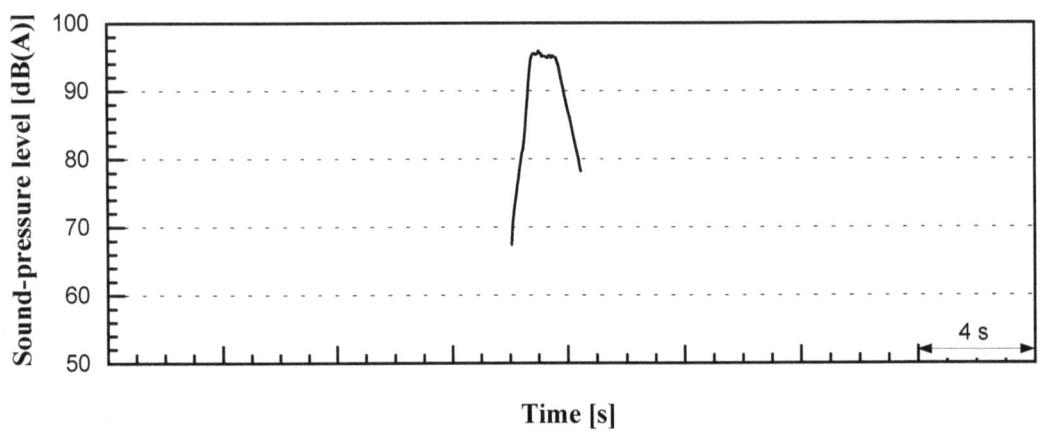

Figure B-83. Time history of the A-weighted SPL during a passby of the TR08 travelling on the prototype concrete guideway at about 300 km/h (186 mph) measured at 6.5 m (21.3 ft) distance from track centerline and 1.5 m (5.0 ft) below the upper surface of the guideway.

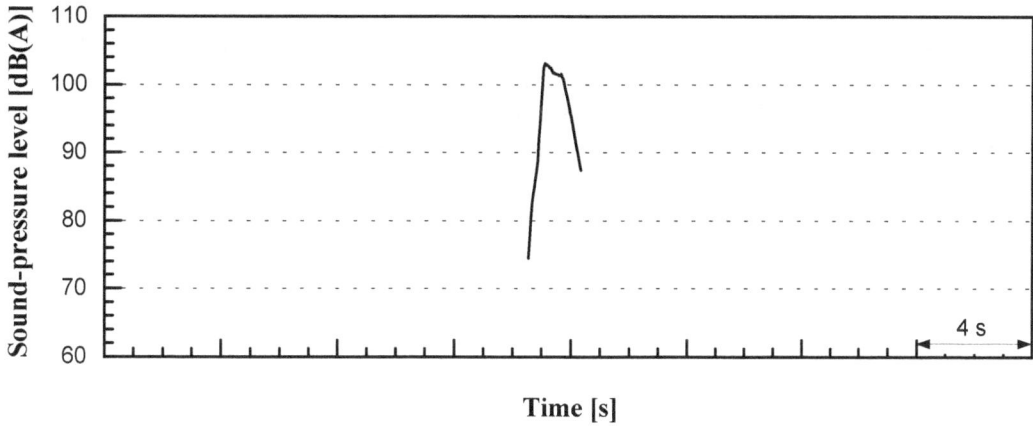

Figure B-84. Time history of the A-weighted SPL during a passby of the TR08 travelling on the prototype concrete guideway at about 400 km/h (249 mph) measured at 6.5 m (21.3 ft) distance from track centerline and 1.5 m (5.0 ft) below the upper surface of the guideway.

Table B-17. Results of the microphone positioned close to the prototype concrete guideway at 6.5 m (21.3 ft) distance from track centerline and 1.5 m (5.0 ft) below the upper surface of the guideway (measuring series L).

Date	Time	Vehicle speed [km/h (mph)]	$L_{Amax,\,fast}$ [dB(A)]	$L_{Aeq,E}$ [dB(A)]	t_E [s]	SEL [dB(A)]	$L_{Aeq,1h}$ [dB(A)]
2002-05-15	13:31	100.0 (62.1)	86.7	84.2	4.03	90.3	54.7
2002-05-15	13:36	99.9 (62.1)	90.5	86.3	3.72	92.0	56.5
2002-05-15	13:42	100.1 (62.2)	86.9	84.1	3.84	90.0	54.4
2002-05-15	13:47	100.1 (62.2)	87.2	84.6	4.03	90.7	55.1
2001-08-24	8:43	149.8 (93.1)	90.7	88.7	2.52	92.7	57.2
2001-08-24	9:26	149.8 (93.1)	88.3	86.2	2.48	90.2	54.6
2001-08-24	08:54	175.6 (109.1)	85.7	83.2	2.64	87.4	51.9
2001-08-24	9:00	199.6 (124.1)	89.5	86.8	2.08	90.0	54.4
2001-08-24	11:30	199.6 (124.1)	88.8	86.6	2.00	89.6	54.1
2001-08-24	13:37	200.0 (124.3)	88.5	86.3	2.04	89.4	53.9
2001-08-24	08:47	299.4 (186.1)	95.8	93.7	1.60	95.7	60.2
2001-08-24	10:30	299.5 (186.1)	95.6	93.5	1.64	95.6	60.1
2001-08-24	09:31	399.1 (248.0)	102.8	100.4	1.32	101.6	66.1
2001-08-24	11:35	399.5 (248.3)	103.1	100.4	1.28	101.5	65.9
2001-08-24	13:41	385.4 (239.5)	102.7	100.2	1.28	101.3	65.7

B.4.6 Microphone beneath guideway centerline

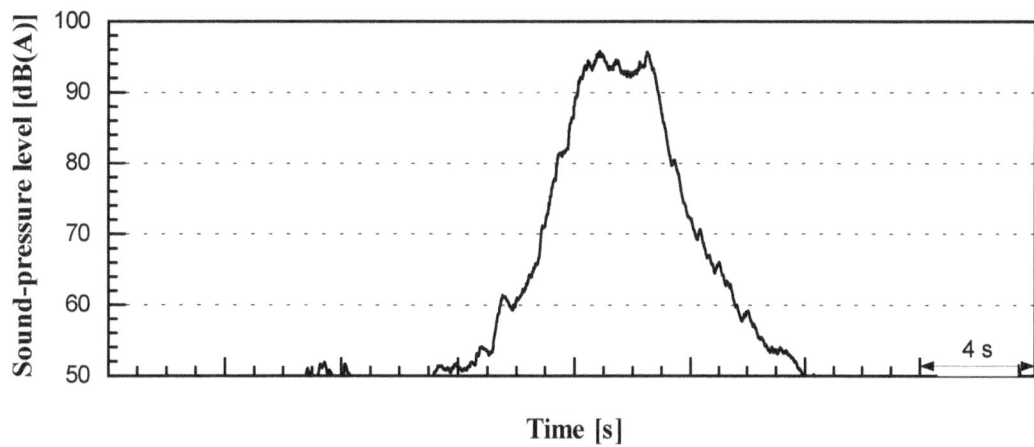

Figure B-85. Time history of the A-weighted SPL during a passby of the TR08 travelling on the prototype concrete guideway at about 100 km/h (62 mph) measured beneath the guideway centerline at a height of 1.5 m (5.0 ft) above the ground.

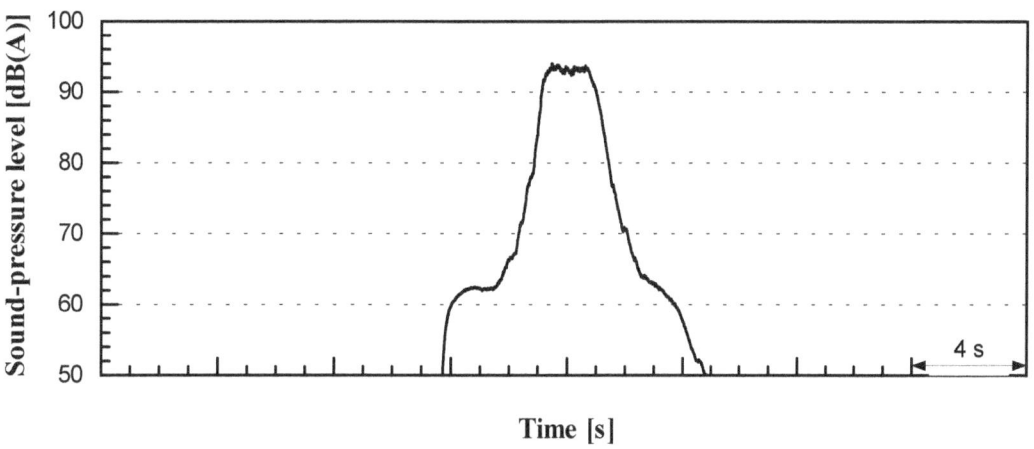

Figure B-86. Time history of the A-weighted SPL during a passby of the TR08 travelling on the prototype concrete guideway at about 150 km/h (93 mph) measured beneath guideway centerline at a height of 1.5 m (5.0 ft) above the ground.

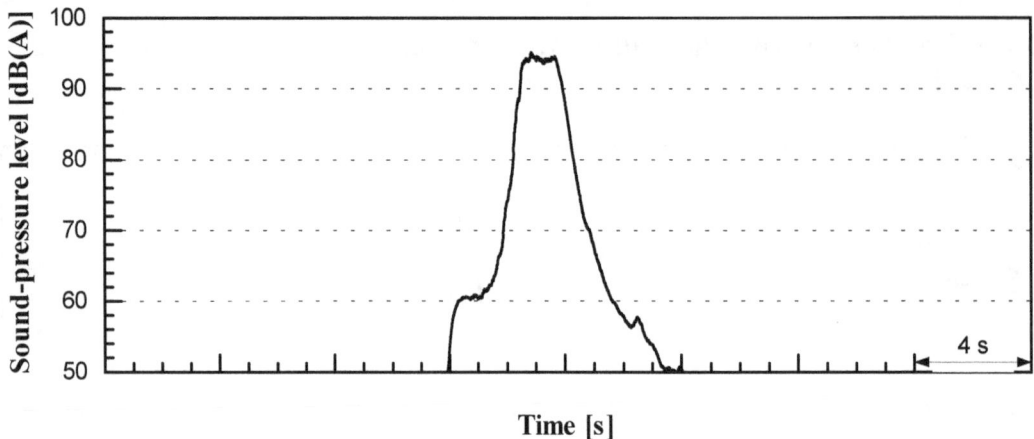

Figure B-87. Time history of the A-weighted SPL during a passby of the TR08 travelling on the prototype concrete guideway at about 200 km/h (124 mph) measured beneath guideway centerline at a height of 1.5 m (5.0 ft) above the ground.

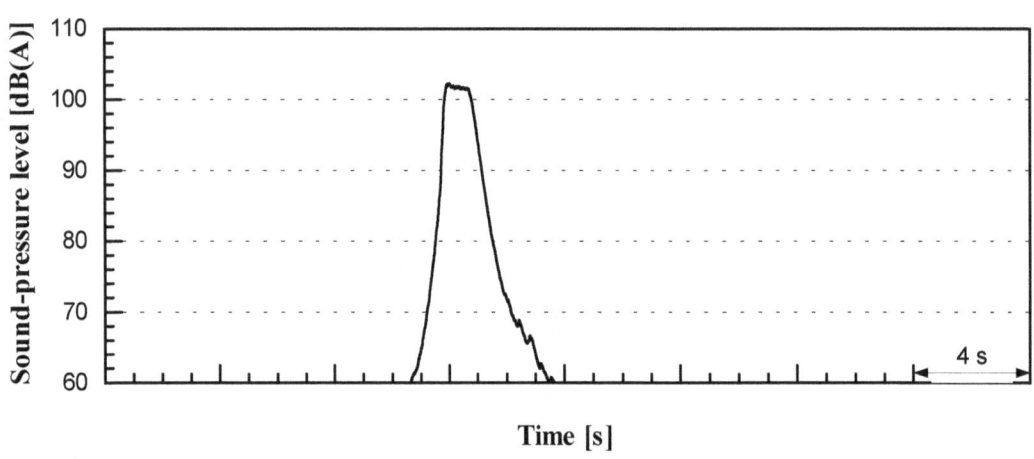

Figure B-88. Time history of the A-weighted SPL during a passby of the TR08 travelling on the prototype concrete guideway at about 300 km/h (186 mph) measured beneath guideway centerline at a height of 1.5 m (5.0 ft) above the ground.

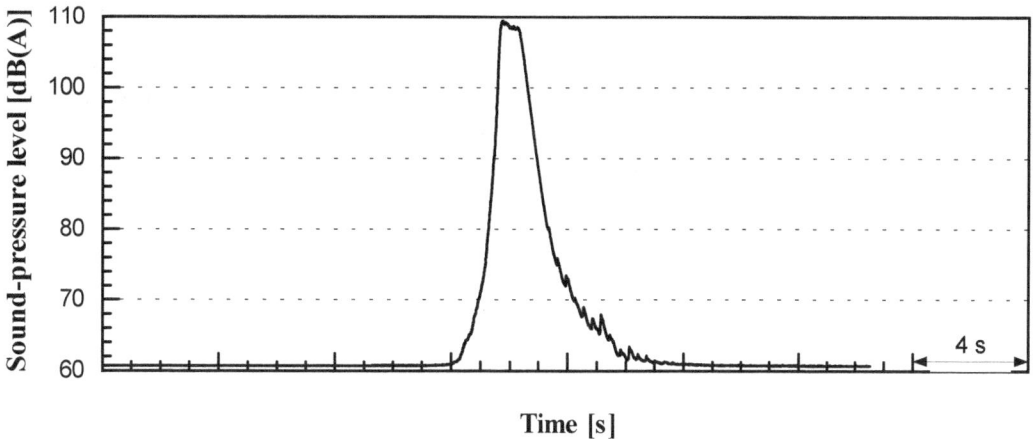

Figure B-89. Time history of the A-weighted SPL during a passby of the TR08 travelling on the prototype concrete guideway at about 400 km/h (249 mph) measured beneath guideway centerline at a height of 1.5 m (5.0 ft) above the ground.

Table B-18. Results of the microphone positioned beneath the centerline of the prototype concrete guideway at a height of 1.5 m (5.0 ft) above the ground (measuring series L).

Date	Time	Vehicle speed [km/h (mph)]	$L_{Amax, fast}$ [dB(A)]	$L_{Aeq,E}$ [dB(A)]	t_E [s]	SEL [dB(A)]	$L_{Aeq,1h}$ [dB(A)]
2002-05-15	13:31	100.0 (62.1)	95.1	91.8	4.29	98.1	62.6
2002-05-15	13:36	99.9 (62.1)	95.3	90.9	4.87	97.8	62.2
2002-05-15	13:42	100.1 (62.2)	95.8	91.8	4.48	98.3	62.8
2002-05-15	13:47	100.1 (62.2)	95.2	91.0	4.48	97.5	62.0
2001-08-21	09:55	149.8 (93.1)	95.5	92.4	3.26	97.5	62.0
2001-08-21	10:35	149.6 (93.0)	95.5	92.4	3.23	97.5	61.9
2001-08-21	13:53	150.0 (93.2)	94.0	90.9	3.17	95.9	60.4
2001-08-21	14:10	149.7 (93.0) (N→S)	93.0	89.5	3.39	94.8	59.2
2001-08-21	11:15	199.6 (124.1)	96.6	93.3	2.34	97.0	61.4
2001-08-21	11:27	199.4 (123.9)	95.8	92.7	2.40	96.5	60.9
2001-08-21	12:23	199.5 (124.0)	96.0	92.5	2.40	96.3	60.7
2001-08-21	13:08	199.6 (124.1)	95.0	91.9	2.46	95.8	60.2
2001-08-21	14:03	199.6 (124.1)	95.7	92.1	2.43	96.0	60.4
2001-08-21	09:25	299.7 (186.3)	102.3	99.2	1.91	102.0	66.4
2001-08-21	10:00	299.3 (186.0)	102.2	99.3	1.89	102.1	66.5
2001-08-21	11:20	299.7 (186.3)	102.1	99.2	1.86	101.9	66.9
2001-08-21	10:41	399.7 (248.4)	109.3	106.1	1.51	107.9	72.3
2001-08-21	12:28	400.1 (248.7)	109.5	106.1	1.51	107.9	72.3
2001-08-21	13:13	399.9 (248.5)	109.4	106.1	1.54	107.9	72.4
2001-08-21	13:58	400.0 (248.6)	109.6	106.2	1.54	108.1	72.5

B.5 Hybrid Guideway

B.5.1 Microphone at 30.5 m (100.0 ft) distance from track centerline

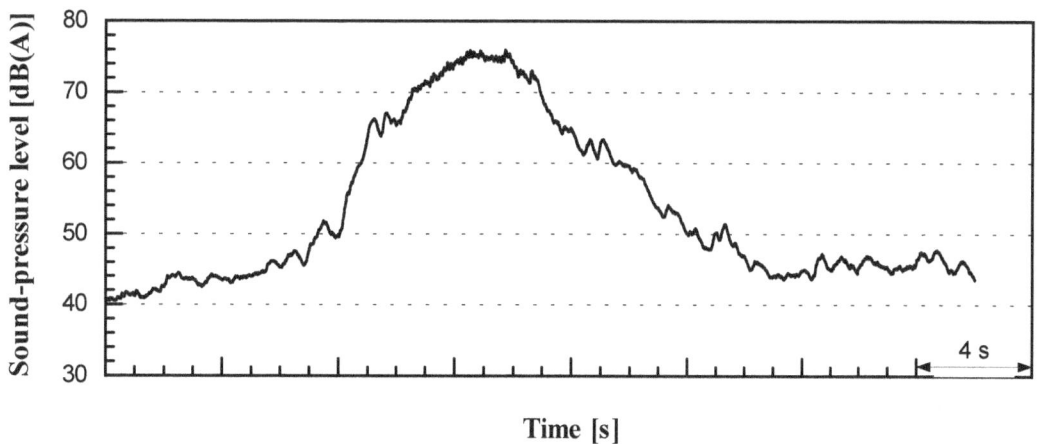

Figure B-90. Time history of the A-weighted SPL during a passby of the TR08 travelling on the hybrid guideway at about 100 km/h (62 mph) measured at 30.5 m (100.0 ft) distance from track centerline and 1.2 m (4.0 ft) above the ground.

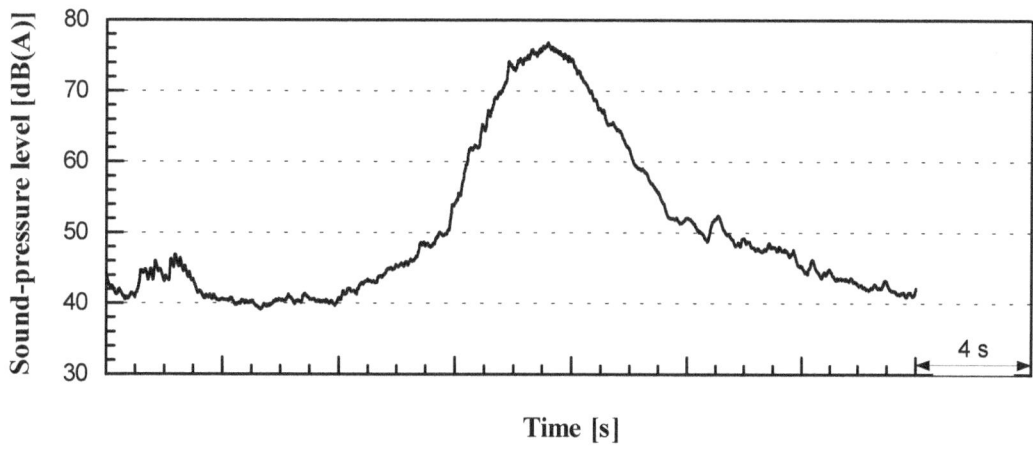

Figure B-91. Time history of the A-weighted SPL during a passby of the TR08 travelling on the hybrid guideway at about 150 km/h (93 mph) measured at 30.5 m (100.0 ft) distance from track centerline and 1.2 m (4.0 ft) above the ground.

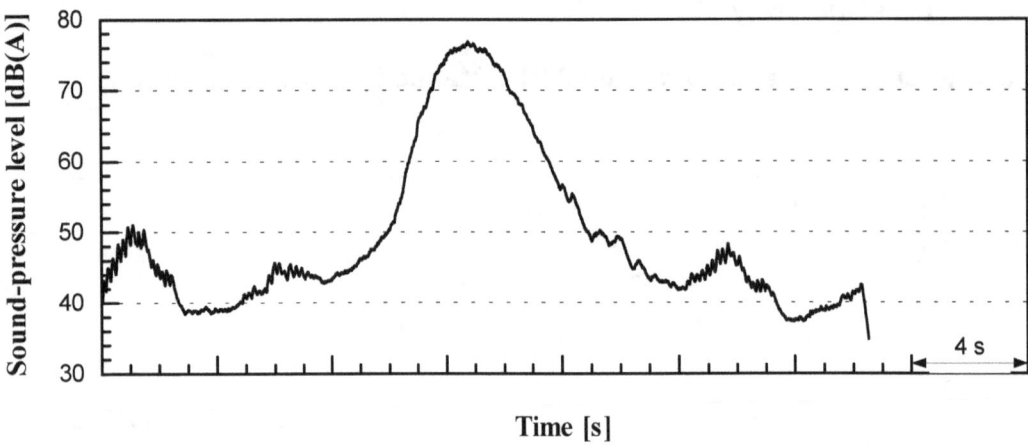

Figure B-92. Time history of the A-weighted SPL during a passby of the TR08 travelling on the hybrid guideway at about 200 km/h (124 mph) measured at 30.5 m (100.0 ft) distance from track centerline and 1.2 m (4.0 ft) above the ground.

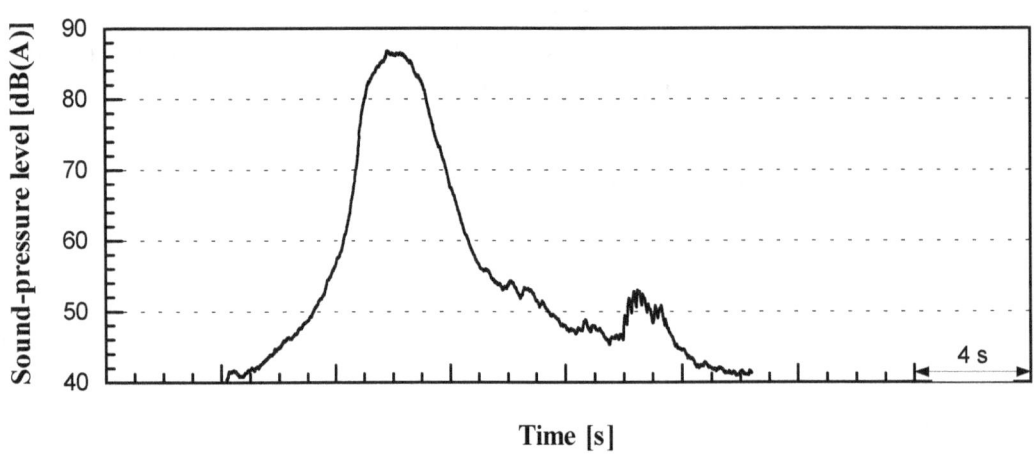

Figure B-93. Time history of the A-weighted SPL during a passby of the TR08 travelling on the hybrid guideway at about 300 km/h (186 mph) measured at 30.5 m (100.0 ft) distance from track centerline and 1.2 m (4.0 ft) above the ground.

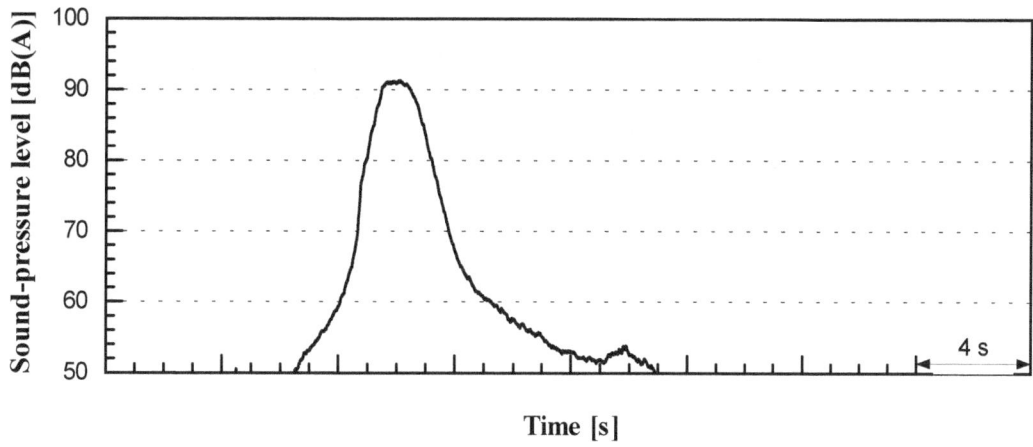

Figure B-94. Time history of the A-weighted SPL during a passby of the TR08 travelling on the hybrid guideway at about 400 km/h (249 mph) measured at 30.5 m (100.0 ft) distance from track centerline and 1.2 m (4.0 ft) above the ground.

Table B-19. Results of the microphone positioned close to the hybrid guideway at 30.5 m (100.0 ft) distance from track centerline and 1.2 m (4.0 ft) above the ground (measuring series J).

Date	Time	Vehicle speed [km/h (mph)]	$L_{Amax,\,fast}$ [dB(A)]	$L_{Aeq,E}$ [dB(A)]	t_E [s]	SEL [dB(A)]	$L_{Aeq,1h}$ [dB(A)]
2002-05-16	09:40	100.5 (62.5)	76.5	71.5	9.73	81.4	45.8
2002-05-16	10:08	100.1 (62.2)	76.0	70.5	10.30	80.7	45.1
2002-05-16	11:14	100.6 (62.5)	76.1	71.0	9.60	80.9	45.3
2002-05-16	09:57	150.1 (93.3)	77.3	71.7	7.20	80.3	44.7
2002-05-16	10:50	150.4 (93.5)	76.8	71.6	6.55	79.8	44.2
2002-05-16	11:10	150.4 (93.5)	77.1	72.1	5.87	79.8	44.2
2002-05-16	10:02	200.3 (124.5)	77.1	72.3	6.07	80.2	44.6
2002-05-16	10:59	200.3 (124.5)	76.8	72.3	5.36	79.6	44.0
2002-05-16	11:49	200.2 (124.4)	77.0	71.7	5.83	79.4	43.8
2002-05-16	10:18	300.3 (186.6)	86.2	81.9	3.86	87.8	52.2
2002-05-16	11:04	300.3 (186.6)	86.7	82.9	3.48	88.4	52.8
2002-05-16	11:53	300.2 (186.6)	86.6	82.8	3.57	88.3	52.8
2002-05-16	09:24	393.6 (244.6)	92.3	87.9	3.25	93.0	57.4
2002-05-16	10:53	392.0 (243.6)	91.3	87.7	3.06	92.5	57.0
2002-05-16	11:43	392.4 (243.9)	91.7	87.7	3.34	92.9	57.3

B.5.2 Microphone at 25.0 m (82.0 ft) distance from track centerline

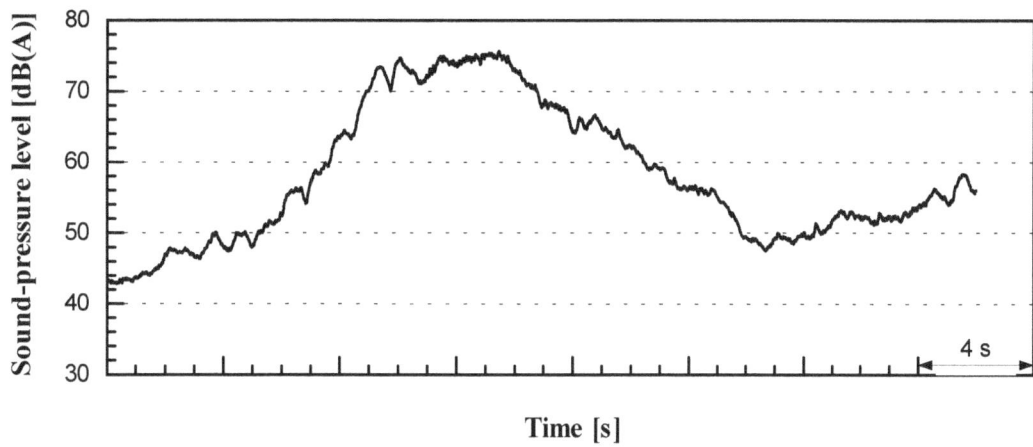

Figure B-95. Time history of the A-weighted SPL during a passby of the TR08 travelling on the hybrid guideway at about 100 km/h (62 mph) measured at 25.0 m (82.0 ft) distance from track centerline and 3.5 m (11.5 ft) above the ground.

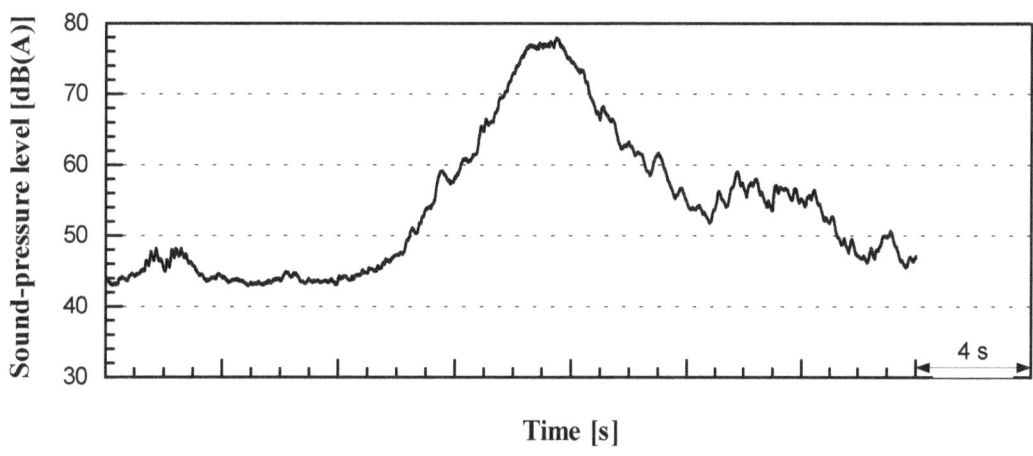

Figure B-96. Time history of the A-weighted SPL during a passby of the TR08 travelling on the hybrid guideway at about 150 km/h (93 mph) measured at 25.0 m (82.0 ft) distance from track centerline and 3.5 m (11.5 ft) above the ground.

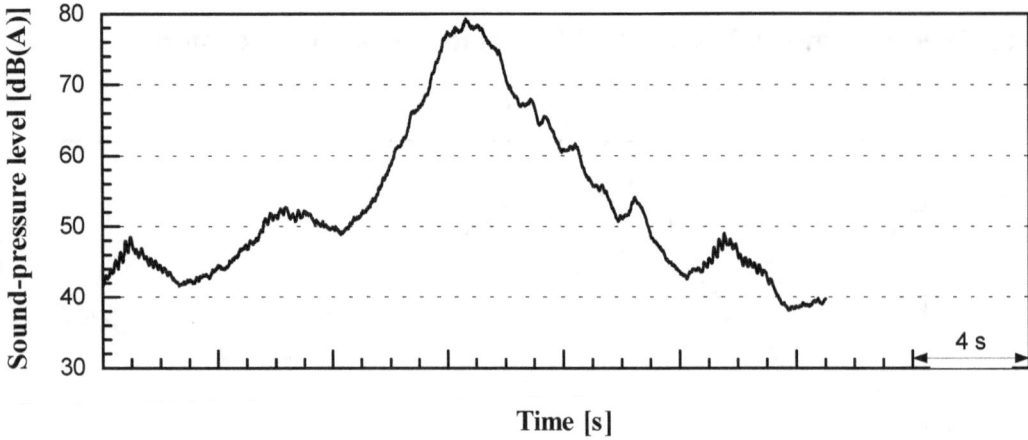

Figure B-97. Time history of the A-weighted SPL during a passby of the TR08 travelling on the hybrid guideway at about 200 km/h (124 mph) measured at 25.0 m (82.0 ft) distance from track centerline and 3.5 m (11.5 ft) above the ground.

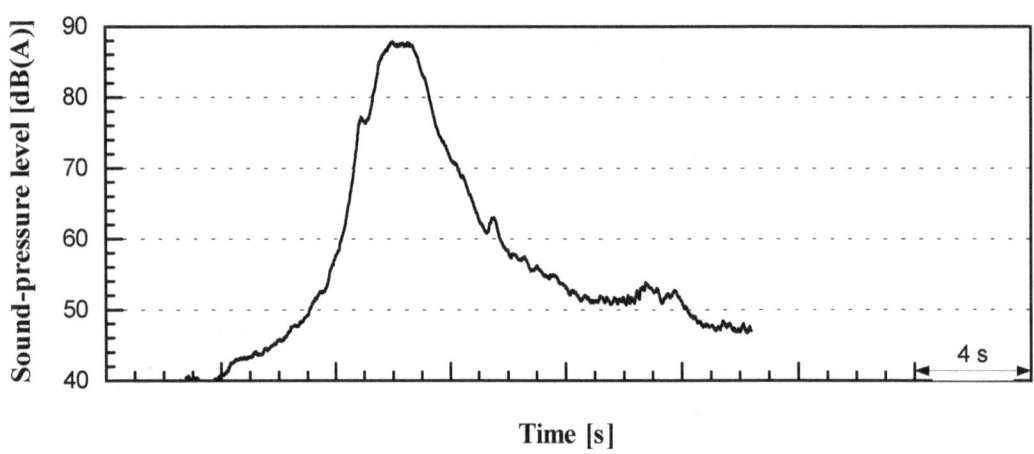

Figure B-98. Time history of the A-weighted SPL during a passby of the TR08 travelling on the hybrid guideway at about 300 km/h (186 mph) measured at 25.0 m (82.0 ft) distance from track centerline and 3.5 m (11.5 ft) above the ground.

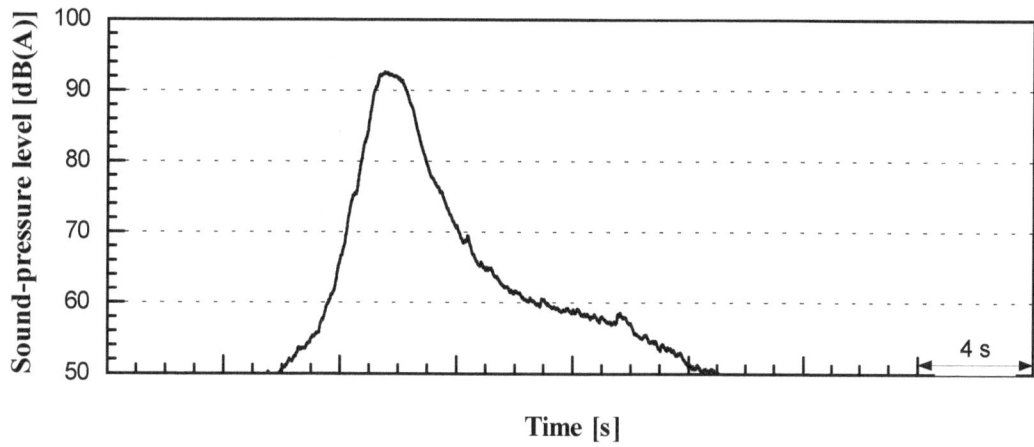

Figure B-99. Time history of the A-weighted SPL during a passby of the TR08 travelling on the hybrid guideway at about 400 km/h (249 mph) measured at 25.0 m (82.0 ft) distance from track centerline and 3.5 m (11.5 ft) above the ground.

Table B-20. Results of the microphone positioned close to the hybrid guideway at 25.0 m (82.0 ft) distance from track centerline and 3.5 m (11.5 ft) above the ground (measuring series J).

Date	Time	Vehicle speed [km/h (mph)]	$L_{Amax, fast}$ [dB(A)]	$L_{Aeq,E}$ [dB(A)]	t_E [s]	SEL [dB(A)]	$L_{Aeq,1h}$ [dB(A)]
2002-05-16	09:40	100.5 (62.5)	77.3	71.8	11.95	82.6	47.0
2002-05-16	10:08	100.1 (62.2)	75.6	70.5	13.39	81.7	46.2
2002-05-16	11:14	100.6 (62.5)	78.1	71.8	11.26	82.3	46.8
2002-05-16	09:57	150.1 (93.3)	77.4	70.9	9.24	80.5	45.0
2002-05-16	10:50	150.4 (93.5)	78.0	71.9	7.36	80.6	45.1
2002-05-16	11:10	150.4 (93.5)	77.8	71.9	7.63	80.8	45.2
2002-05-16	10:02	200.3 (124.5)	79.6	74.0	6.34	82.0	46.5
2002-05-16	10:59	200.3 (124.5)	79.2	73.1	6.56	81.3	45.7
2002-05-16	11:49	200.2 (124.4)	78.5	72.6	6.86	80.9	45.4
2002-05-16	10:18	300.3 (186.6)	86.9	81.9	4.33	88.2	52.7
2002-05-16	11:04	300.3 (186.6)	87.8	83.3	3.97	89.3	53.7
2002-05-16	11:53	300.2 (186.6)	87.4	82.5	4.46	89.0	53.4
2002-05-16	09:24	393.6 (244.6)	93.3	88.4	3.50	93.8	58.3
2002-05-16	10:53	392.0 (243.6)	92.6	87.9	3.44	93.3	57.7
2002-05-16	11:43	392.4 (243.9)	93.0	88.1	3.75	93.9	58.3

B.5.3 Microphone at 15.2 m (50.0 ft) distance from track centerline

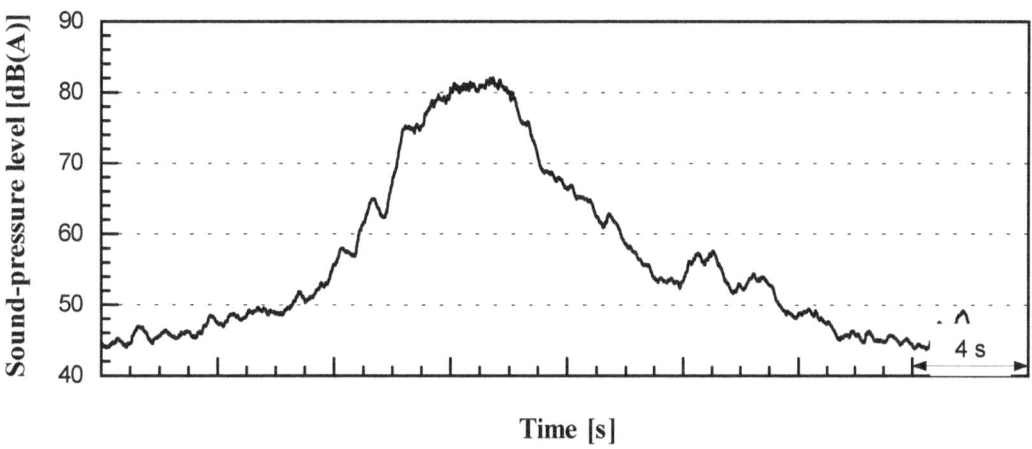

Figure B-100. Time history of the A-weighted SPL during a passby of the TR08 travelling on the hybrid guideway at about 100 km/h (62 mph) measured at 15.2 m (50.0 ft) distance from track centerline and 1.5 m (5.0 ft) above the ground.

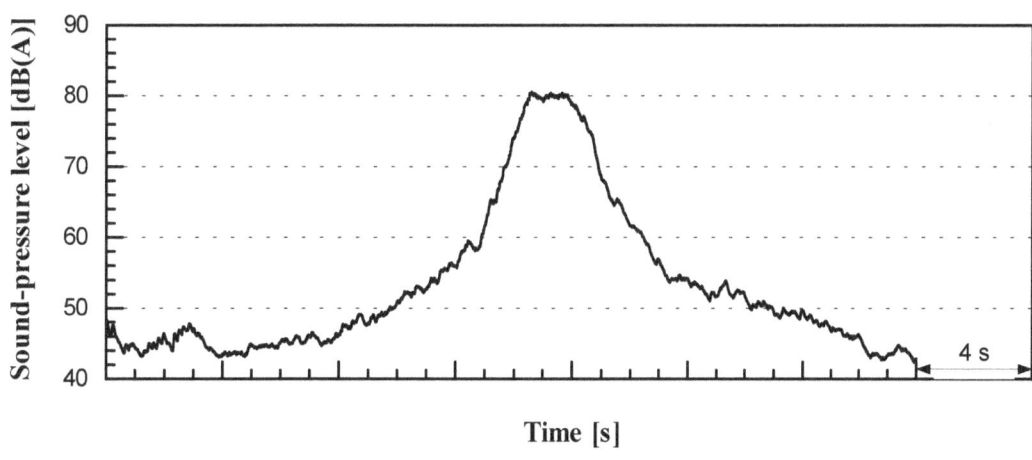

Figure B-101. Time history of the A-weighted SPL during a passby of the TR08 travelling on the hybrid guideway at about 150 km/h (93 mph) measured at 15.2 m (50.0 ft) distance from track centerline and 1.5 m (5.0 ft) above the ground.

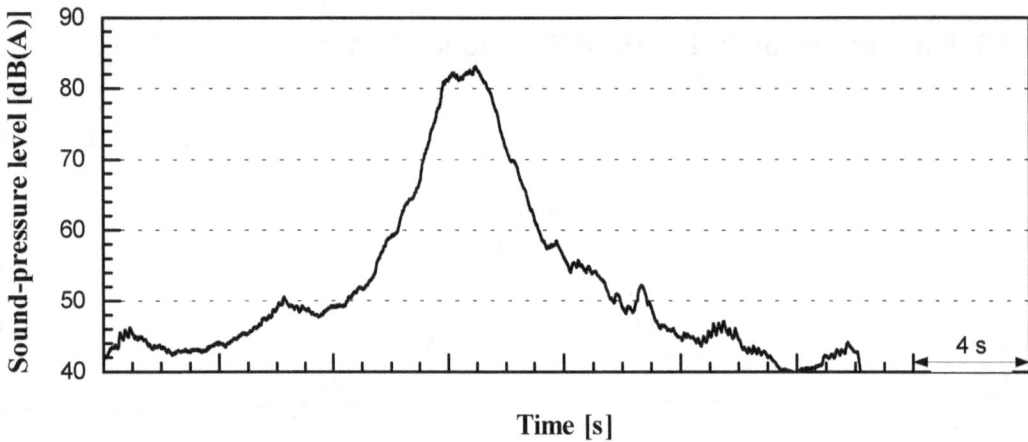

Figure B-102. Time history of the A-weighted SPL during a passby of the TR08 travelling on the hybrid guideway at about 200 km/h (124 mph) measured at 15.2 m (50.0 ft) distance from track centerline and 1.5 m (5.0 ft) above the ground.

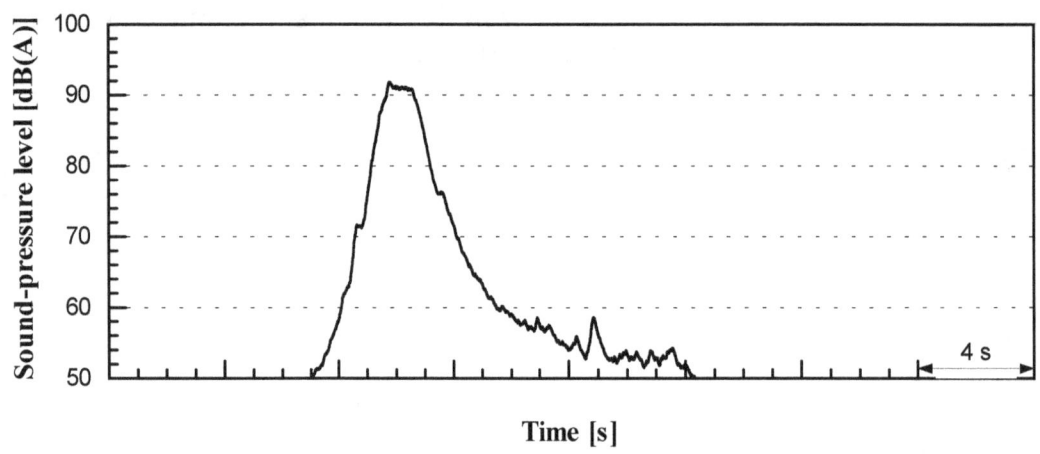

Figure B-103. Time history of the A-weighted SPL during a passby of the TR08 travelling on the hybrid guideway at about 300 km/h (186 mph) measured at 15.2 m (50.0 ft) distance from track centerline and 1.5 m (5.0 ft) above the ground.

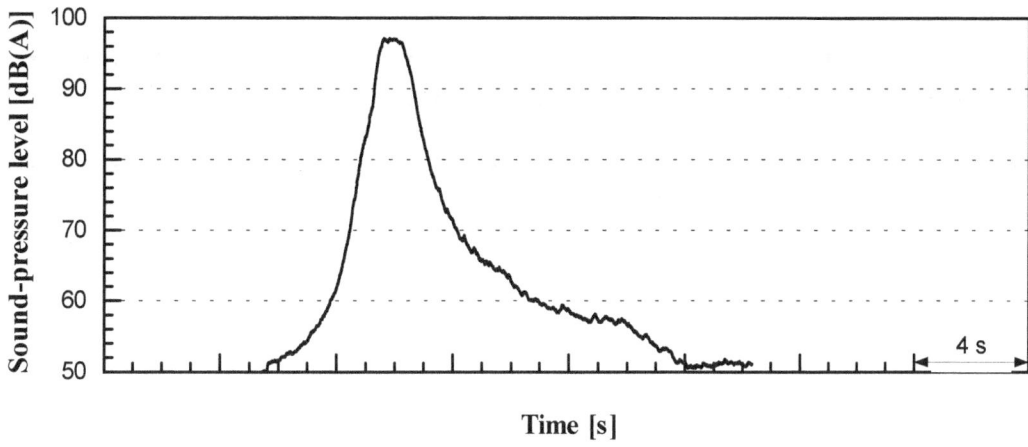

Figure B-1. Time history of the A-weighted SPL during a passby of the TR08 travelling on the hybrid guideway at about 400 km/h (249 mph) measured at 15.2 m (50.0 ft) distance from track centerline and 1.5 m (5.0 ft) above the ground.

Table B-21. Results of the microphone positioned close to the hybrid guideway at 15.2 m (50.0 ft) distance from track centerline and 1.5 m (5.0 ft) above the ground (measuring series J).

Date	Time	Vehicle speed [km/h (mph)]	$L_{Amax, fast}$ [dB(A)]	$L_{Aeq,E}$ [dB(A)]	t_E [s]	SEL [dB(A)]	$L_{Aeq,1h}$ [dB(A)]
2002-05-16	09:40	100.5 (62.5)	82.4	77.1	7.82	86.0	50.5
2002-05-16	10:08	100.1 (62.2)	82.0	76.9	8.00	86.0	50.4
2002-05-16	11:14	100.6 (62.5)	81.7	76.6	7.78	85.5	49.9
2002-05-16	09:57	150.1 (93.3)	80.6	76.4	5.34	83.7	48.1
2002-05-16	10:50	150.4 (93.5)	80.5	76.1	5.46	83.5	47.9
2002-05-16	11:10	150.4 (93.5)	81.7	77.5	4.50	84.0	48.5
2002-05-16	10:02	200.3 (124.5)	83.5	78.6	4.80	85.4	49.8
2002-05-16	10:59	200.3 (124.5)	83.1	78.3	4.40	84.7	49.2
2002-05-16	11:49	200.2 (124.4)	83.1	78.1	4.38	84.5	49.0
2002-05-16	10:18	300.3 (186.6)	92.5	87.8	3.11	92.8	57.2
2002-05-16	11:04	300.3 (186.6)	91.8	87.4	3.13	92.3	56.8
2002-05-16	11:53	300.2 (186.6)	92.0	87.7	2.98	92.5	56.9
2002-05-16	09:24	393.6 (244.6)	97.6	93.2	2.57	97.3	61.7
2002-05-16	10:53	392.0 (243.6)	97.1	93.0	2.67	97.3	61.7
2002-05-16	11:43	392.4 (243.9)	97.3	92.9	2.57	97.0	61.4

B.5.4 Microphone at 6.5 m (21.3 ft) distance from track centerline (high position)

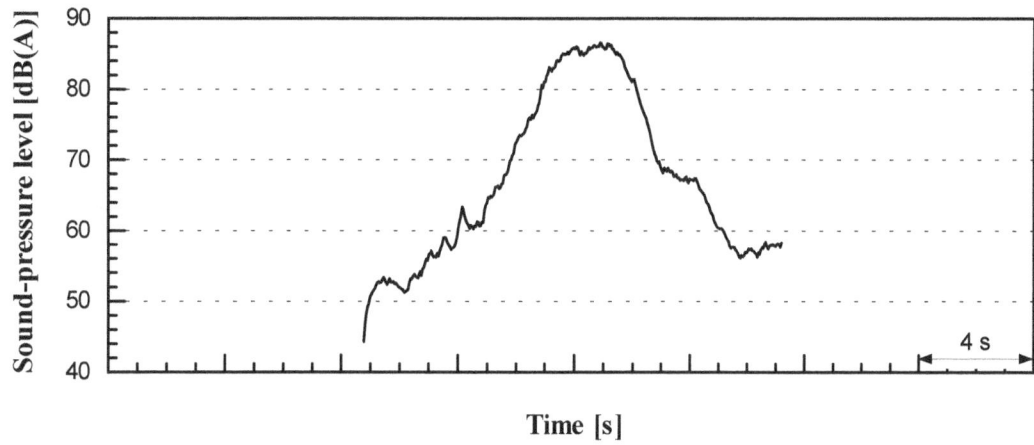

Figure B-105. Time history of the A-weighted SPL during a passby of the TR08 travelling on the hybrid guideway at about 100 km/h (62 mph) measured at 6.5 m (21.3 ft) distance from track centerline and the height of the upper surface of the guideway.

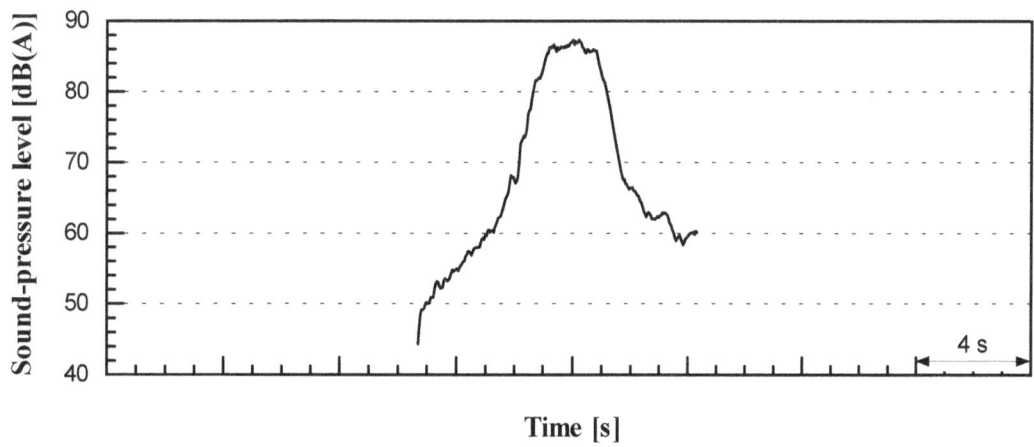

Figure B-106. Time history of the A-weighted SPL during a passby of the TR08 travelling on the hybrid guideway at about 150 km/h (93 mph) measured at 6.5 m (21.3 ft) distance from track centerline and the height of the upper surface of the guideway.

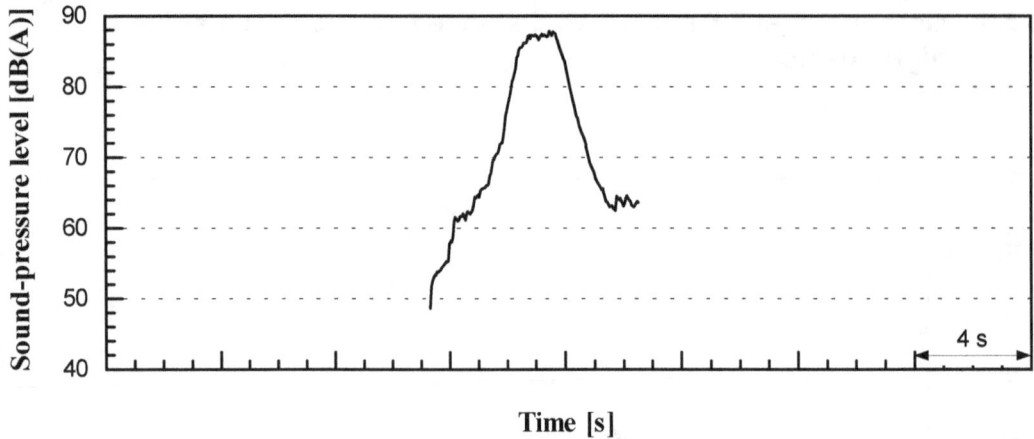

Figure B-107. Time history of the A-weighted SPL during a passby of the TR08 travelling on the hybrid guideway at about 200 km/h (124 mph) measured at 6.5 m (21.3 ft) distance from track centerline and the height of the upper surface of the guideway.

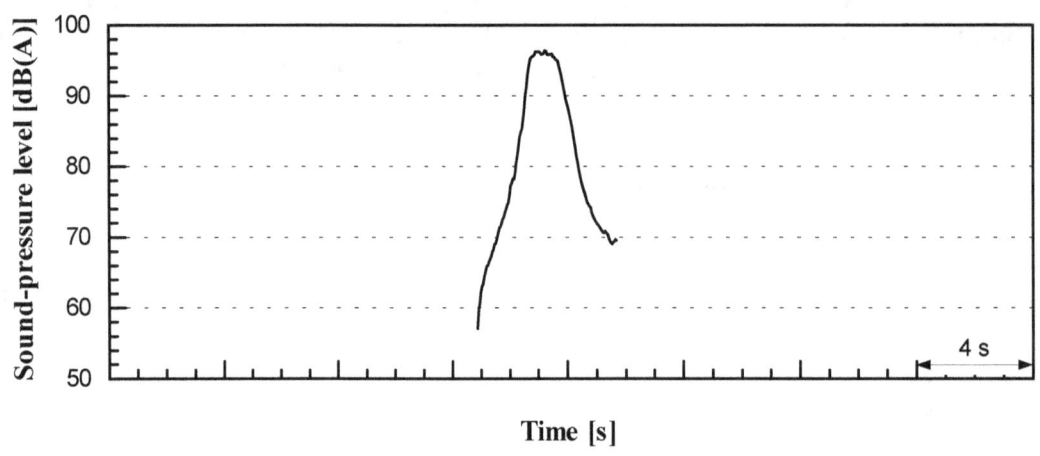

Figure B-108. Time history of the A-weighted SPL during a passby of the TR08 travelling on the hybrid guideway at about 300 km/h (186 mph) measured at 6.5 m (21.3 ft) distance from track centerline and the height of the upper surface of the guideway.

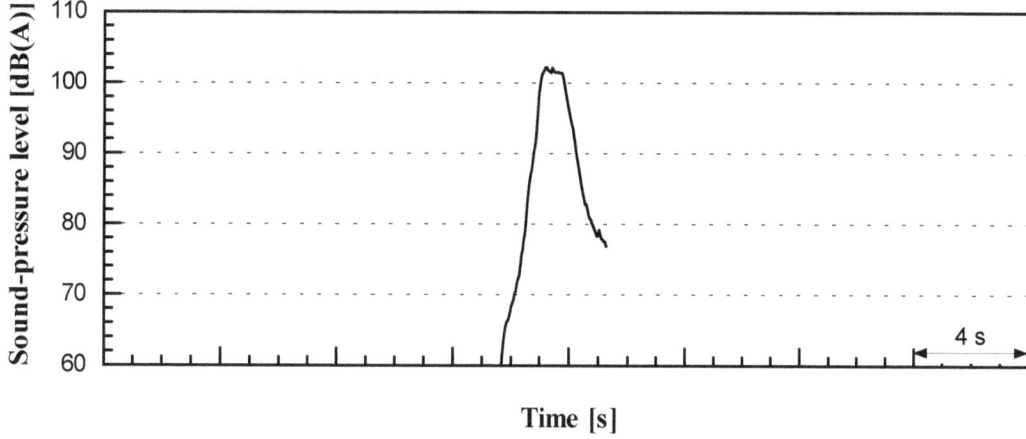

Figure B-109. Time history of the A-weighted SPL during a passby of the TR08 travelling on the hybrid guideway at about 400 km/h (249 mph) measured at 6.5 m (21.3 ft) distance from track centerline and the height of the upper surface of the guideway.

Table B-22. Results of the microphone positioned close to the hybrid guideway at 6.5 m (21.3 ft) distance from track centerline and the height of the upper surface of the guideway (measuring series I/J).

Date	Time	Vehicle speed [km/h (mph)]	$L_{Amax, fast}$ [dB(A)]	$L_{Aeq,E}$ [dB(A)]	t_E [s]	SEL [dB(A)]	$L_{Aeq,1h}$ [dB(A)]
2002-05-16	09:40	100.5 (62.5)	86.6	84.1	3.64	89.7	54.2
2002-05-16	09:45	99.6 (61.9)	86.3	84.1	4.16	90.3	54.7
2002-05-16	13:37	150.6 (93.6)	87.3	85.0	2.84	89.5	54.0
2002-05-16	13:57	149.8 (93.1)	86.1	84.1	2.92	88.8	53.2
2002-05-16	12:08	150.3 (93.4)	87.7	85.4	2.76	89.8	54.3
2002-05-16	13:11	150.3 (93.4)	87.1	84.9	2.84	89.4	53.9
2002-05-16	14:14	159.8 (99.3)	87.6	85.7	2.52	89.7	54.1
2002-05-16	13:42	173.7 (108.0)	87.4	85.0	2.60	89.1	53.6
2002-05-16	12:18	200.2 (124.4)	87.8	85.5	2.32	89.2	53.6
2002-05-16	13:27	200.3 (124.5)	87.7	85.5	2.32	89.2	53.6
2002-05-16	14:18	250.4 (155.6)	92.8	90.2	2.08	93.4	57.8
2002-05-16	14:01	291.2 (181.0)	95.5	93.6	1.72	95.9	60.4
2002-05-16	12:23	300.4 (186.7)	96.4	94.3	1.76	96.7	61.1
2002-05-16	13:30	300.4 (186.7)	96.2	94.2	1.72	96.6	61.0
2002-05-16	12:12	392.1 (243.7)	102.2	100.0	1.44	101.6	66.0
2002-05-16	13:14	392.0 (243.6)	101.8	99.6	1.48	101.3	65.8

B.5.5 Microphone at 6.5 m (21.3 ft) distance from track centerline (low position)

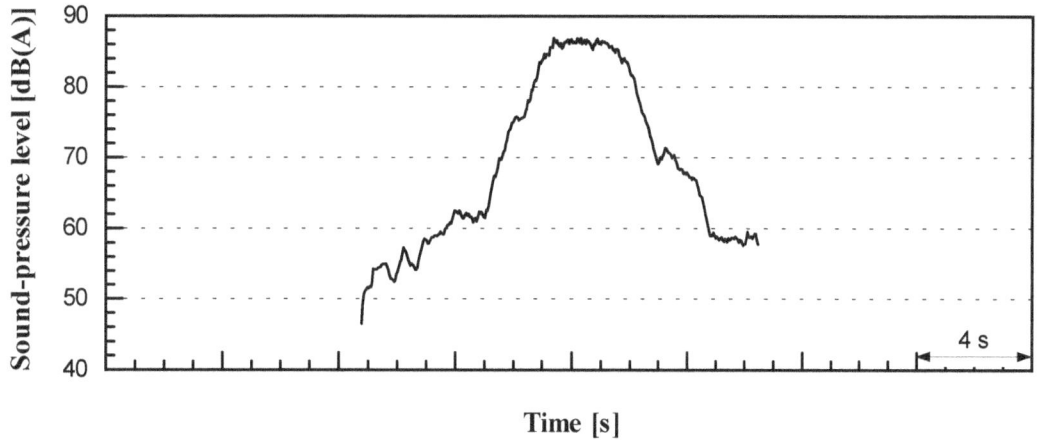

Figure B-110. Time history of the A-weighted SPL during a passby of the TR08 travelling on the hybrid guideway at about 100 km/h (62 mph) measured at 6.5 m (21.3 ft) distance from track centerline and 1.5 m (5.0 ft) below the upper surface of the guideway.

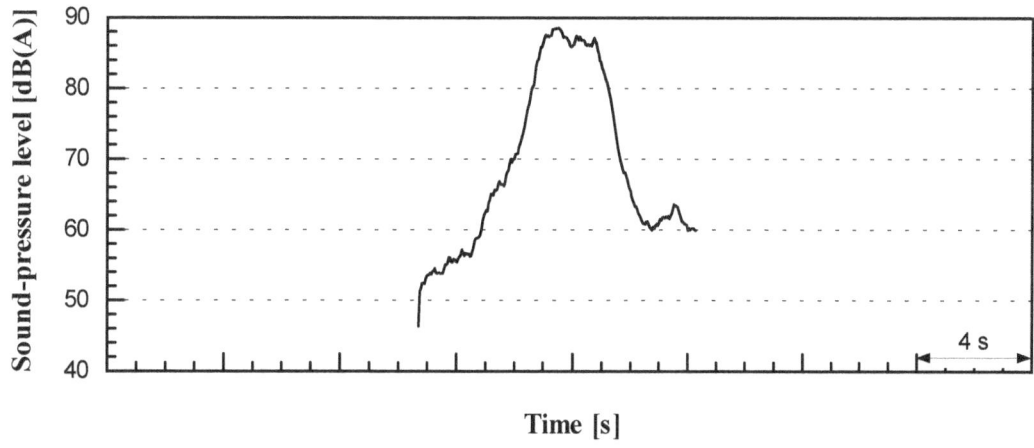

Figure B-111. Time history of the A-weighted SPL during a passby of the TR08 travelling on the hybrid guideway at about 150 km/h (93 mph) measured at 6.5 m (21.3 ft) distance from track centerline and 1.5 m (5.0 ft) below the upper surface of the guideway.

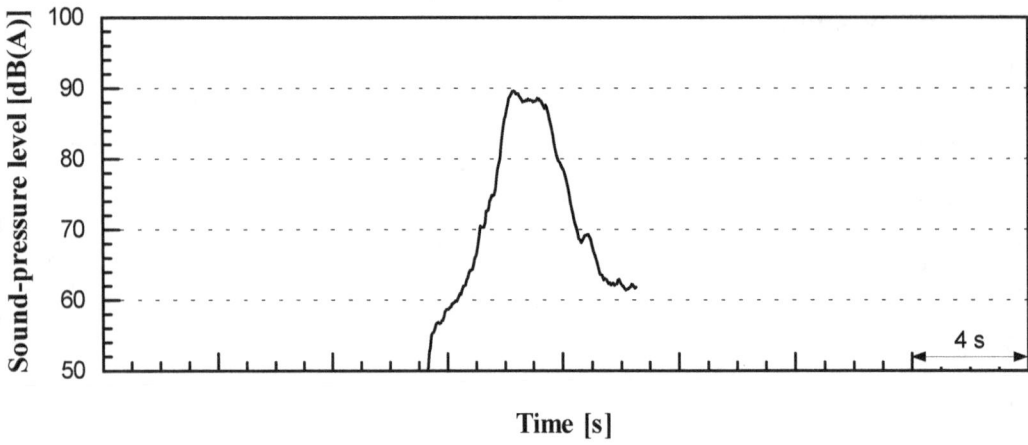

Figure B-112. Time history of the A-weighted SPL during a passby of the TR08 travelling on the hybrid guideway at about 200 km/h (124 mph) measured at 6.5 m (21.3 ft) distance from track centerline and 1.5 m (5.0 ft) below the upper surface of the guideway.

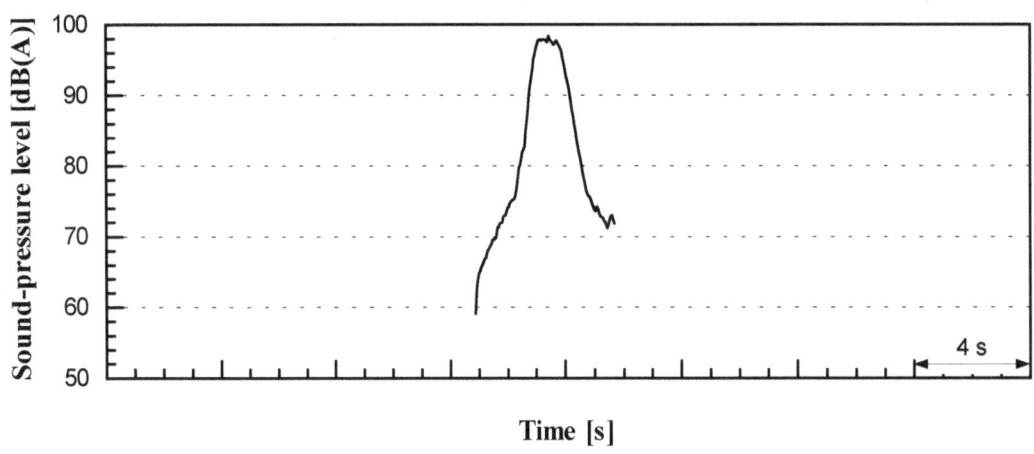

Figure B-113. Time history of the A-weighted SPL during a passby of the TR08 travelling on the hybrid guideway at about 300 km/h (186 mph) measured at 6.5 m (21.3 ft) distance from track centerline and 1.5 m (5.0 ft) below the upper surface of the guideway.

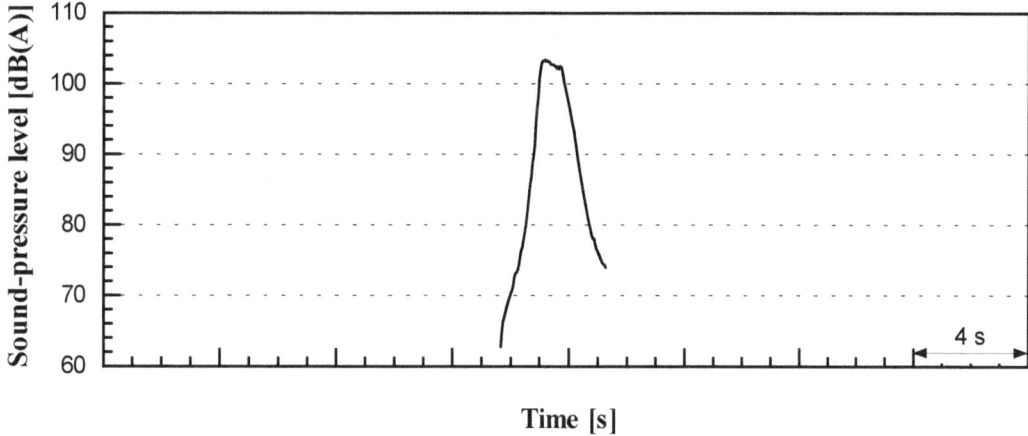

Figure B-114. Time history of the A-weighted SPL during a passby of the TR08 travelling on the hybrid guideway at about 400 km/h (249 mph) measured at 6.5 m (21.3 ft) distance from track centerline and 1.5 m (5.0 ft) below the upper surface of the guideway.

Table B-23. Results of the microphone positioned close to the hybrid guideway at 6.5 m (21.3 ft) distance from track centerline and 1.5 m (5.0 ft) below the upper surface of the guideway (measuring series J).

Date	Time	Vehicle speed [km/h (mph)]	$L_{Amax, fast}$ [dB(A)]	$L_{Aeq,E}$ [dB(A)]	t_E [s]	SEL [dB(A)]	$L_{Aeq,1h}$ [dB(A)]
2002-05-16	10:08	100.1 (62.2)	86.9	84.9	4.00	90.9	55.3
2002-05-16	10:22	100.4 (62.4)	88.7	85.8	3.72	91.5	55.9
2002-05-16	11:14	100.6 (62.5)	88.3	85.6	3.76	91.4	55.8
2002-05-16	10:00	150.1 (93.3)	88.5	86.0	2.84	90.5	55.0
2002-05-16	10:22	150.4 (93.5)	87.6	85.6	2.80	90.1	54.5
2002-05-16	11:10	150.4 (93.5)	87.5	85.1	2.92	89.8	54.2
2002-05-16	11:39	150.6 (93.6)	87.6	85.7	2.76	90.1	54.5
2002-05-16	10:02	200.3 (124.5)	89.6	87.3	2.16	90.6	55.1
2002-05-16	10:59	200.3 (124.5)	88.7	86.6	2.32	90.2	54.7
2002-05-16	11:49	200.2 (124.4)	88.3	86.0	2.36	89.8	54.2
2002-05-16	10:18	300.3 (186.6)	98.3	96.1	1.56	98.0	62.5
2002-05-16	11:04	300.3 (186.6)	97.9	95.8	1.68	98.0	62.4
2002-05-16	11:53	300.2 (186.6)	97.6	95.5	1.68	97.8	62.2
2002-05-16	10:13	395.6 (245.9)	103.7	101.4	1.40	102.9	67.3
2002-05-16	10:53	392.0 (243.6)	103.3	101.1	1.44	102.7	67.1
2002-05-16	11:43	392.4 (243.9)	103.1	100.8	1.44	102.4	66.8

B.5.6 Microphone beneath guideway centerline

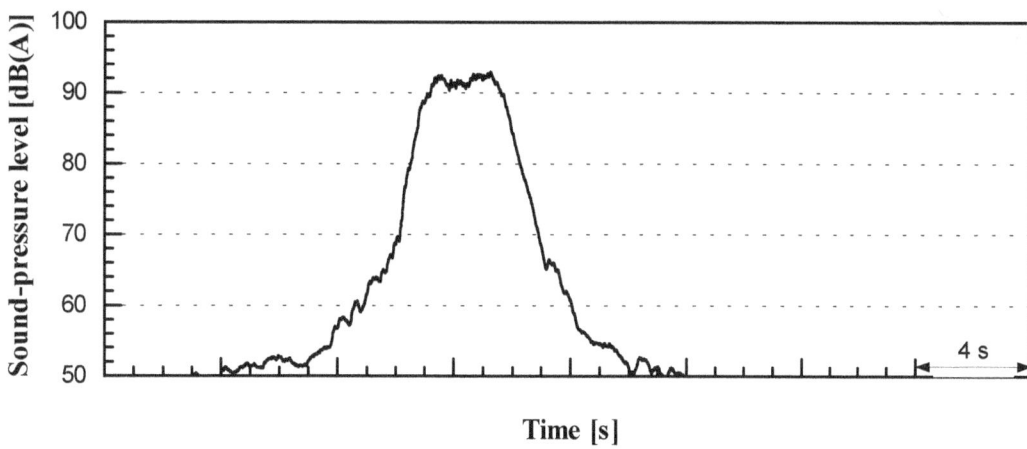

Figure B-115. Time history of the A-weighted SPL during a passby of the TR08 travelling on the hybrid guideway at about 100 km/h (62 mph) measured beneath guideway centerline at a height of 1.5 m (5.0 ft) above the ground.

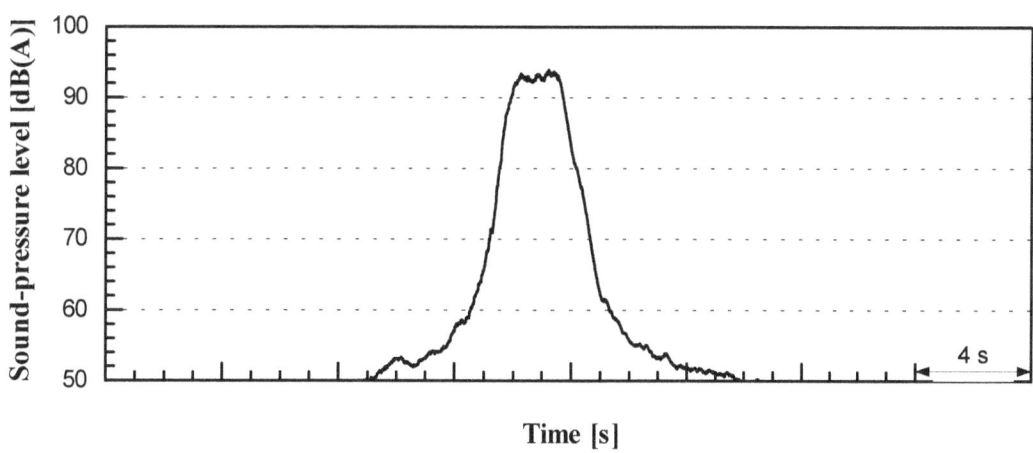

Figure B-116. Time history of the A-weighted SPL during a passby of the TR08 travelling on the hybrid guideway at about 150 km/h (93 mph) measured beneath guideway centerline at a height of 1.5 m (5.0 ft) above the ground.

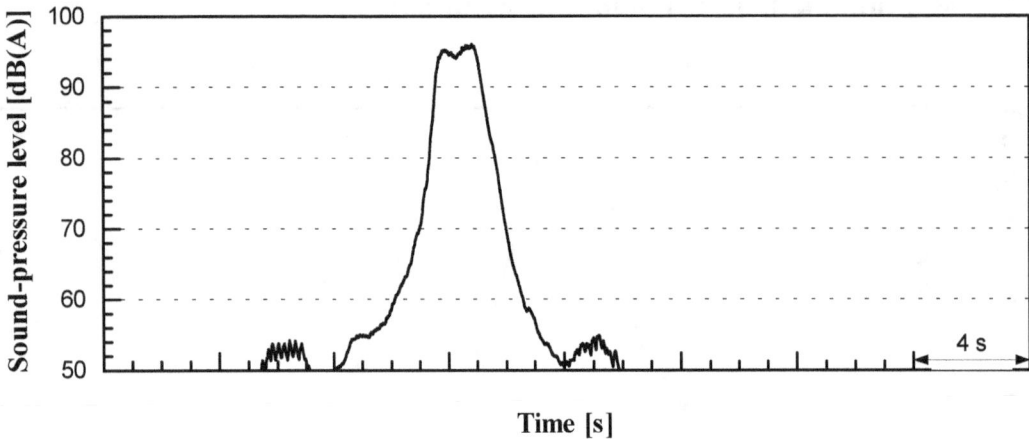

Figure B-117. Time history of the A-weighted SPL during a passby of the TR08 travelling on the hybrid guideway at about 200 km/h (124 mph) measured beneath guideway centerline at a height of 1.5 m (5.0 ft) above the ground.

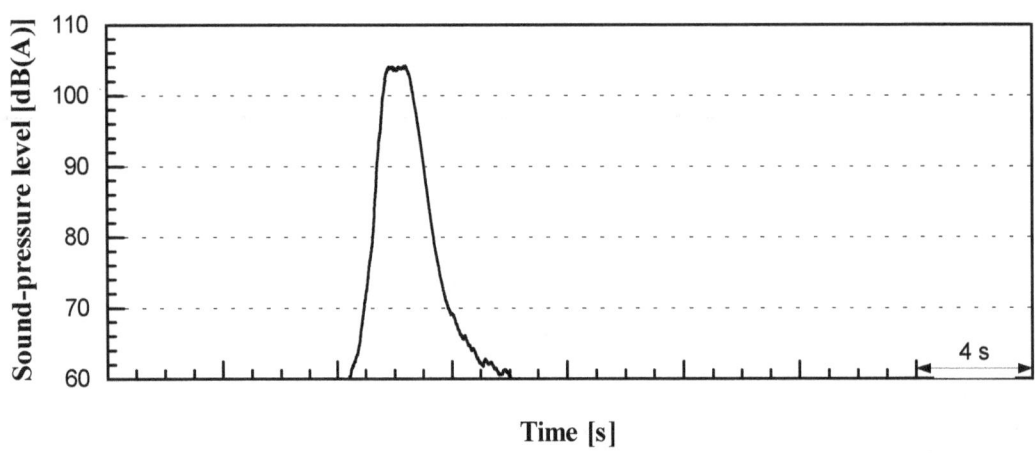

Figure B-118. Time history of the A-weighted SPL during a passby of the TR08 travelling on the hybrid guideway at about 300 km/h (186 mph) measured beneath guideway centerline at a height of 1.5 m (5.0 ft) above the ground.

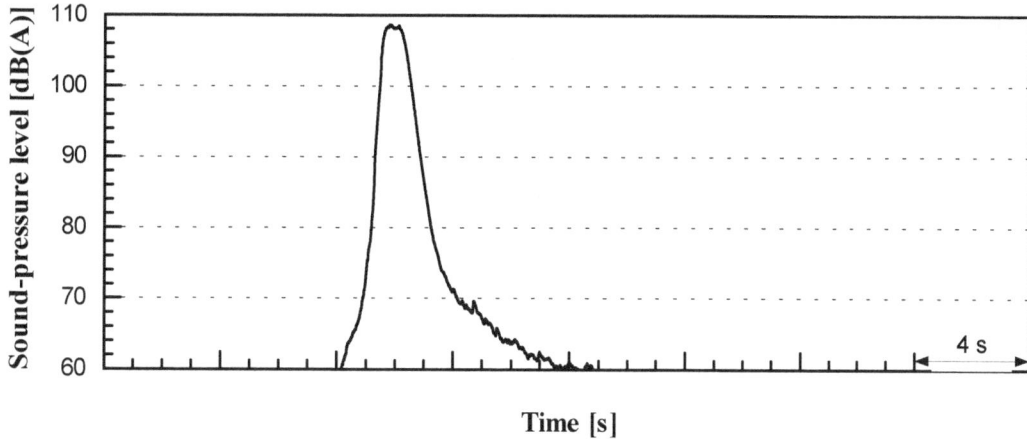

Figure B-119. Time history of the A-weighted SPL during a passby of the TR08 travelling on the hybrid guideway at about 400 km/h (249 mph) measured beneath guideway centerline at a height of 1.5 m (5.0 ft) above the ground.

Table B-24. Results of the microphone positioned beneath the centerline of the hybrid guideway at a height of 1.5 m (5.0 ft) above the ground (measuring series J).

Date	Time	Vehicle speed [km/h (mph)]	$L_{Amax,\,fast}$ [dB(A)]	$L_{Aeq,E}$ [dB(A)]	t_E [s]	SEL [dB(A)]	$L_{Aeq,1h}$ [dB(A)]
2002-05-16	10:08	100.1 (62.2)	92.9	89.7	4.52	96.3	60.7
2002-05-16	11:14	100.6 (62.5)	92.2	89.0	4.73	95.7	60.1
2002-05-16	09:57	150.1 (93.3)	93.6	90.4	3.26	95.6	60.0
2002-05-16	10:50	150.4 (93.5)	93.9	90.7	3.14	95.7	60.1
2002-05-16	11:10	150.4 (93.5)	94.1	90.5	3.17	95.6	60.0
2002-05-16	10:02	200.3 (124.5)	95.7	91.6	2.64	95.8	60.2
2002-05-16	10:59	200.3 (124.5)	96.0	92.9	2.53	96.9	61.4
2002-05-16	11:49	200.2 (124.4)	95.8	92.8	2.55	96.8	61.3
2002-05-16	10:18	300.3 (186.6)	104.0	101.2	1.85	103.8	68.3
2002-05-16	11:04	300.3 (186.6)	104.1	101.2	1.90	103.9	68.4
2002-05-16	11:53	300.2 (186.6)	103.9	100.8	1.93	103.7	68.1
2002-05-16	10:53	392.0 (243.6)	108.6	105.4	1.63	107.6	72.0
2002-05-16	11:43	392.4 (243.9)	108.8	105.8	1.62	107.9	72.3

B.6 North switch

B.6.1 Microphone at 30.5 m (100.0 ft) distance from track centerline

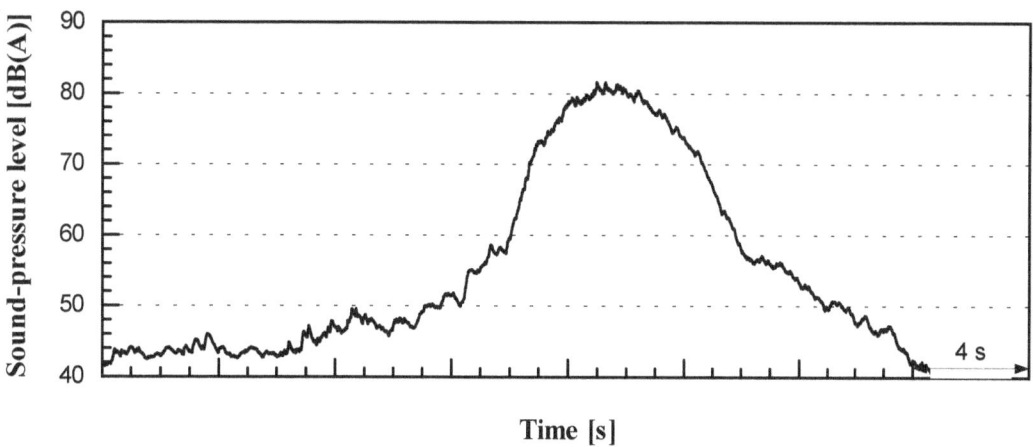

Figure B-120. Time history of the A-weighted SPL during a passby of the TR08 travelling on the North switch at about 100 km/h (62 mph) measured at 30.5 m (100.0 ft) distance from track centerline and 1.2 m (4.0 ft) above the ground.

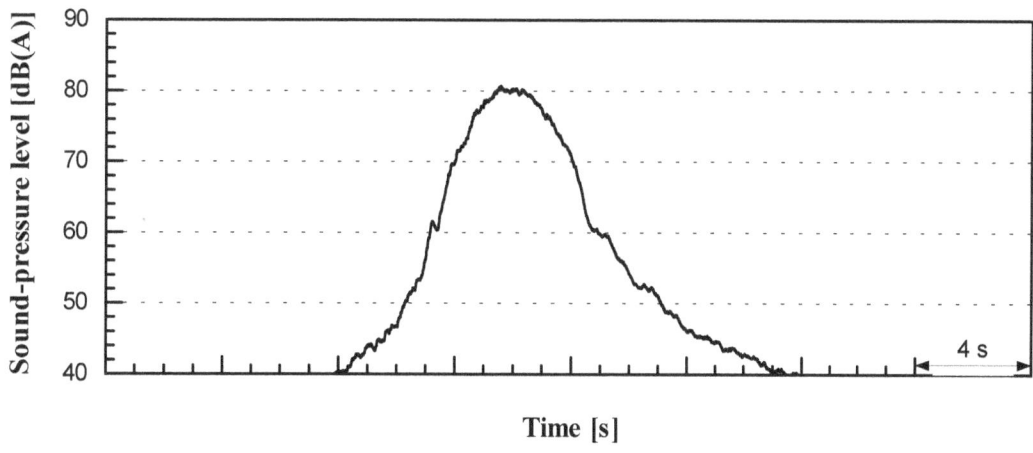

Figure B-121. Time history of the A-weighted SPL during a passby of the TR08 travelling on the North switch at about 150 km/h (93 mph) measured at 30.5 m (100.0 ft) distance from track centerline and 1.2 m (4.0 ft) above the ground.

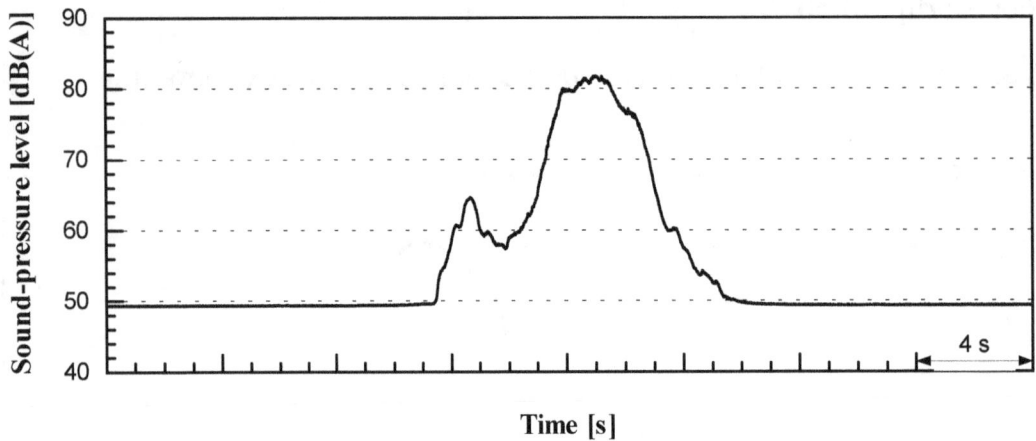

Figure B-122. Time history of the A-weighted SPL during a passby of the TR08 travelling on the North switch at about 200 km/h (124 mph) measured at 30.5 m (100.0 ft) distance from track centerline and 1.2 m (4.0 ft) above the ground.

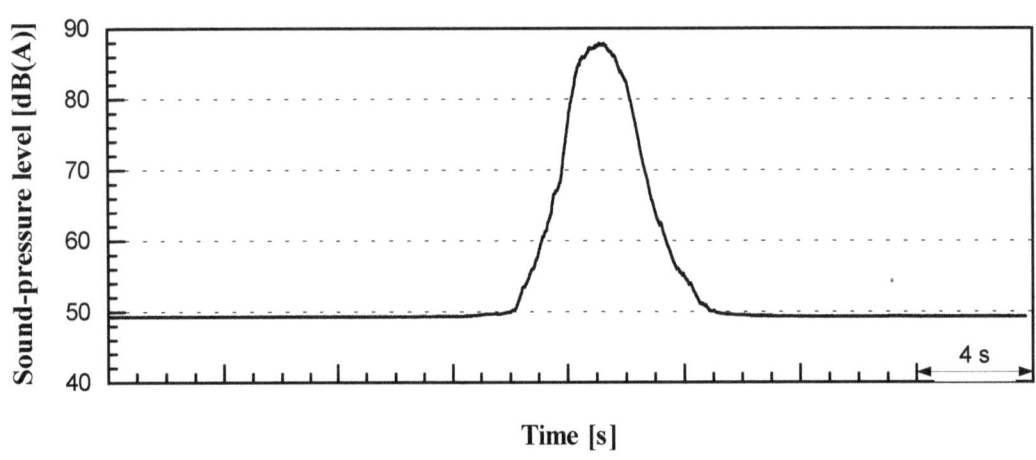

Figure B-123. Time history of the A-weighted SPL during a passby of the TR08 travelling on the North switch at about 300 km/h (186 mph) measured at 30.5 m (100.0 ft) distance from track centerline and 1.2 m (4.0 ft) above the ground.

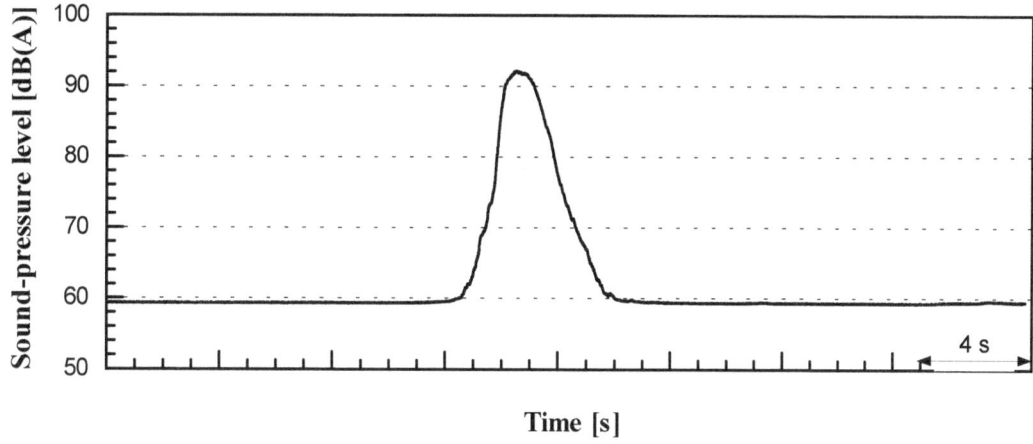

Figure B-124. Time history of the A-weighted SPL during a passby of the TR08 travelling on the North switch at about 385 km/h (239 mph) measured at 30.5 m (100.0 ft) distance from track centerline and 1.2 m (4.0 ft) above the ground.

Table B-25. Results of the microphone positioned close to the North switch at 30.5 m (100.0 ft) distance from track centerline and 1.2 m (4.0 ft) above the ground (measuring series Z)

Date	Time	Vehicle speed [km/h (mph)]	$L_{Amax, fast}$ [dB(A)]	$L_{Aeq,E}$ [dB(A)]	t_E [s]	SEL [dB(A)]	$L_{Aeq,1h}$ [dB(A)]
2002-05-17	09:52	100.2 (62.3)	81.6	77.0	7.54	85.8	50.2
2002-05-17	10:03	100.2 (62.3)	81.5	77.2	7.23	85.8	50.3
2002-05-17	10:14	100.3 (62.3)	81.9	77.5	7.12	86.1	50.5
2002-05-17	10:49	99.9 (62.1)	80.8	77.0	7.33	85.6	50.1
2001-08-24	09:27	149.8 (93.1)	81.1	77.0	5.36	84.3	48.7
2001-08-24	10:14	149.8 (93.1)	81.4	77.2	5.14	84.3	48.8
2001-08-24	10:26	150.0 (93.2)	80.7	76.7	5.25	83.9	48.3
2001-08-24	12:35	199.7 (124.1)	82.5	78.5	4.44	85.0	49.4
2001-08-24	12:46	199.9 (124.2)	81.8	77.8	4.64	84.5	48.9
2001-08-24	13:39	199.9 (124.2)	82.8	78.9	4.30	85.3	49.7
2001-08-24	10:31	299.4 (186.1)	87.9	84.2	3.06	89.0	53.5
2001-08-24	12:50	300.0 (186.5)	87.9	84.4	3.19	89.4	53.8
2001-08-24	09:32	379.1 (235.6)	92.3	88.1	3.07	93.0	57.4
2001-08-24	10:18	385.3 (239.5)	92.1	88.2	2.90	92.8	57.2
2001-08-24	11:35	384.9 (239.2)	92.8	88.6	2.78	93.0	57.5

B.6.2 Microphone at 25.0 m (82.0 ft) distance from track centerline

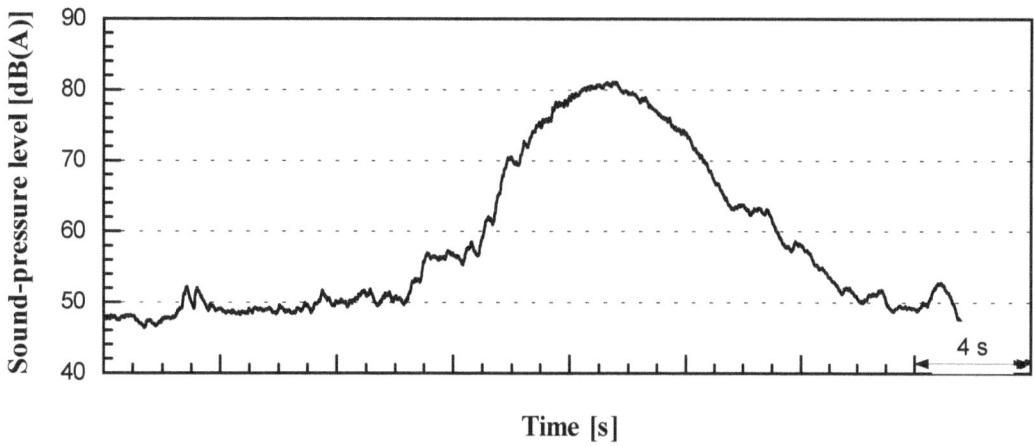

Figure B-125. Time history of the A-weighted SPL during a passby of the TR08 travelling on the North switch at about 100 km/h (62 mph) measured at 25.0 m (82.0 ft) distance from track centerline and 3.5 m (11.5 ft) above the ground.

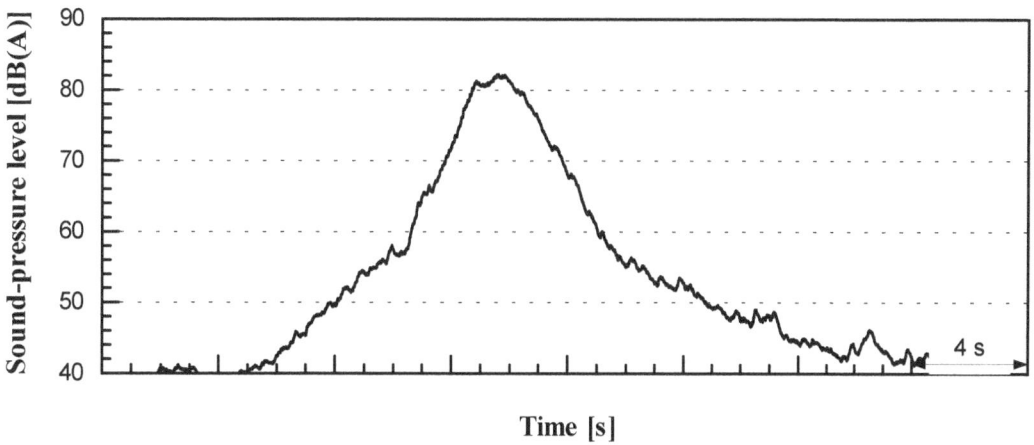

Figure B-126. Time history of the A-weighted SPL during a passby of the TR08 travelling on the North switch at about 150 km/h (93 mph) measured at 25.0 m (82.0 ft) distance from track centerline and 3.5 m (11.5 ft) above the ground.

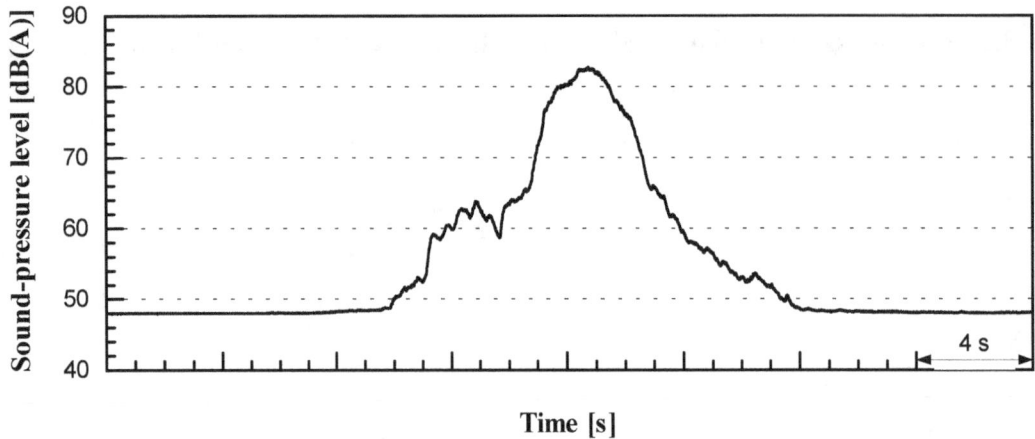

Figure B-127. Time history of the A-weighted SPL during a passby of the TR08 travelling on the North switch at about 200 km/h (124 mph) measured at 25.0 m (82.0 ft) distance from track centerline and 3.5 m (11.5 ft) above the ground.

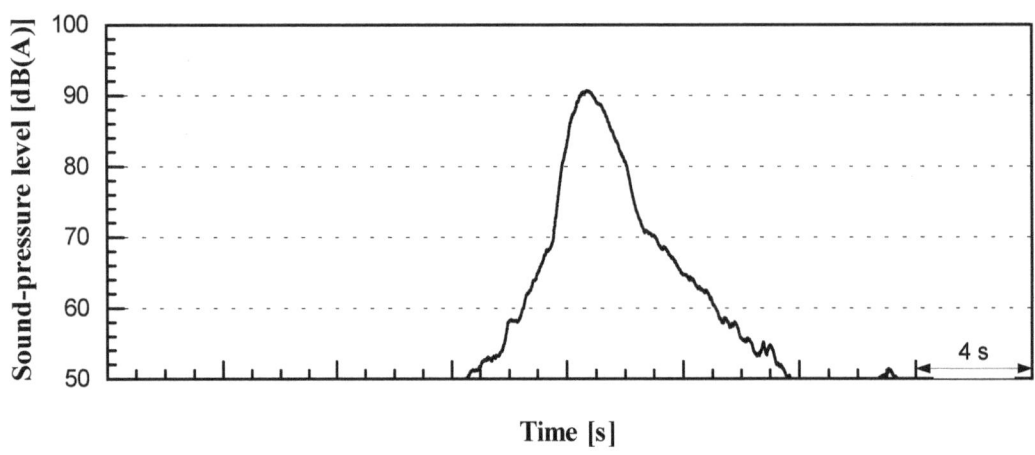

Figure B-128. Time history of the A-weighted SPL during a passby of the TR08 travelling on the North switch at about 300 km/h (186 mph) measured at 25.0 m (82.0 ft) distance from track centerline and 3.5 m (11.5 ft) above the ground.

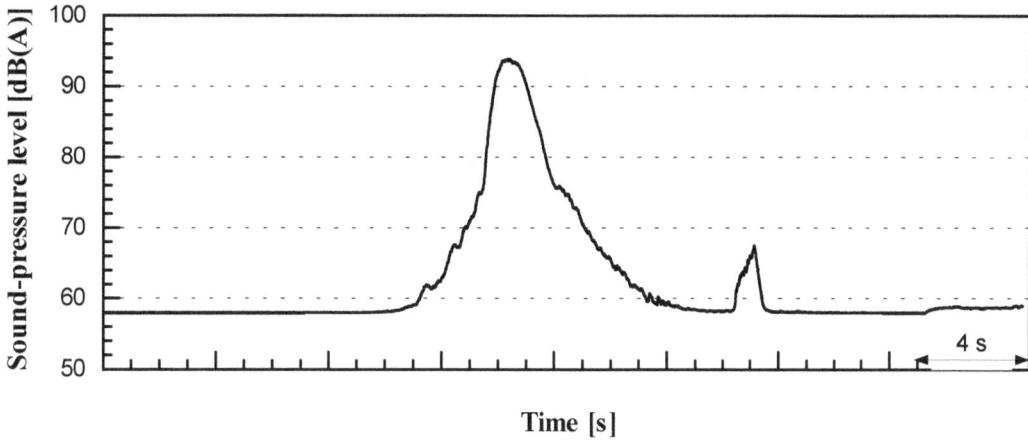

Figure B-129. Time history of the A-weighted SPL during a passby of the TR08 travelling on the North switch at about 385 km/h (239 mph) measured at 25.0 m (82.0 ft) distance from track centerline and 3.5 m (11.5 ft) above the ground.

Table B-26. Results of the microphone positioned close to the North switch at 25.0 m (82.0 ft) distance from track centerline and 3.5 m (11.5 ft) above the ground (measuring series Z).

Date	Time	Vehicle speed [km/h (mph)]	$L_{Amax, fast}$ [dB(A)]	$L_{Aeq,E}$ [dB(A)]	t_E [s]	SEL [dB(A)]	$L_{Aeq,1h}$ [dB(A)]
2002-05-17	09:52	100.2 (62.3)	81.1	76.2	9.69	86.1	50.5
2002-05-17	10:03	100.2 (62.3)	81.0	76.7	8.54	86.0	50.4
2002-05-17	10:14	100.3 (62.3)	82.4	77.4	8.14	86.5	50.9
2002-05-17	10:49	99.9 (62.1)	81.3	77.0	8.34	86.2	50.6
2001-08-24	09:27	149.8 (93.1)	81.9	77.7	5.69	85.2	49.7
2001-08-24	10:14	149.8 (93.1)	82.2	77.3	6.01	85.1	49.6
2001-08-24	10:26	150.0 (93.2)	82.2	77.1	6.02	84.9	49.3
2001-08-24	12:35	199.7 (124.1)	83.2	78.2	5.59	85.7	50.2
2001-08-24	12:46	199.9 (124.2)	82.7	77.7	5.63	85.2	49.7
2001-08-24	13:39	199.9 (124.2)	82.7	78.0	5.65	85.5	49.9
2001-08-24	10:31	299.4 (186.1)	90.7	86.2	3.12	91.1	55.6
2001-08-24	12:50	300.0 (186.5)	89.9	85.7	3.12	90.6	55.1
2001-08-24	09:32	379.1 (235.6)	93.5	89.0	3.09	93.9	58.4
2001-08-24	10:18	385.3 (239.5)	93.9	88.9	3.35	94.2	58.6
2001-08-24	11:35	384.9 (239.2)	93.7	89.2	3.15	94.2	58.6

B.6.3 Microphone at 15.2 m (50.0 ft) distance from track centerline

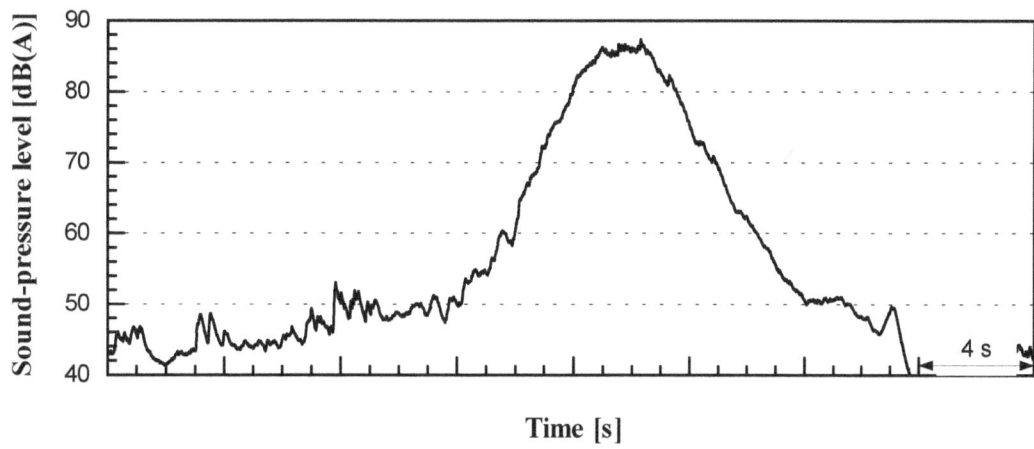

Figure B-130. Time history of the A-weighted SPL during a passby of the TR08 travelling on the North switch at about 100 km/h (62 mph) measured at 15.2 m (50.0 ft) distance from track centerline and 1.5 m (5.0 ft) above the ground.

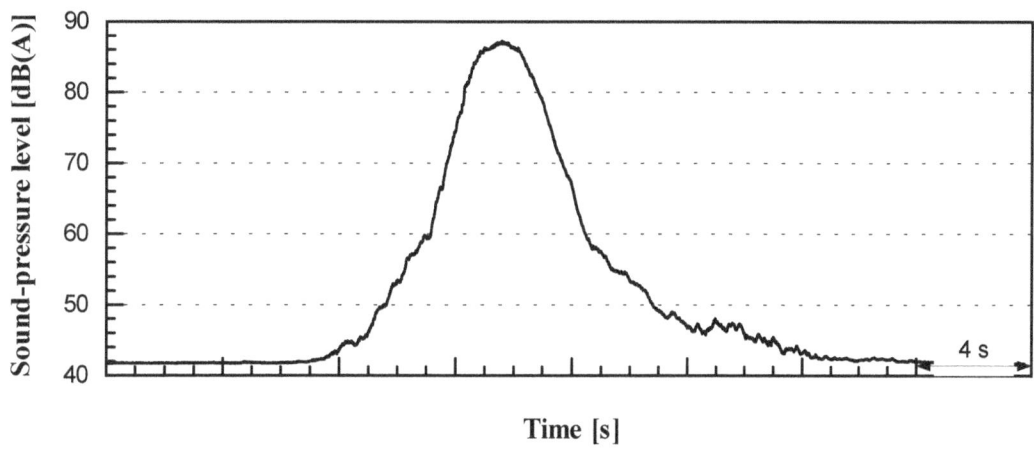

Figure B-131. Time history of the A-weighted SPL during a passby of the TR08 travelling on the North switch at about 150 km/h (93 mph) measured at 15.2 m (50.0 ft) distance from track centerline and 1.5 m (5.0 ft) above the ground.

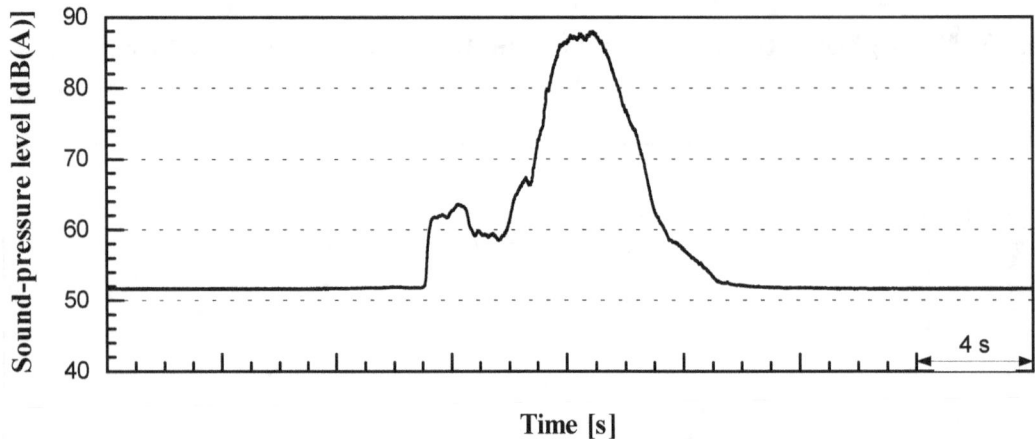

Figure B-132. Time history of the A-weighted SPL during a passby of the TR08 travelling on the North switch at about 200 km/h (124 mph) measured at 15.2 m (50.0 ft) distance from track centerline and 1.5 m (5.0 ft) above the ground.

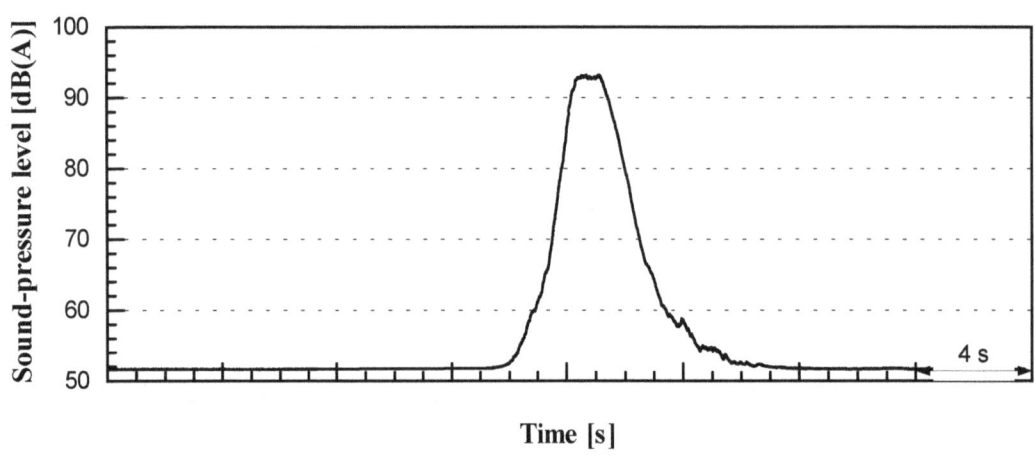

Figure B-133. Time history of the A-weighted SPL during a passby of the TR08 travelling on the North switch at about 300 km/h (186 mph) measured at 15.2 m (50.0 ft) distance from track centerline and 1.5 m (5.0 ft) above the ground.

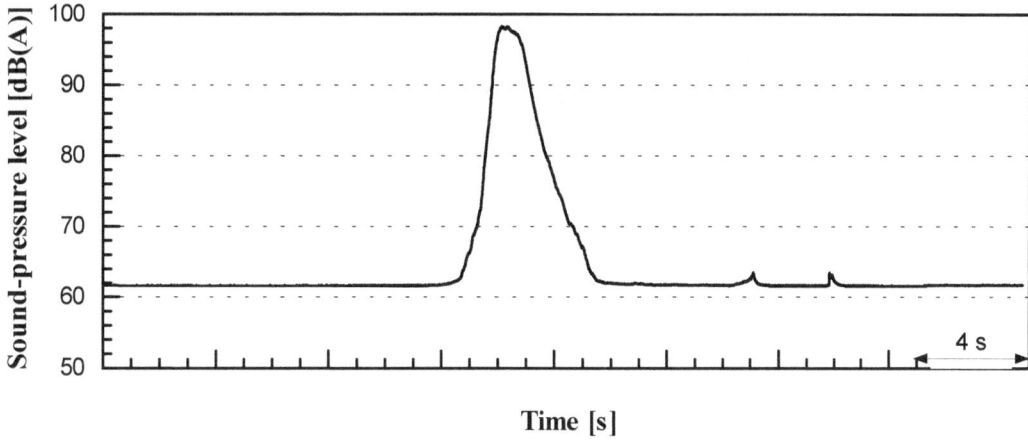

Figure B-134. Time history of the A-weighted SPL during a passby of the TR08 travelling on the North switch at about 385 km/h (239 mph) measured at 15.2 m (50.0 ft) distance from track centerline and 1.5 m (5.0 ft) above the ground.

Table B-27. Results of the microphone positioned close to the North switch at 15.2 m (50.0 ft) distance from track centerline and 1.5 m (5.0 ft) above the ground (measuring series Z).

Date	Time	Vehicle speed [km/h (mph)]	$L_{Amax, fast}$ [dB(A)]	$L_{Aeq,E}$ [dB(A)]	t_E [s]	SEL [dB(A)]	$L_{Aeq,1h}$ [dB(A)]
2002-05-17	09:52	100.2 (62.3)	87.3	82.3	6.67	90.5	55.0
2002-05-17	10:03	100.2 (62.3)	86.2	81.9	6.88	90.2	54.7
2002-05-17	10:14	100.3 (62.3)	87.1	82.2	6.75	90.5	54.9
2002-05-17	10:49	99.9 (62.1)	86.2	81.6	6.79	89.9	54.4
2001-08-24	09:27	149.8 (93.1)	86.9	83.2	4.45	89.7	54.1
2001-08-24	10:14	149.8 (93.1)	86.6	82.9	4.57	89.5	54.0
2001-08-24	10:26	150.0 (93.2)	87.2	83.2	4.38	89.6	54.0
2001-08-24	12:35	199.7 (124.1)	88.5	84.4	3.82	90.2	54.6
2001-08-24	12:46	199.9 (124.2)	88.0	84.0	3.88	89.9	54.4
2001-08-24	13:39	199.9 (124.2)	88.0	84.1	3.86	90.0	54.4
2001-08-24	10:31	299.4 (186.1)	97.4	90.7	0.68	89.0	53.4
2001-08-24	12:50	300.0 (186.5)	93.6	89.8	2.67	94.1	58.5
2001-08-24	09:32	379.1 (235.6)	98.4	94.4	2.25	97.9	62.4
2001-08-24	10:18	385.3 (239.5)	98.3	94.3	2.35	98.0	62.4
2001-08-24	11:35	384.9 (239.2)	98.6	94.4	2.20	97.8	62.3

B.6.4 Microphone at 6.5 m (21.3 ft) distance from track centerline (high position)

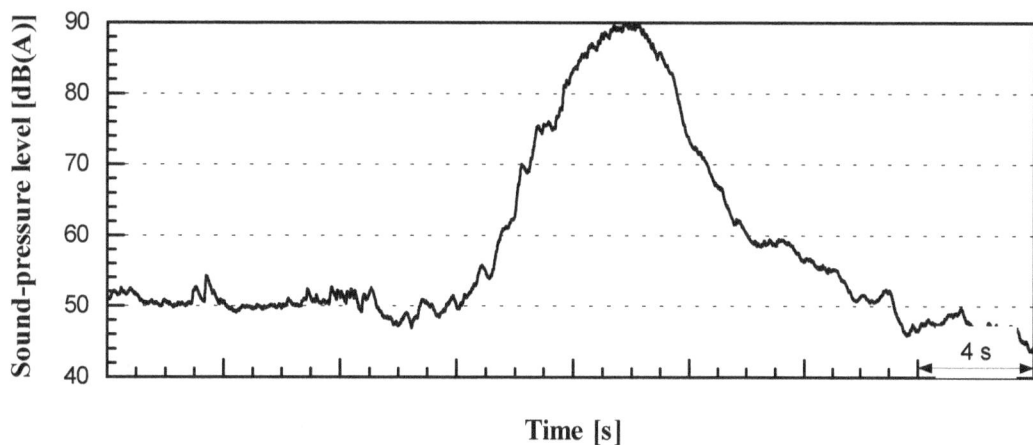

Figure B-135. Time history of the A-weighted SPL during a passby of the TR08 travelling on the North switch at about 100 km/h (62 mph) measured at 6.5 m (21.3 ft) distance from track centerline and the height of the upper surface of the guideway.

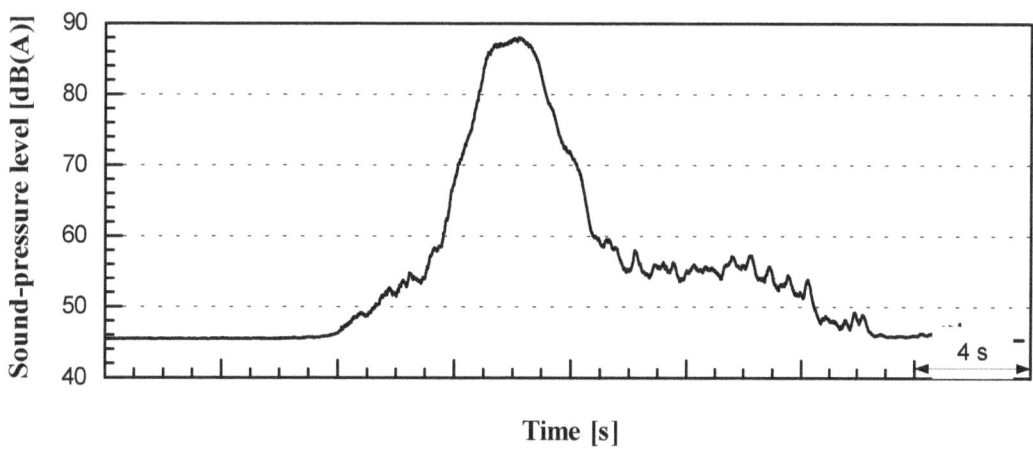

Figure B-136. Time history of the A-weighted SPL during a passby of the TR08 travelling on the North switch at about 150 km/h (93 mph) measured at 6.5 m (21.3 ft) distance from track centerline and the height of the upper surface of the guideway.

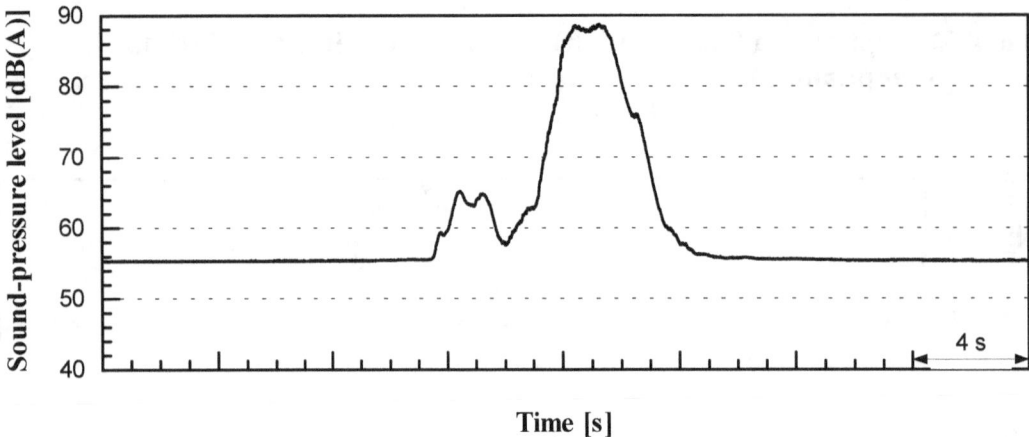

Figure B-137. Time history of the A-weighted SPL during a passby of the TR08 travelling on the North switch at about 200 km/h (124 mph) measured at 6.5 m (21.3 ft) distance from track centerline and the height of the upper surface of the guideway.

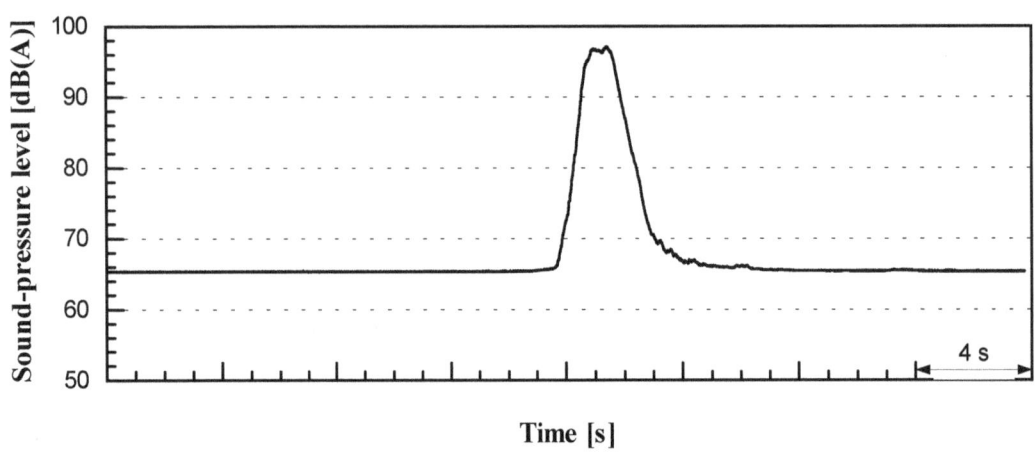

Figure B-138. Time history of the A-weighted SPL during a passby of the TR08 travelling on the North switch at about 300 km/h (186 mph) measured at 6.5 m (21.3 ft) distance from track centerline and the height of the upper surface of the guideway.

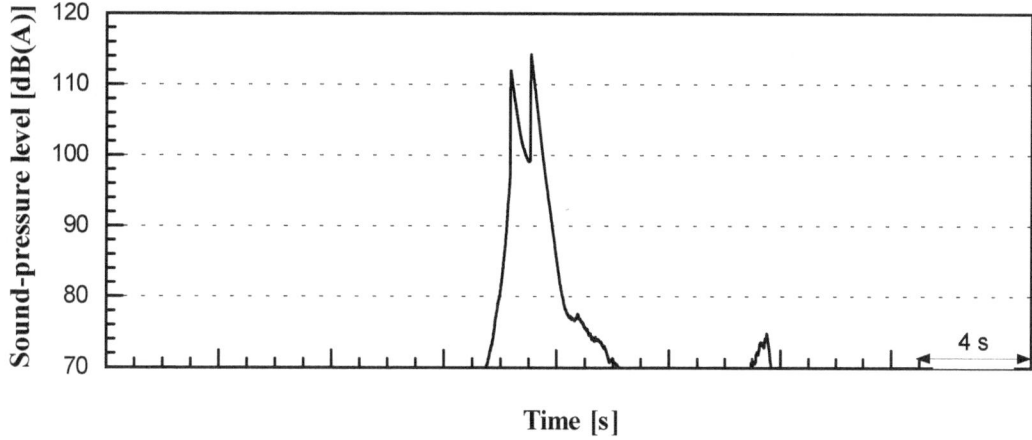

Figure B-139. Time history of the A-weighted SPL during a passby of the TR08 travelling on the North switch at about 385 km/h (239 mph) measured at 6.5 m (21.3 ft) distance from track centerline and the height of the upper surface of the guideway.

Table B-28. Results of the microphone positioned close to the North switch at 6.5 m (21.3 ft) distance from track centerline and the height of the upper surface of the guideway (measuring series Z).

Date	Time	Vehicle speed [km/h (mph)]	$L_{Amax, fast}$ [dB(A)]	$L_{Aeq,E}$ [dB(A)]	t_E [s]	SEL [dB(A)]	$L_{Aeq,1h}$ [dB(A)]
2002-05-17	09:52	100.2 (62.3)	90.0	87.1	3.92	93.1	57.5
2002-05-17	10:03	100.2 (62.3)	89.7	86.9	3.94	92.9	57.3
2002-05-17	10:14	100.3 (62.3)	90.2	87.3	3.81	93.1	57.6
2002-05-17	10:49	99.9 (62.1)	89.0	86.4	3.97	92.4	56.8
2001-08-24	09:27	149.8 (93.1)	88.6	86.4	2.58	90.5	55.0
2001-08-24	10:14	149.8 (93.1)	88.4	86.0	2.63	90.2	54.6
2001-08-24	10:26	150.0 (93.2)	88.0	85.8	2.62	90.0	54.4
2001-08-24	12:35	199.7 (124.1)	89.4	87.2	2.31	90.8	55.2
2001-08-24	12:46	199.9 (124.2)	88.7	86.8	2.32	90.5	54.9
2001-08-24	13:39	199.9 (124.2)	88.6	86.7	2.36	90.4	54.8
2001-08-24	10:31	299.4 (186.1)	97.1	95.1	1.54	96.9	61.4
2001-08-24	12:50	300.0 (186.5)	96.6	94.5	1.55	96.4	60.9
2001-08-24	09:32	379.1 (235.6)	114.6	110.6	0.32	105.7	70.1
2001-08-24	10:18	385.3 (239.5)	114.2	110.2	0.33	105.4	69.8
2001-08-24	11:35	384.9 (239.2)	113.0	109.0	0.32	104.1	68.5

B.6.5 Microphone at 6.5 m (21.3 ft) distance from track centerline (low position)

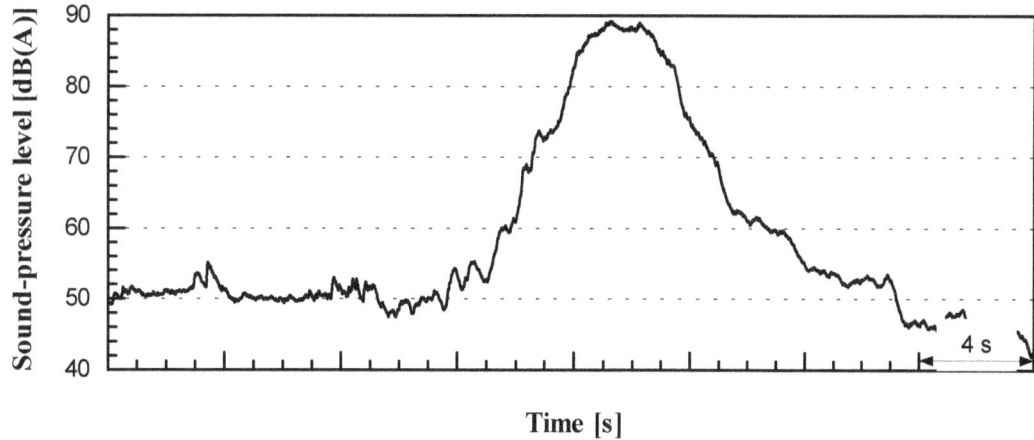

Figure B-140. Time history of the A-weighted SPL during a passby of the TR08 travelling on the North switch at about 100 km/h (62 mph) measured at 6.5 m (21.3 ft) distance from track centerline and 1.5 m (5.0 ft) below the upper surface of the guideway.

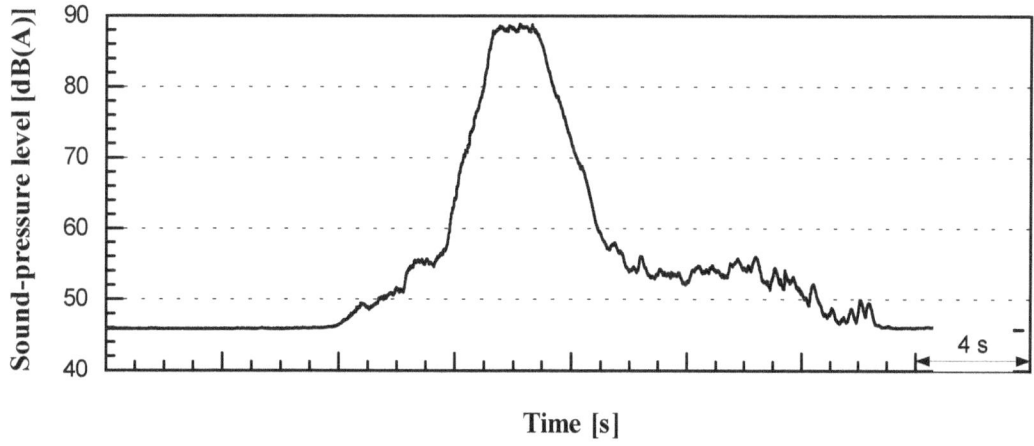

Figure B-141. Time history of the A-weighted SPL during a passby of the TR08 travelling on the North switch at about 150 km/h (93 mph) measured at 6.5 m (21.3 ft) distance from track centerline and 1.5 m (5.0 ft) below the upper surface of the guideway.

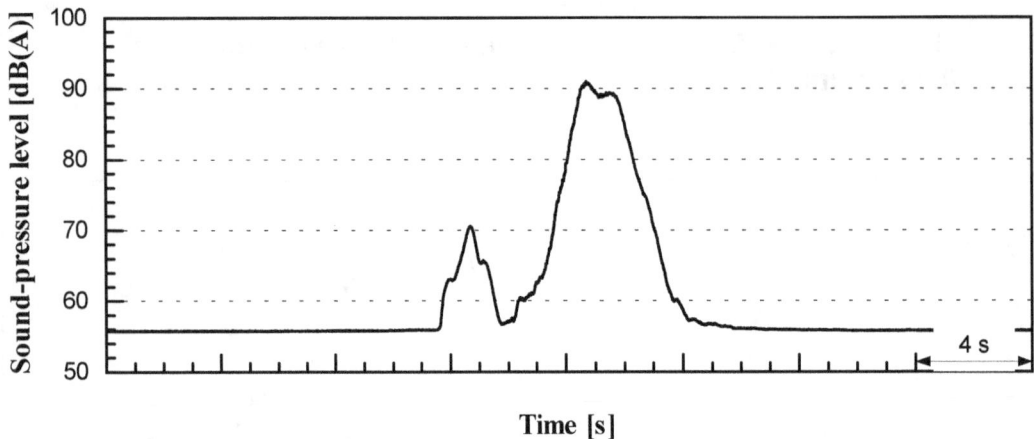

Figure B-142. Time history of the A-weighted SPL during a passby of the TR08 travelling on the North switch at about 200 km/h (124 mph) measured at 6.5 m (21.3 ft) distance from track centerline and 1.5 m (5.0 ft) below the upper surface of the guideway.

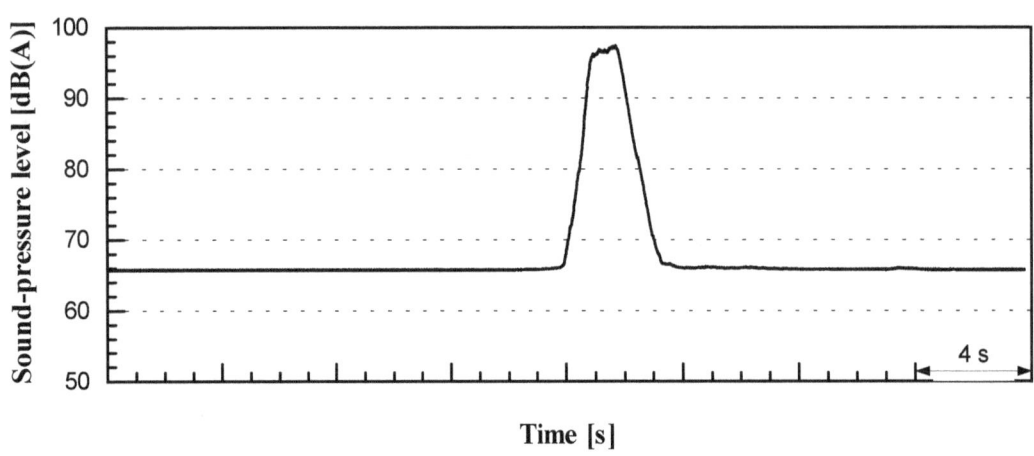

Figure B-143. Time history of the A-weighted SPL during a passby of the TR08 travelling on the North switch at about 300 km/h (186 mph) measured at 6.5 m (21.3 ft) distance from track centerline and 1.5 m (5.0 ft) below the upper surface of the guideway.

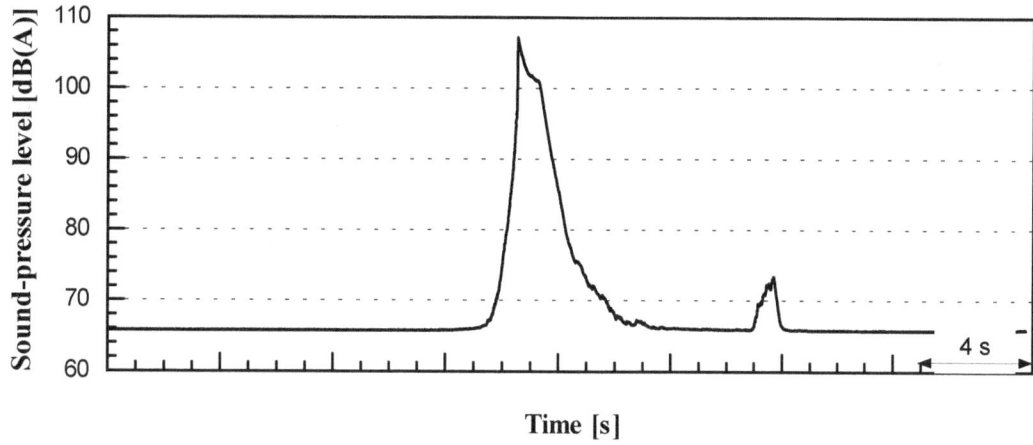

Figure B-144. Time history of the A-weighted SPL during a passby of the TR08 travelling on the North switch at about 385 km/h (239 mph) measured at 6.5 m (21.3 ft) distance from track centerline and 1.5 m (5.0 ft) below the upper surface of the guideway.

Table B-29. Results of the microphone positioned close to the North switch at 6.5 m (21.3 ft) distance from track centerline and 1.5 m (5.0 ft) below the upper surface of the guideway (measuring series Z).

Date	Time	Vehicle speed [km/h (mph)]	$L_{Amax, fast}$ [dB(A)]	$L_{Aeq,E}$ [dB(A)]	t_E [s]	SEL [dB(A)]	$L_{Aeq,1h}$ [dB(A)]
2002-05-17	09:52	100.2 (62.3)	89.2	86.8	3.84	92.7	57.1
2002-05-17	10:03	100.2 (62.3)	88.7	86.7	3.95	92.7	57.2
2002-05-17	10:14	100.3 (62.3)	89.3	87.0	3.80	92.8	57.2
2002-05-17	10:49	99.9 (62.1)	89.1	86.8	3.82	92.6	57.0
2001-08-24	09:27	149.8 (93.1)	89.8	87.5	2.45	91.4	55.9
2001-08-24	10:14	149.8 (93.1)	89.5	87.4	2.51	91.4	55.8
2001-08-24	10:26	150.0 (93.2)	88.9	87.0	2.54	91.0	55.5
2001-08-24	12:35	199.7 (124.1)	90.7	88.5	2.22	92.0	56.4
2001-08-24	12:46	199.9 (124.2)	91.0	88.4	2.16	91.7	56.2
2001-08-24	13:39	199.9 (124.2)	91.6	88.9	2.14	92.2	56.7
2001-08-24	10:31	299.4 (186.1)	97.4	95.5	1.54	97.4	61.8
2001-08-24	12:50	300.0 (186.5)	97.7	95.6	1.56	97.6	62.0
2001-08-24	09:32	379.1 (235.6)	107.7	102.5	0.95	102.3	66.7
2001-08-24	10:18	385.3 (239.5)	107.2	102.6	0.96	102.4	66.9
2001-08-24	11:35	384.9 (239.2)	106.8	102.1	0.97	102.0	66.4

B.6.6 Microphone beneath guideway centerline

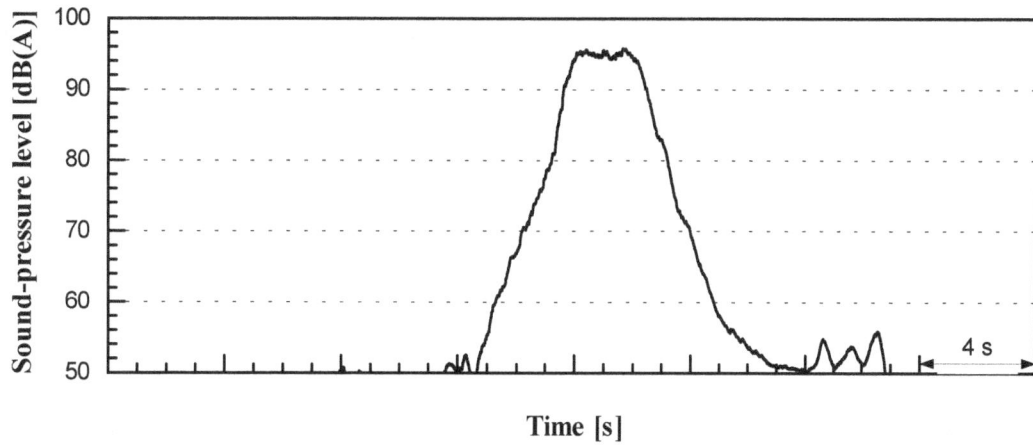

Figure B-145. Time history of the A-weighted SPL during a passby of the TR08 travelling on the North switch at about 100 km/h (62 mph) measured beneath guideway centerline at a height of 1.5 m (5.0 ft) above the ground.

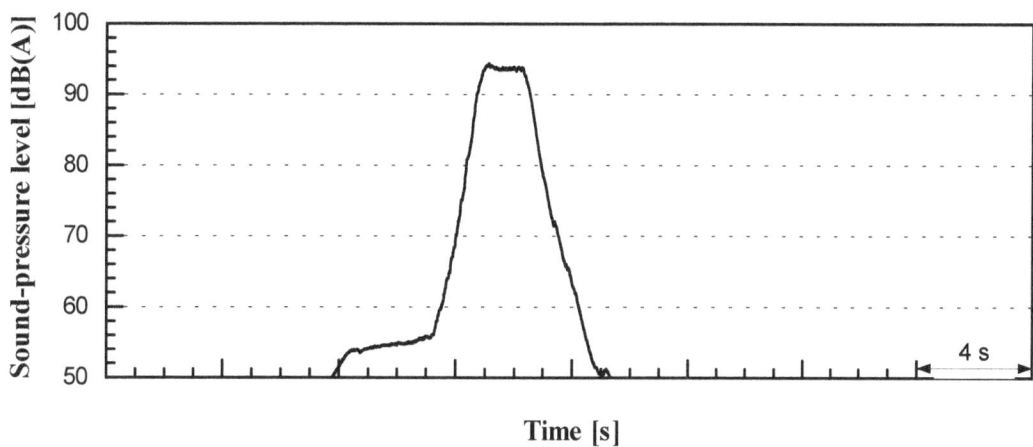

Figure B-146. Time history of the A-weighted SPL during a passby of the TR08 travelling on the North switch at about 150 km/h (93 mph) measured beneath guideway centerline at a height of 1.5 m (5.0 ft) above the ground.

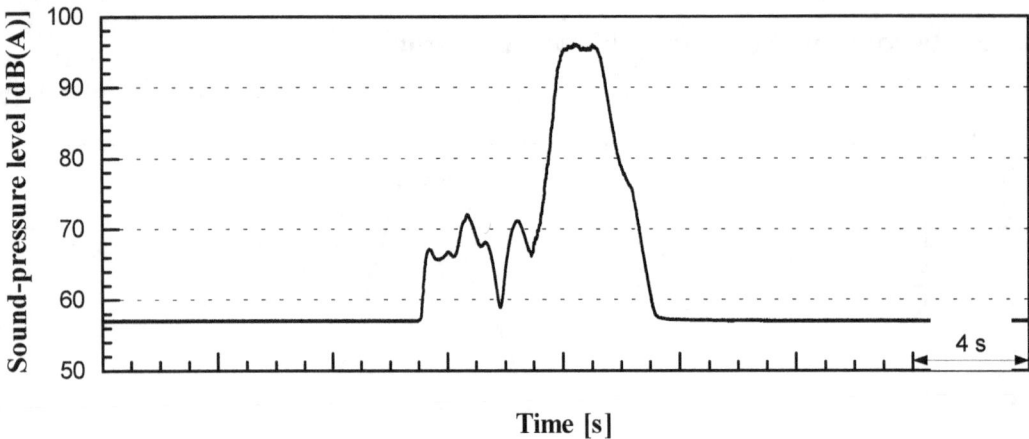

Figure B-147. Time history of the A-weighted SPL during a passby of the TR08 travelling on the North switch at about 200 km/h (124 mph) measured beneath guideway centerline at a height of 1.5 m (5.0 ft) above the ground.

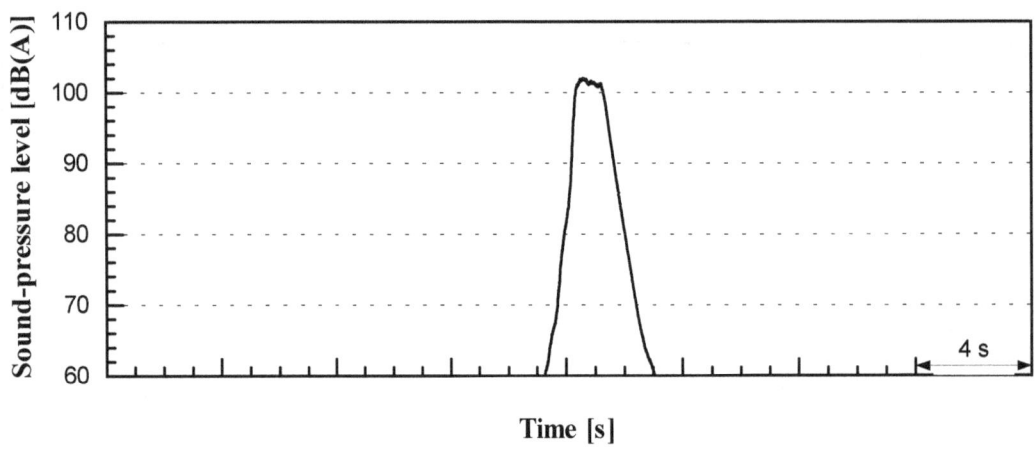

Figure B-148. Time history of the A-weighted SPL during a passby of the TR08 travelling on the North switch at about 300 km/h (186 mph) measured beneath guideway centerline at a height of 1.5 m (5.0 ft) above the ground.

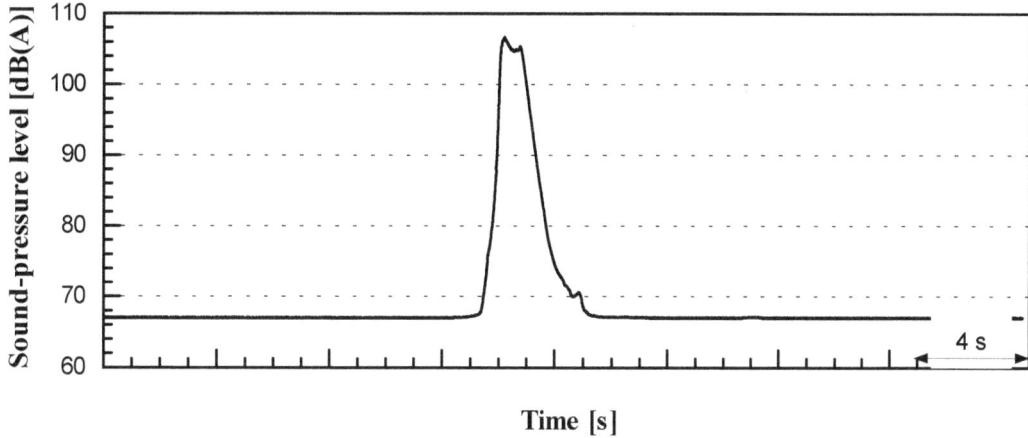

Figure B-149. Time history of the A-weighted SPL during a passby of the TR08 travelling on the North switch at about 385 km/h (239 mph) measured beneath guideway centerline at a height of 1.5 m (5.0 ft) above the ground.

Table B-30. Results of the microphone positioned beneath the centerline of the North switch at a height of 1.5 m (5.0 ft) above the ground (measuring series Z).

Date	Time	Vehicle speed [km/h (mph)]	$L_{Amax, fast}$ [dB(A)]	$L_{Aeq,E}$ [dB(A)]	t_E [s]	SEL [dB(A)]	$L_{Aeq,1h}$ [dB(A)]
2002-05-17	09:52	100.2 (62.3)	95.7	92.6	4.54	99.2	63.6
2002-05-17	10:03	100.2 (62.3)	96.1	93.0	4.45	99.5	63.9
2002-05-17	10:14	100.3 (62.3)	95.7	92.7	4.42	99.1	63.6
2002-05-17	10:49	99.9 (62.1)	96.0	92.9	4.42	99.4	63.8
2001-08-24	09:27	149.8 (93.1)	96.0	92.9	3.08	97.8	62.3
2001-08-24	10:14	149.8 (93.1)	96.3	93.1	3.11	98.0	62.5
2001-08-24	10:26	150.0 (93.2)	96.0	93.1	3.05	97.9	62.4
2001-08-24	12:35	199.7 (124.1)	96.4	93.1	2.84	97.6	62.0
2001-08-24	12:46	199.9 (124.2)	96.1	92.9	2.93	97.5	62.0
2001-08-24	13:39	199.9 (124.2)	96.6	93.3	2.76	97.7	62.2
2001-08-24	10:31	299.4 (186.1)	102.0	98.9	1.92	101.8	66.2
2001-08-24	12:50	300.0 (186.5)	101.9	98.7	1.94	101.6	66.0
2001-08-24	09:32	379.1 (235.6)	105.8	102.7	1.62	104.8	69.2
2001-08-24	10:18	385.3 (239.5)	106.7	103.1	1.56	105.0	69.5
2001-08-24	11:35	384.9 (239.2)	106.6	103.3	1.56	105.2	69.6

B.7 At-Grade Guideway

B.7.1 Beam 340 (Steel)

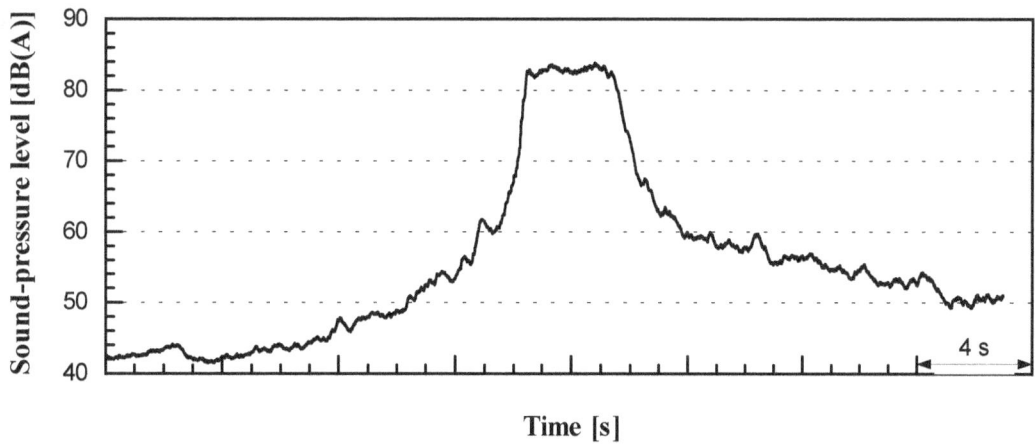

Figure B-150. Time history of the A-weighted SPL during a passby of the TR08 travelling on the at-grade steel guideway at about 100 km/h (62 mph) measured at 6.5 m (21.3 ft) distance from track centerline and the height of the upper surface of the guideway.

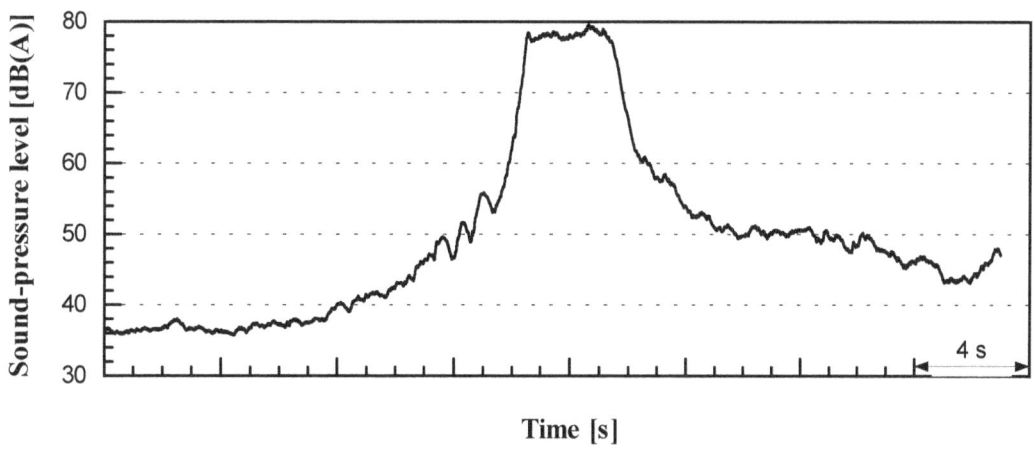

Figure B-151. Time history of the A-weighted SPL during a passby of the TR08 travelling on the at-grade steel guideway at about 100 km/h (62 mph) measured at 6.5 m (21.3 ft) distance from track centerline and 1.5 m (5.0 ft) below the upper surface of the guideway.

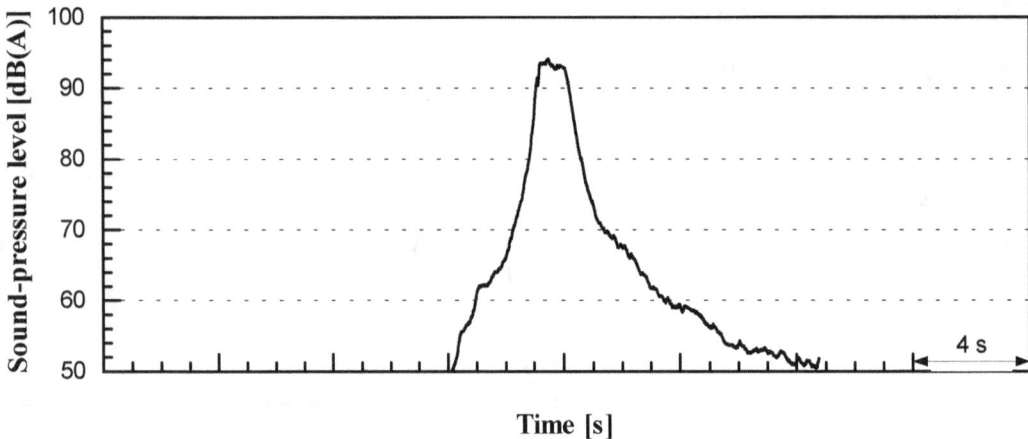

Figure B-152. Time history of the A-weighted SPL during a passby of the TR08 travelling on the at-grade steel guideway at about 300 km/h (186 mph) measured at 6.5 m (21.3 ft) distance from track centerline and the height of the upper surface of the guideway.

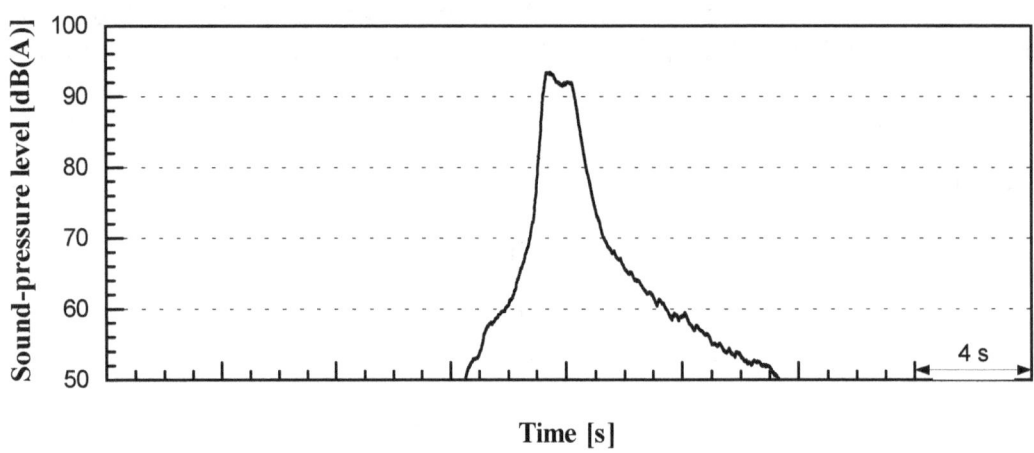

Figure B-153. Time history of the A-weighted SPL during a passby of the TR08 travelling on the at-grade steel guideway at about 300 km/h (186 mph) measured at 6.5 m (21.3 ft) distance from track centerline and 1.5 m (5.0 ft) below the upper surface of the guideway.

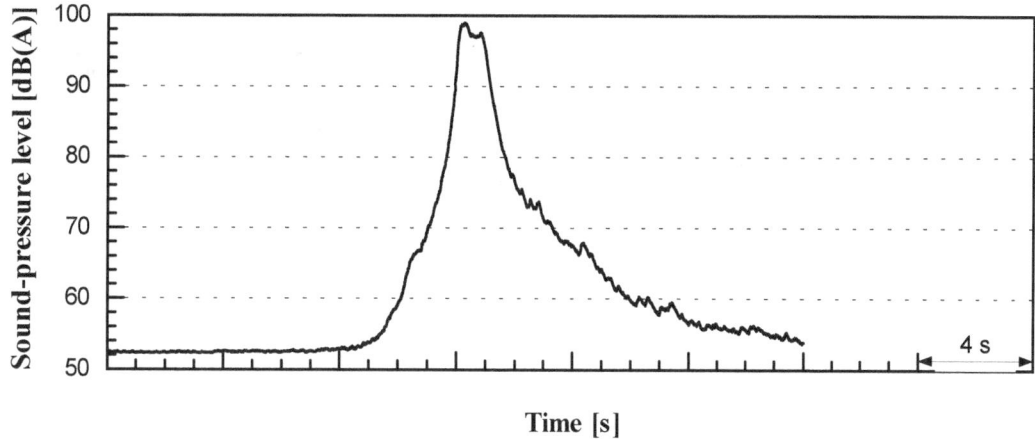

Figure B-154. Time history of the A-weighted SPL during a passby of the TR08 travelling on the at-grade steel guideway at about 370 km/h (230 mph) measured at 6.5 m (21.3 ft) distance from track centerline and the height of the upper surface of the guideway.

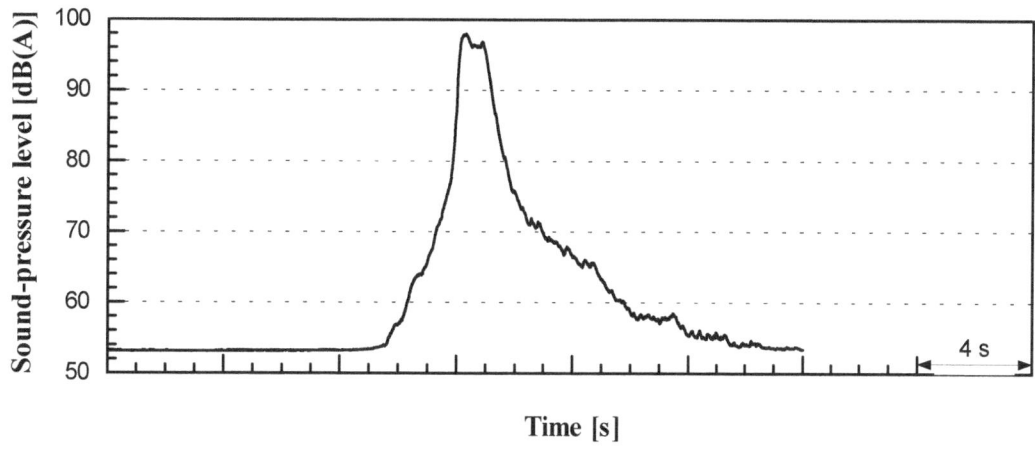

Figure B-155. Time history of the A-weighted SPL during a passby of the TR08 travelling on the at-grade steel guideway at about 370 km/h (230 mph) measured at 6.5 m (21.3 ft) distance from track centerline and 1.5 m (5.0 ft) below the upper surface of the guideway.

Table B-31. Results of the microphone positioned close to the at-grade steel guideway at 6.5 m (21.3 ft) distance from track centerline and the height of the upper surface of the guideway (measuring series Y).

Date	Time	Vehicle speed [km/h (mph)]	$L_{Amax, fast}$ [dB(A)]	$L_{Aeq,E}$ [dB(A)]	t_E [s]	SEL [dB(A)]	$L_{Aeq,1h}$ [dB(A)]
2002-05-16	13:42	100.3 (62.3)	83.8	82.3	3.67	87.9	52.3
2002-05-16	15:47	100.2 (62.3)	83.6	81.6	3.57	87.1	51.5
2001-08-17	12:43	368.0 (228.7)	98.9	96.5	1.26	97.6	62.0
2001-08-17	14:15	259.6 (161.3)	91.0	88.8	1.63	90.9	55.3
2001-08-22	09:05	349.2 (217.0)	97.0	94.7	1.36	96.0	60.5
2001-08-22	10:16	315.7 (196.2)	95.0	93.3	1.45	94.9	59.3
2001-08-22	10:38	299.1 (185.9)	93.8	92.0	1.47	93.7	58.1
2001-08-22	11:16	299.6 (186.2)	94.2	92.2	1.46	93.8	58.3
2001-08-17	10:50	367.1 (228.2)	98.4	96.0	1.31	97.2	61.6
2001-08-17	11:21	366.8 (228.0)	99.1	96.8	1.26	97.8	62.3
2001-08-17	12:29	366.5 (227.8)	99.1	96.7	1.26	97.8	62.2
2001-08-17	13:18	367.1 (228.2)	99.1	96.6	1.26	97.6	62.1
2001-08-17	14:07	368.3 (228.9)	98.8	96.4	1.26	97.4	61.9
2001-08-22	11:48	372.0 (231.2)	99.1	96.8	1.26	97.8	62.3

Table B-32. Results of the microphone positioned close to the at-grade steel guideway at 6.5 m (21.3 ft) distance from track centerline and 1.5 m (5.0 ft) below the upper surface of the guideway (measuring series Y).

Date	Time	Vehicle speed [km/h (mph)]	$L_{Amax, fast}$ [dB(A)]	$L_{Aeq,E}$ [dB(A)]	t_E [s]	SEL [dB(A)]	$L_{Aeq,1h}$ [dB(A)]
2002-05-16	13:42	100.3 (62.3)	79.6	77.7	3.52	83.1	47.6
2002-05-16	15:47	100.2 (62.3)	78.9	77.2	3.50	82.7	47.1
2001-08-17	12:43	368.0 (228.7)	97.5	95.3	1.28	96.4	60.8
2001-08-17	14:15	259.6 (161.3)	90.0	87.7	1.65	89.9	54.3
2001-08-22	09:05	349.2 (217.0)	96.2	93.9	1.34	95.2	59.6
2001-08-22	10:16	315.7 (196.2)	94.3	92.6	1.44	94.2	58.6
2001-08-22	10:38	299.1 (185.9)	93.0	90.8	1.47	92.5	56.9
2001-08-22	11:16	299.6 (186.2)	93.4	91.3	1.44	92.9	57.3
2001-08-17	10:50	367.1 (228.2)	96.7	94.5	1.33	95.8	60.2
2001-08-17	11:21	366.8 (228.0)	97.5	95.5	1.29	96.6	61.0
2001-08-17	12:29	366.5 (227.8)	97.6	95.3	1.25	96.3	60.7
2001-08-17	13:18	367.1 (228.2)	97.6	95.4	1.26	96.4	60.8
2001-08-17	14:07	368.3 (228.9)	97.6	95.3	1.27	96.3	60.8
2001-08-22	11:48	372.0 (231.2)	97.9	95.8	1.25	96.8	61.2

B.7.2 Beam 341 (Concrete)

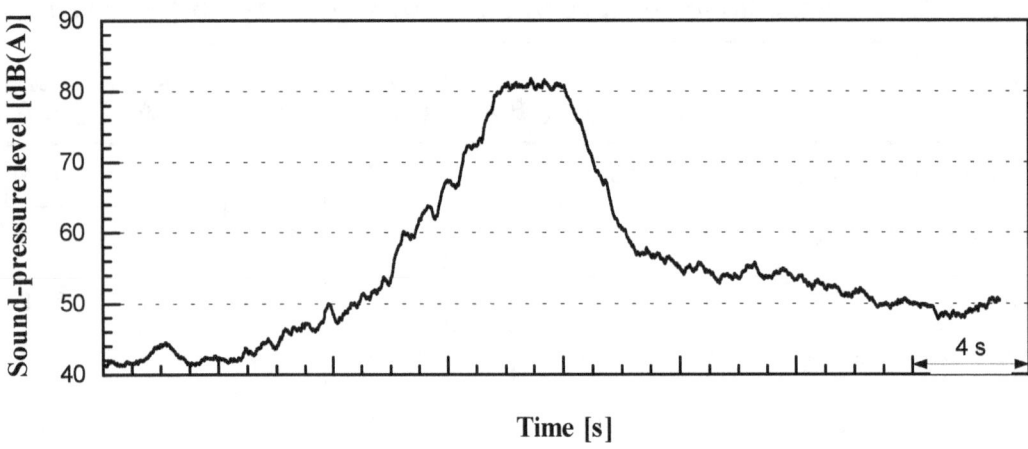

Figure B-156. Time history of the A-weighted SPL during a passby of the TR08 travelling on the at-grade concrete guideway at about 100 km/h (62 mph) measured at 6.5 m (21.3 ft) distance from track centerline and the height of the upper surface of the guideway.

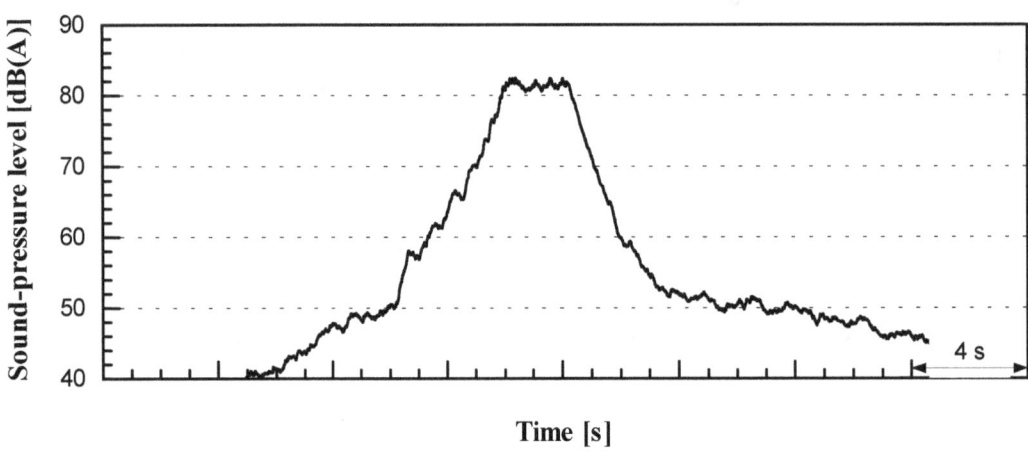

Figure B-157. Time history of the A-weighted SPL during a passby of the TR08 travelling on the at-grade concrete guideway at about 100 km/h (62 mph) measured at 6.5 m (21.3 ft) distance from track centerline and 1.5 m (5.0 ft) below the upper surface of the guideway.

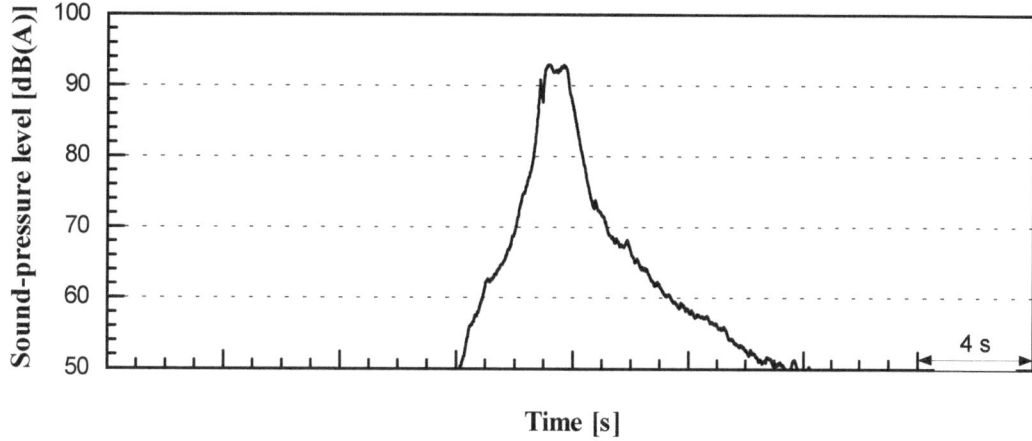

Figure B-158. Time history of the A-weighted SPL during a passby of the TR08 travelling on the at-grade concrete guideway at about 300 km/h (186 mph) measured at 6.5 m (21.3 ft) distance from track centerline and the height of the upper surface of the guideway.

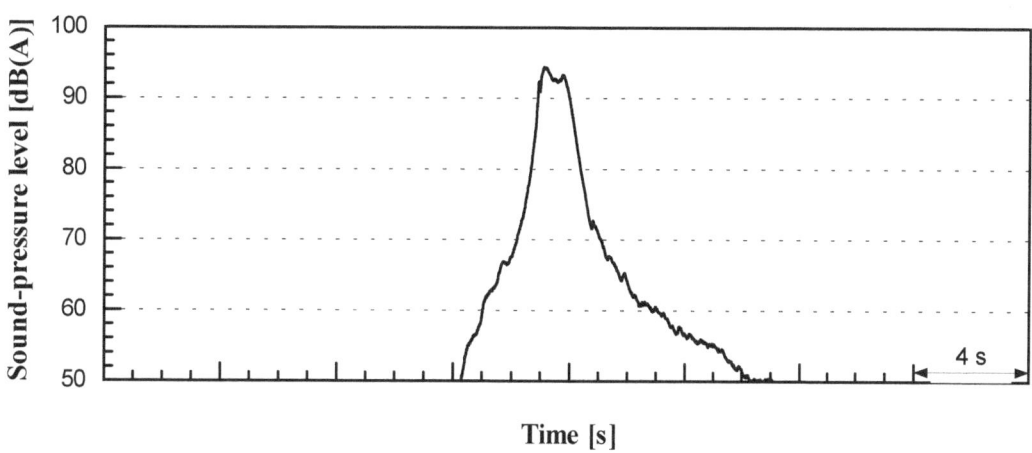

Figure B-159. Time history of the A-weighted SPL during a passby of the TR08 travelling on the at-grade concrete guideway at about 300 km/h (186 mph) measured at 6.5 m (21.3 ft) distance from track centerline and 1.5 m (5.0 ft) below the upper surface of the guideway.

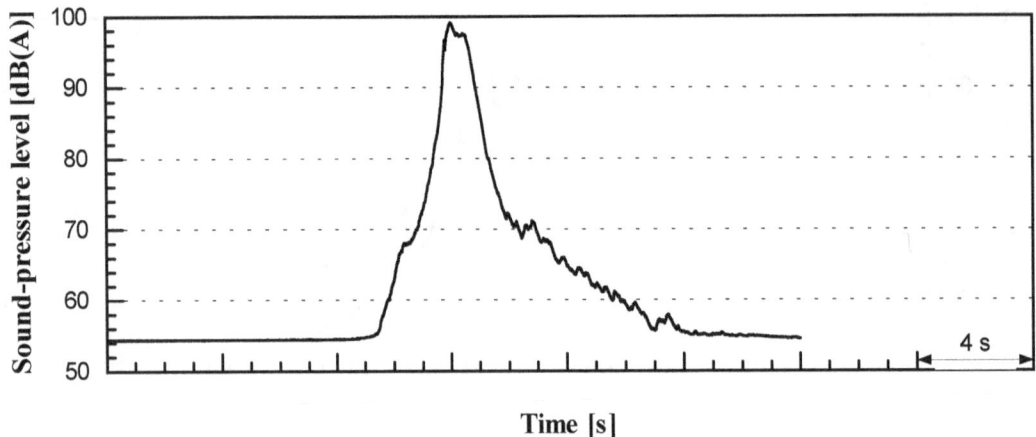

Figure B-160. Time history of the A-weighted SPL during a passby of the TR08 travelling on the at-grade concrete guideway at about 370 km/h (230 mph) measured at 6.5 m (21.3 ft) distance from track centerline and the height of the upper surface of the guideway.

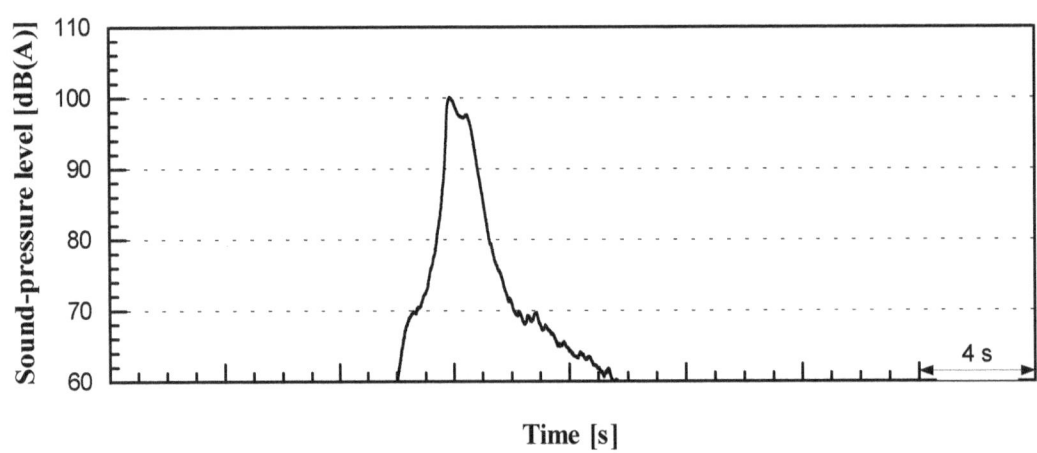

Figure B-161. Time history of the A-weighted SPL during a passby of the TR08 travelling on the at-grade concrete guideway at about 370 km/h (230 mph) measured at 6.5 m (21.3 ft) distance from track centerline and 1.5 m (5.0 ft) below the upper surface of the guideway.

Table B-33. Results of the microphone positioned close to the at-grade concrete guideway at 6.5 m (21.3 ft) distance from track centerline and the height of the upper surface of the guideway (measuring series Y).

Date	Time	Vehicle speed [km/h (mph)]	$L_{Amax, fast}$ [dB(A)]	$L_{Aeq,E}$ [dB(A)]	t_E [s]	SEL [dB(A)]	$L_{Aeq,1h}$ [dB(A)]
2002-05-16	13:42	100.3 (62.3)	81.8	79.4	4.01	85.4	49.9
2002-05-16	15:47	100.2 (62.3)	82.0	79.7	3.79	85.5	49.9
2001-08-17	12:43	368.0 (228.7)	99.2	96.9	1.25	97.9	62.4
2001-08-17	14:15	259.6 (161.3)	90.3	87.9	1.70	90.2	54.6
2001-08-22	09:05	349.2 (217.0)	97.6	95.2	1.30	96.3	60.8
2001-08-22	10:16	315.7 (196.2)	94.3	91.8	1.45	93.4	57.9
2001-08-22	10:38	299.1 (185.9)	92.8	90.8	1.47	92.5	57.0
2001-08-22	11:16	299.6 (186.2)	92.9	90.9	1.46	92.5	57.0
2001-08-17	10:50	367.1 (228.2)	98.0	95.9	1.28	97.0	61.4
2001-08-17	11:21	366.8 (228.0)	99.3	97.0	1.26	98.0	62.4
2001-08-17	12:29	366.5 (227.8)	99.3	96.9	1.25	97.9	62.3
2001-08-17	13:18	367.1 (228.2)	99.1	96.8	1.25	97.8	62.2
2001-08-17	14:07	368.3 (228.9)	99.0	96.9	1.27	97.9	62.3
2001-08-22	11:48	372.0 (231.2)	99.1	96.6	1.22	97.5	61.9

Table B-34. Results of the microphone positioned close to the at-grade concrete guideway at 6.5 m (21.3 ft) distance from track centerline and 1.5 m (5.0 ft) below the upper surface of the guideway (measuring series Y).

Date	Time	Vehicle speed [km/h (mph)]	$L_{Amax,\,fast}$ [dB(A)]	$L_{Aeq,E}$ [dB(A)]	t_E [s]	SEL [dB(A)]	$L_{Aeq,1h}$ [dB(A)]
2002-05-16	13:42	100.3 (62.3)	82.4	80.3	3.63	85.9	50.4
2002-05-16	15:47	100.2 (62.3)	82.8	80.5	3.54	86.0	50.4
2001-08-17	12:43	368.0 (228.7)	99.3	96.8	1.20	97.6	62.0
2001-08-17	14:15	259.6 (161.3)	91.8	88.7	1.56	90.6	55.0
2001-08-22	09:05	349.2 (217.0)	98.4	95.8	1.24	96.7	61.1
2001-08-22	10:16	315.7 (196.2)	94.9	92.6	1.40	94.0	58.5
2001-08-22	10:38	299.1 (185.9)	94.3	92.0	1.38	93.4	57.8
2001-08-22	11:16	299.6 (186.2)	94.3	92.0	1.41	93.5	57.9
2001-08-17	10:50	367.1 (228.2)	97.5	95.4	1.25	96.3	60.7
2001-08-17	11:21	366.8 (228.0)	99.8	96.9	1.18	97.6	62.0
2001-08-17	12:29	366.5 (227.8)	99.2	96.7	1.20	97.5	61.9
2001-08-17	13:18	367.1 (228.2)	99.5	96.8	1.17	97.4	61.9
2001-08-17	14:07	368.3 (228.9)	99.3	96.5	1.19	97.3	61.7
2001-08-22	11:48	372.0 (231.2)	100.1	97.3	1.17	98.0	62.4

APPENDIX C. SOUND SOURCE DISTRIBUTION GRAPHICS

Appendix C presents additional sound-source distribution graphics for the TR08 Maglev System. These graphics present the same averaged, sound-source data measured using the WV microphone array as are presented in Figures 2-13 through 2-16 in the main body of the report. Whereas consistent color codes are used in Figures 2-13 through 2-16, the Appendix C graphics do not use consistent colors throughout the range of speeds for each graphic. Essentially, Figures 2-13 through 2-16 present a clear trend in the location of sound sources as a function of speed, and Figures C-1 through C-4 present each speed's sound sources in more detail.

Figure C-1 Averaged sound-source distributions measured with the WV array during passbys of the TR08 on the reference concrete guideway at speeds between 150 and 400 km/h (93 and 249 mph).

Figure C-2 Averaged sound-source distributions measured with the WV array during passbys of the TR08 on the prototype steel guideway at speeds between 150 and 400 km/h (93 and 249 mph).

Figure C-3 Averaged sound-source distributions measured with the WV array during passbys of the TR08 on the prototype concrete guideway at speeds between 150 and 400 km/h (93 and 249 mph).

Figure C-4 Averaged sound-source distributions measured with the WV array during passbys of the TR08 on the hybrid guideway at speeds between 150 and 400 km/h (93 and 249 mph).

APPENDIX D. TABULAR ONE-THIRD OCTAVE BAND DATA

Appendix D presents tabular one-third octave-band data for the TR08. These data are simply the tabular format of the data presented graphically in Figures 2-26, 2-28, 2-30, and 2-32.

Table D-1. Sound-pressure levels of the unweighted one-third octave-band spectra measured with the WV array during passbys of the TR08 on the reference concrete guideway at speeds between 150 and 400 km/h (93 and 249 mph) at a height of 2.1 m (6.9 ft) with reference to the upper surface of the guideway.

One-third octave-band center frequency [Hz]	Sound-pressure level [dB]			
	150 km/h	200 km/h	300 km/h	400 km/h
315	58.9	65.3	73.5	81.6
400	61.6	62.9	72.8	78.9
500	66.7	61.1	71.8	78.3
630	58.2	63.4	72.0	76.7
800	54.9	63.4	72.7	78.9
1000	61.2	60.0	76.3	80.4
1250	52.8	60.3	73.4	82.9
1600	51.3	58.9	71.5	83.8
2000	51.9	57.5	71.1	80.2
2500	50.1	56.2	70.6	81.9
3150	48.2	53.2	66.2	79.2
4000	47.9	52.6	62.5	72.4

Table D-2. Sound-pressure levels of the unweighted one-third octave-band spectra measured with the WV array during passbys of the TR08 on the reference concrete guideway at speeds between 150 and 400 km/h (93 and 249 mph) at a height of 0.7 m (2.3 ft) with reference to the upper surface of the guideway.

One-third octave-band center frequency [Hz]	Sound-pressure level [dB]			
	150 km/h	200 km/h	300 km/h	400 km/h
315	64.5	69.9	78.0	86.4
400	65.0	67.8	76.9	82.9
500	69.1	65.4	76.4	82.9
630	60.8	67.5	74.9	80.6
800	59.5	66.1	75.5	83.2
1000	63.9	62.8	77.5	81.4
1250	55.7	60.0	72.0	81.8
1600	56.3	59.8	70.5	81.5
2000	56.3	59.4	70.6	79.4
2500	53.9	56.7	69.2	79.3
3150	50.9	53.3	65.2	77.3
4000	52.1	53.6	62.4	72.7

Table D-3. Sound-pressure levels of the unweighted one-third octave-band spectra measured with the WV array during passbys of the TR08 on the reference concrete guideway at speeds between 150 and 400 km/h (93 and 249 mph) at a height of 0 m (0 ft) with reference to the upper surface of the guideway.

One-third octave-band center frequency [Hz]	Sound-pressure level [dB]			
	150 km/h	200 km/h	300 km/h	400 km/h
315	64.2	71.6	79.1	86.8
400	67.8	67.6	77.7	84.3
500	73.4	65.8	75.8	83.6
630	62.4	69.3	78.0	82.3
800	58.1	68.0	75.3	82.0
1000	63.4	61.8	76.5	82.2
1250	56.1	62.5	75.2	81.9
1600	54.9	61.6	71.5	80.6
2000	53.5	58.8	70.2	79.0
2500	52.8	58.9	70.6	78.6
3150	49.7	56.1	67.5	76.5
4000	48.4	52.7	64.4	75.5

Table D-4. Sound-pressure levels of the unweighted one-third octave-band spectra measured with the WV array during passbys of the TR08 on the reference concrete guideway at speeds between 150 and 400 km/h (93 and 249 mph) at a height of -0.7 m (-2.3 ft) with reference to the upper surface of the guideway.

One-third octave-band center frequency [Hz]	Sound-pressure level [dB]			
	150 km/h	200 km/h	300 km/h	400 km/h
315	64.1	74.5	79.8	85.6
400	69.1	68.1	78.1	84.6
500	76.0	65.2	78.7	84.1
630	65.9	76.1	78.0	83.0
800	60.1	71.1	76.9	83.6
1000	66.3	62.5	78.4	82.0
1250	59.6	66.4	78.2	87.0
1600	60.4	64.6	72.1	83.7
2000	57.0	60.9	71.3	79.1
2500	58.0	59.0	69.2	77.8
3150	55.2	56.6	65.4	75.0
4000	52.5	52.9	60.8	71.4

Table D-5. Sound-pressure levels of the unweighted one-third octave-band spectra measured with the WV array during passbys of the TR08 on the reference concrete guideway at speeds between 150 and 400 km/h (93 and 249 mph) at a height of -1.4 m (-4.6 ft) with reference to the upper surface of the guideway.

One-third octave-band center frequency [Hz]	Sound-pressure level [dB]			
	150 km/h	200 km/h	300 km/h	400 km/h
315	62.2	71.6	79.7	85.3
400	68.0	67.3	77.8	82.6
500	77.8	64.2	79.8	83.1
630	65.6	71.8	78.0	81.6
800	59.2	69.5	74.9	82.0
1000	63.2	60.2	74.5	80.1
1250	55.2	61.4	73.8	80.4
1600	55.7	59.7	66.7	79.5
2000	51.7	54.9	65.3	74.7
2500	53.1	54.7	64.3	72.9
3150	50.3	51.7	60.7	69.9
4000	48.1	48.7	57.1	67.1

Table D-6. Sound-pressure levels of the unweighted one-third octave-band spectra measured with the WV array during passbys of the TR08 on the prototype steel guideway at speeds between 150 and 400 km/h (93 and 249 mph) at a height of 2.1 m (6.9 ft) with reference to the upper surface of the guideway.

One-third octave-band center frequency [Hz]	Sound-pressure level [dB]			
	150 km/h	200 km/h	300 km/h	400 km/h
315	60.9	67.4	76.4	82.7
400	66.3	65.9	74.8	80.8
500	70.7	64.2	74.1	81.0
630	66.5	67.2	73.4	79.5
800	62.2	67.5	75.2	80.0
1000	66.3	63.4	77.8	83.1
1250	55.6	63.4	75.0	84.1
1600	54.2	61.5	73.1	86.0
2000	57.9	60.6	72.5	81.2
2500	51.8	58.6	71.6	82.1
3150	49.8	55.7	67.0	78.8
4000	50.8	54.7	63.6	72.5

Table D-7. Sound-pressure levels of the unweighted one-third octave-band spectra measured with the WV array during passbys of the TR08 on the prototype steel guideway at speeds between 150 and 400 km/h (93 and 249 mph) at a height of 0.7 m (2.3 ft) with reference to the upper surface of the guideway.

One-third octave-band center frequency [Hz]	Sound-pressure level [dB]			
	150 km/h	200 km/h	300 km/h	400 km/h
315	62.8	69.0	78.7	86.6
400	66.4	67.7	77.4	84.6
500	71.3	66.0	77.1	82.7
630	68.9	70.2	75.3	80.8
800	67.8	69.4	78.0	83.1
1000	68.6	64.5	78.8	83.7
1250	55.0	60.8	72.6	81.0
1600	57.7	61.3	71.6	82.5
2000	58.2	60.6	72.2	80.9
2500	47.9	56.1	69.3	79.4
3150	48.9	53.0	65.4	77.3
4000	53.9	53.7	63.9	73.7

Table D-8. Sound-pressure levels of the unweighted one-third octave-band spectra measured with the WV array during passbys of the TR08 on the prototype steel guideway at speeds between 150 and 400 km/h (93 and 249 mph) at a height of 0 m (0 ft) with reference to the upper surface of the guideway.

One-third octave-band center frequency [Hz]	Sound-pressure level [dB]			
	150 km/h	200 km/h	300 km/h	400 km/h
315	64.7	70.4	80.6	87.3
400	70.7	69.6	79.0	85.9
500	76.2	67.1	79.0	84.0
630	72.0	71.8	76.9	82.9
800	66.7	69.7	77.4	83.1
1000	69.9	64.3	78.1	85.7
1250	59.7	64.7	76.7	82.7
1600	57.1	62.0	71.9	81.3
2000	56.1	59.4	71.8	80.0
2500	53.9	59.1	71.1	79.7
3150	50.8	55.6	68.6	77.2
4000	51.5	53.3	65.8	75.9

Table D-9. Sound-pressure levels of the unweighted one-third octave-band spectra measured with the WV array during passbys of the TR08 on the prototype steel guideway at speeds between 150 and 400 km/h (93 and 249 mph) at a height of -0.7 m (-2.3 ft) with reference to the upper surface of the guideway.

One-third octave-band center frequency [Hz]	Sound-pressure level [dB]			
	150 km/h	200 km/h	300 km/h	400 km/h
315	67.2	72.3	82.5	88.5
400	72.7	70.6	81.0	86.8
500	79.6	68.5	80.3	86.4
630	76.1	77.8	79.6	85.3
800	71.0	73.7	80.3	84.8
1000	71.1	66.0	80.4	85.2
1250	65.6	71.0	79.7	86.3
1600	63.3	65.9	73.3	83.4
2000	59.9	61.1	71.5	80.0
2500	59.8	59.6	70.0	78.6
3150	57.4	56.0	66.5	76.2
4000	56.7	54.5	63.0	72.3

Table D-10. Sound-pressure levels of the unweighted one-third octave-band spectra measured with the WV array during passbys of the TR08 on the prototype steel guideway at speeds between 150 and 400 km/h (93 and 249 mph) at a height of -1.4 m (-4.6 ft) with reference to the upper surface of the guideway.

One-third octave-band center frequency [Hz]	Sound-pressure level [dB]			
	150 km/h	200 km/h	300 km/h	400 km/h
315	67.5	72.5	82.3	87.7
400	72.1	71.5	80.8	87.7
500	79.8	68.9	80.5	86.3
630	75.5	78.8	81.5	85.6
800	71.8	74.4	81.6	85.2
1000	72.3	66.4	81.5	85.4
1250	63.1	69.1	79.4	85.5
1600	61.1	63.7	71.5	82.9
2000	57.2	59.2	70.3	80.3
2500	57.9	58.8	69.8	77.6
3150	55.6	55.1	65.8	75.4
4000	54.8	53.0	62.4	71.8

Table D-11. Sound-pressure levels of the unweighted one-third octave-band spectra measured with the WV array during passbys of the TR08 on the prototype concrete guideway at speeds between 150 and 400 km/h (93 and 249 mph) at a height of 2.1 m (6.9 ft) with reference to the upper surface of the guideway.

One-third octave-band center frequency [Hz]	Sound-pressure level [dB]			
	150 km/h	200 km/h	300 km/h	400 km/h
315	61.0	65.9	76.4	80.7
400	64.4	64.5	73.2	79.6
500	68.9	62.5	71.3	78.1
630	66.9	66.9	72.0	77.3
800	58.9	65.8	74.1	79.7
1000	62.8	61.1	76.2	82.1
1250	54.9	61.3	72.9	81.7
1600	53.1	59.9	71.8	83.4
2000	53.4	58.6	71.1	79.7
2500	51.5	57.2	70.4	81.6
3150	49.4	54.4	65.9	78.9
4000	49.5	53.6	62.2	72.1

Table D-12. Sound-pressure levels of the unweighted one-third octave-band spectra measured with the WV array during passbys of the TR08 on the prototype concrete guideway at speeds between 150 and 400 km/h (93 and 249 mph) at a height of 0.7 m (2.3 ft) with reference to the upper surface of the guideway.

One-third octave-band center frequency [Hz]	Sound-pressure level [dB]			
	150 km/h	200 km/h	300 km/h	400 km/h
315	64.5	70.0	79.4	85.9
400	65.5	68.4	76.9	84.0
500	69.7	65.8	75.7	82.1
630	67.4	70.2	76.2	81.0
800	61.0	67.5	76.4	82.9
1000	65.2	63.0	78.4	82.8
1250	56.0	59.8	72.4	80.9
1600	57.4	59.4	71.0	82.4
2000	56.4	59.4	70.7	79.3
2500	53.2	56.6	69.5	79.3
3150	51.4	53.2	65.5	77.1
4000	51.9	52.1	62.6	73.2

Table D-13. Sound-pressure levels of the unweighted one-third octave-band spectra measured with the WV array during passbys of the TR08 on the prototype concrete guideway at speeds between 150 and 400 km/h (93 and 249 mph) at a height of 0 m (0 ft) with reference to the upper surface of the guideway.

One-third octave-band center frequency [Hz]	Sound-pressure level [dB]			
	150 km/h	200 km/h	300 km/h	400 km/h
315	64.7	70.0	81.7	87.0
400	68.9	68.5	78.8	86.2
500	74.3	66.8	77.2	84.0
630	69.3	71.8	76.9	83.0
800	61.6	69.0	77.0	83.1
1000	64.7	62.7	77.5	83.8
1250	57.7	62.4	76.4	81.4
1600	55.9	61.2	72.5	81.8
2000	54.5	59.8	71.3	79.8
2500	52.7	59.6	71.1	79.5
3150	49.5	56.0	68.3	76.9
4000	48.2	53.2	65.2	74.8

Table D-14. Sound-pressure levels of the unweighted one-third octave-band spectra measured with the WV array during passbys of the TR08 on the prototype concrete guideway at speeds between 150 and 400 km/h (93 and 249 mph) at a height of -0.7 m (-2.3 ft) with reference to the upper surface of the guideway.

One-third octave-band center frequency [Hz]	Sound-pressure level [dB]			
	150 km/h	200 km/h	300 km/h	400 km/h
315	65.8	71.0	81.5	86.1
400	71.0	69.3	79.1	85.1
500	77.4	68.1	79.4	86.1
630	74.5	75.6	79.0	84.4
800	62.9	70.8	77.9	83.3
1000	66.2	62.2	79.7	82.4
1250	61.9	65.1	78.8	85.0
1600	61.1	62.5	71.9	83.3
2000	57.1	60.4	70.3	79.1
2500	58.1	58.9	70.6	78.9
3150	55.4	55.4	66.7	76.7
4000	52.6	51.7	62.3	72.5

Table D-15. Sound-pressure levels of the unweighted one-third octave-band spectra measured with the WV array during passbys of the TR08 on the prototype concrete guideway at speeds between 150 and 400 km/h (93 and 249 mph) at a height of -1.4 m (-4.6 ft) with reference to the upper surface of the guideway.

One-third octave-band center frequency [Hz]	Sound-pressure level [dB]			
	150 km/h	200 km/h	300 km/h	400 km/h
315	66.0	70.8	81.5	85.2
400	68.7	68.4	78.5	84.7
500	76.7	66.9	77.4	85.1
630	76.0	76.9	77.7	82.8
800	63.6	71.6	77.3	83.3
1000	65.1	60.9	76.1	82.2
1250	59.6	61.8	75.2	80.3
1600	57.9	59.7	67.7	79.6
2000	53.4	56.2	67.0	76.8
2500	53.5	54.4	65.3	73.5
3150	50.6	50.7	61.4	71.5
4000	48.8	49.1	59.4	69.2

Table D-16. Sound-pressure levels of the unweighted one-third octave-band spectra measured with the WV array during passbys of the TR08 on the hybrid guideway at speeds between 150 and 400 km/h (93 and 249 mph) at a height of 2.1 m (6.9 ft) with reference to the upper surface of the guideway.

One-third octave-band center frequency [Hz]	Sound-pressure level [dB]			
	150 km/h	200 km/h	300 km/h	400 km/h
315	61.1	66.6	76.5	83.0
400	62.2	64.7	74.8	80.8
500	67.4	63.4	72.8	79.4
630	65.0	66.6	74.0	77.9
800	59.4	67.0	76.1	79.7
1000	62.9	62.2	78.5	83.7
1250	54.6	61.3	73.9	82.8
1600	52.6	60.0	72.8	83.4
2000	52.4	58.9	71.3	79.9
2500	51.1	57.3	70.8	81.6
3150	49.5	54.5	65.9	79.0
4000	48.0	53.7	62.8	72.4

Table D-17. Sound-pressure levels of the unweighted one-third octave-band spectra measured with the WV array during passbys of the TR08 on the hybrid guideway at speeds between 150 and 400 km/h (93 and 249 mph) at a height of 0.7 m (2.3 ft) with reference to the upper surface of the guideway.

One-third octave-band center frequency [Hz]	Sound-pressure level [dB]			
	150 km/h	200 km/h	300 km/h	400 km/h
315	64.8	70.6	81.0	86.6
400	65.1	68.8	79.1	85.1
500	70.3	66.4	78.1	82.8
630	70.5	68.9	77.0	81.8
800	63.9	70.3	79.5	82.4
1000	64.5	64.5	79.5	83.6
1250	57.1	61.2	74.3	81.1
1600	57.0	61.1	72.4	80.9
2000	56.4	60.9	72.3	79.1
2500	53.8	57.4	69.9	78.6
3150	51.3	54.0	66.4	76.6
4000	49.0	51.8	64.6	72.9

Table D-18. Sound-pressure levels of the unweighted one-third octave-band spectra measured with the WV array during passbys of the TR08 on the hybrid guideway at speeds between 150 and 400 km/h (93 and 249 mph) at a height of 0 m (0 ft) with reference to the upper surface of the guideway.

One-third octave-band center frequency [Hz]	Sound-pressure level [dB]			
	150 km/h	200 km/h	300 km/h	400 km/h
315	65.4	71.5	82.8	88.2
400	66.7	69.4	80.5	87.1
500	74.3	67.6	78.3	85.1
630	72.8	70.3	77.1	82.5
800	64.8	69.7	77.6	82.3
1000	64.3	63.4	78.5	83.9
1250	58.2	63.0	77.4	81.6
1600	55.1	62.1	73.0	80.5
2000	52.6	59.7	71.0	78.4
2500	50.2	59.2	71.3	78.5
3150	46.4	53.4	66.9	76.1
4000	43.7	50.2	64.0	74.1

Table D-19. Sound-pressure levels of the unweighted one-third octave-band spectra measured with the WV array during passbys of the TR08 on the hybrid guideway at speeds between 150 and 400 km/h (93 and 249 mph) at a height of -0.7 m (-2.3 ft) with reference to the upper surface of the guideway.

One-third octave-band center frequency [Hz]	Sound-pressure level [dB]			
	150 km/h	200 km/h	300 km/h	400 km/h
315	66.3	72.5	82.2	88.8
400	69.5	72.8	83.8	88.8
500	78.6	70.8	80.3	87.6
630	80.6	78.6	81.0	87.6
800	68.0	74.1	82.4	85.2
1000	69.4	66.9	83.0	87.4
1250	62.6	67.9	81.8	88.0
1600	61.7	67.1	77.6	85.8
2000	57.0	63.6	75.6	82.1
2500	54.0	63.4	75.0	82.3
3150	50.2	57.0	71.5	80.5
4000	47.7	52.6	68.1	77.2

Table D-20. Sound-pressure levels of the unweighted one-third octave-band spectra measured with the WV array during passbys of the TR08 on the hybrid guideway at speeds between 150 and 400 km/h (93 and 249 mph) at a height of -1.4 m (-4.6 ft) with reference to the upper surface of the guideway.

One-third octave-band center frequency [Hz]	Sound-pressure level [dB]			
	150 km/h	200 km/h	300 km/h	400 km/h
315	67.9	71.7	80.9	88.2
400	68.8	70.5	82.4	88.4
500	77.1	68.6	78.6	85.6
630	74.0	75.3	80.0	84.6
800	66.7	72.9	80.9	84.0
1000	65.7	63.9	81.0	84.1
1250	58.6	63.8	77.8	81.8
1600	56.6	63.2	71.5	80.4
2000	51.7	58.2	69.1	76.8
2500	48.4	58.0	68.5	75.6
3150	44.9	52.1	64.8	73.7
4000	42.7	47.8	61.6	70.4

www.ingramcontent.com/pod-product-compliance
Lightning Source LLC
Chambersburg PA
CBHW080234180526

45167CB00006B/2275